NONLINEAR EQUATIONS IN ABSTRACT SPACES

ACADEMIC PRESS RAPID MANUSCRIPT REPRODUCTION

Proceedings of an International Symposium
on Nonlinear Equations in Abstract Spaces,
held at The University of Texas at Arlington, Arlington, Texas,
June 8–10, 1977

NONLINEAR EQUATIONS IN ABSTRACT SPACES

Edited by
V. Lakshmikantham

Department of Mathematics
University of Texas at Arlington
Arlington, Texas

ACADEMIC PRESS New York San Francisco London 1978

A Subsidiary of Harcourt Brace Jovanovich, Publishers

ACADEMIC PRESS, INC.
111 Fifth Avenue, New York, New York 10003

United Kingdom Edition published by
ACADEMIC PRESS, INC. (LONDON) LTD.
24/28 Oval Road, London NW1 7DX

Library of Congress Cataloging in Publication Data

International Symposium on Nonlinear Equations in Abstract Spaces, University of Texas at Arlington, 1977.
Nonlinear equations in abstract spaces.

"Proceedings of an International Symposium on Nonlinear Equations in Abstract Spaces, held at the University of Texas at Arlington, Arlington, Texas, June 8-10, 1977."
1. Differential equations, Nonlinear—Numerical solutions—Congresses. 2. Volterra equations—Numerical solutions—Congresses. 3. Banach spaces—Congresses.
I. Lakshmikantham, V. II. Title.
QA371.I553 1977 515'.35 78-8412
ISBN 0-12-434160-8

PRINTED IN THE UNITED STATES OF AMERICA

Contents

List of Contributors vii
Preface ix

INVITED ADDRESSES AND RESEARCH REPORTS

New Results in Stochastic Equations—The Nonlinear Case 3
G. Adomian

Positive Operators and Sturmian Theory of Nonselfadjoint Second-Order Systems 25
Shair Ahmad and Alan C. Lazer

Nonlinear Superposition for Operator Equations 43
W. F. Ames

Random Fixed Point Theorems 67
Heinz W. Engl

Delay Equations of Parabolic Type in Banach Space 81
W. E. Fitzgibbon

The Exact Amount of Nonuniqueness for Singular Ordinary Differential Equations in Banach Spaces with an Application to the Euler-Poisson-Darboux Equation 95
Jerome A. Goldstein

On the Equation $Tx = y$ in Banach Spaces with Weakly Continuous Duality Maps 105
Athanassios G. Kartsatos

Nonlinear Evolution Operators in Banach Spaces 113
Yoshikazu Kobayashi

Abstract Boundary Value Problems 117
V. Lakshmikantham

Existence Theory of Delay Differential Equations in Banach Spaces 125
V. Lakshmikantham, S. Leela, and V. Moauro

Invariant Sets and a Mathematical Model Involving Semilinear Differential Equations 135
Robert H. Martin, Jr.

Total Stability and Classical Hamiltonian Theory 149
V. Moauro, L. Salvadori, and M. Scalia

On Some Mathematical Models of Social Phenomena 161
Elliott W. Montroll

Generalized Inverse Mapping Theorems and Related Applications of Generalized Inverses in Nonlinear Analysis 217
M. Z. Nashed

Iteration for Systems of Nonlinear Partial Differential Equations 253
J. W. Neuberger

Existence Theorems and Approximations in Nonlinear Elasticity 265
J. T. Oden

Existence Theorems for Semilinear Abstract and Differential Equations with Noninvertible Linear Parts and Noncompact Perturbations 275
W. V. Petryshyn

Iterative Methods for Accretive Sets 317
Simeon Reich

Model Equations for Nonlinear Dispersive Systems 327
R. E. Showalter

Second Order Differential Equations in Banach Space 331
C. C. Travis and G. F. Webb

CONTRIBUTED PAPERS

A Characterization of the Range of a Nonlinear Volterra Integral Operator 365
Thomas Kiffe and Michael Stecher

Discontinuous Perturbations of Elliptic Boundary Value Problems at Resonance 375
P. J. McKenna

An Existence Theorem for Weak Solutions of Differential Equations in Banach Spaces 387
A. R. Mitchell and Chris Smith

Monotonicity and Alternative Methods 405
Kent Nagle

The OLP Method of Nonlinear Stability Analysis of Turbulence in Newtonian Fluids 417
Fred R. Payne

Generalized Contractions and Sequence of Iterates 439
K. L. Singh

Criteria for the Existence and Comparison of Solutions to Nonlinear Volterra Integral Equations in Banach Space 463
R. L. Vaughn

Semilinear Boundary Value Problems in Banach Space 469
James R. Ward

Polynomial Perturbations to the Laplacian L^p 479
Fred B. Weissler

LIST OF CONTRIBUTORS

Numbers in parentheses indicate the page on which authors' contributions begin.
*Indicates the author who presented the paper at the conference.

ADOMIAN, G. (3), Center for Applied Mathematics, The University of Georgia, Athens, Georgia 30602

AHMAD, SHAIR (25), Department of Mathematics, Oklahoma State University, Stillwater, Oklahoma 74074

AMES, W. F. (43), Center for Applied Mathematics, The University of Georgia, Athens, Georgia 30602

ENGL, HEINZ W. (67), Institut für Mathematik, Kepler-Universität, Linz, Austria

FITZGIBBON, W. E. (81), Department of Mathematics, University of Houston, Houston, Texas 77004

GOLDSTEIN, JEROME A. (95), Department of Mathematics, Tulane University, New Orleans, Louisiana 70118

KARTSATOS, ATHANASSIOS G. (105), Department of Mathematics, University of South Florida, Tampa, Florida 33620

KIFFE, THOMAS (365), Department of Mathematics, Texas A&M University, College Station, Texas 77843

KOBAYASHI, YOSHIKAZU (113), Faculty of Engineering, Niigata University, Nagaoka, Japan

LAKSHMIKANTHAM, V. (117, 125), Department of Mathematics, The University of Texas at Arlington, Arlington, Texas 76019

LAZER, ALAN C.* (25), Department of Mathematical Sciences, University of Cincinnati, Cincinnati, Ohio 45221

LEELA, S.* (125), Department of Mathematics, SUNY College at Genesco, Genesco, New York 14454

McKENNA, P. J. (375), Department of Mathematics, The University of Wyoming, Laramie, Wyoming 82071

MARTIN, ROBERT H., JR. (135), Department of Mathematics, North Carolina State University, Raleigh, North Carolina 27607

MITCHELL A. R. (387), Department of Mathematics, The University of Texas at Arlington, Arlington, Texas 76019

MOAURO, V.* (125, 149), Istituto Matematico "R. Caccioppoli" dell'Università di Napoli, Via Mezzocannone, Naples, Italy, and The University of Texas at Arlington, Arlington, Texas 76019

MONTROLL, ELLIOTT W. (161), Physics Department, University of Rochester, Rochester, New York 10021

NAGLE, KENT (405), Department of Mathematics, University of South Florida, Tampa, Florida 33620

NASHED, M. Z. (217), Department of Mathematics, University of Delaware, Newark, Delaware 19711

NEUBERGER, J. W. (253), Department of Mathematics, North Texas State University, Denton, Texas 76203

ODEN, J. T. (265), Texas Institute for Computational Mechanics, The University of Texas at Austin, Austin, Texas 78712

PAYNE, FRED R. (417), Aerospace Engineering, The University of Texas at Arlington, Arlington, Texas 76019

PETRYSHYN, W. V. (275), Department of Mathematics, Rutgers University, New Brunswick, New Jersey 08903

REICH, SIMEON (317), Department of Mathematics, University of Southern California, Los Angeles, California 90007

SALVADORI, L. (149), Istituto Matematico "G. Castelnuovo" dell'Università di Roma, Rome, Italy, and The University of Texas at Arlington, Arlington, Texas 76019

SCALIA, M. (149), Istituto Matematico "G. Castelnuovo" dell'Università di Roma, Rome, Italy

SHOWALTER, R. E. (327), Department of Mathematics, The University of Texas at Austin, Austin, Texas 78712

SINGH, K. L. (439), Department of Mathematics, Texas A&M University, College Station, Texas 77843

SMITH, CHRIS* (387), Department of Mathematics, The University of Texas at Arlington, Arlington, Texas 76019

STECHER, MICHAEL* (365), Department of Mathematics, Texas A&M University, College Station, Texas 77843

TRAVIS, C. C. (331), Health and Safety Research Division, Oak Ridge National Laboratory, Oak Ridge, Tennessee 37830

VAUGHN, R. L. (463), Department of Mathematics, The University of Texas at Arlington, Arlington, Texas 76019

WARD, JAMES R. (469), Department of Mathematics, Pan American University, Edinburg, Texas 78539

WEBB, G. F.* (331), Department of Mathematics, Vanderbilt University, Box 1085, Nashville, Tennessee 37235

WEISSLER, FRED B. (479), Department of Mathematics, The University of Texas at Austin, Austin, Texas 78712

Preface

An International Symposium on Nonlinear Equations in Abstract Spaces was held at The University of Texas at Arlington, June 8–10, 1977. The purposes of the symposium were to highlight some of the recent advances in abstract nonlinear equations, to stimulate discussions, and to share ideas for future research in this area.

The present volume consists of the proceedings of the symposium. It includes papers that were delivered as invited talks and research reports as well as contributed papers.

The theme of the symposium was the solvability of nonlinear equations, such as Volterra integral equations, ordinary differential equations, and differential equations with retarded argument, in Banach spaces.

There is a group of papers dealing with boundary value problems using such techniques as nonlinear superposition, alternative methods, and fixed points of monotone mappings in ordered Banach spaces. Another group of papers is concerned with the application of nonlinear semigroup theory in solving Volterra integral equations and semilinear partial differential equations. A third group of papers deals with existence theorems for nonlinear evolution equations in Banach spaces. There are also some applications of the previous results to problems arising in nonlinear elasticity, turbulence in Newtonian fluids, classical mechanics, and mathematical models in social phenomena.

I wish to express my special thanks to my colleagues and friends Bill Beeman, Steve Bernfeld, Dorothy Chestnut, Jerome Eisenfeld, Mike Lord, A. R. Mitchell, R. W. Mitchell, and Bennie Williams for assisting me in planning and organizing the symposium; to my secretaries Mrs. Gloria Brown, Debbie Green, and Mrs. Kandy Dyer for their assistance during the conference; and to Mrs. Mary Ann Crain and Mrs. Sandra Weber for excellent typing of the proceedings.

NEW RESULTS IN STOCHASTIC EQUATIONS - THE NONLINEAR CASE

G. Adomian
Center for Applied Mathematics
University of Georgia

Although basic laws generally lead to nonlinear differential and integral equations in many areas, linear approximations are usually employed for mathematical tractability and the use of superposition. Where nonlinear systems must be solved, perturbation theory is classically used which can be a severe restriction. A further substantial difficulty is that real systems generally involve uncertainties and fluctuations and are consequently to be considered as stochastic systems.[1-6] The objective then must be to consider *nonlinear stochastic* systems to allow more realistic modelling.

Previous papers[7-18] have studied objectives, limitations, and necessary restrictive assumptions for the various methods for solution of linear stochastic differential equations arising in physical problems. An iterative procedure was developed using the concept of stochastic Green's functions[4] and allowing treatment of a wide class of problems without many of the limitations inherent in other methods or restrictive assumptions involving highly special processes. It has been shown[9] that the method is applicable to partial differential equations and wave propagation in randomly varying media[6,9] without an assumption of monochromaticity, which ignores the spectral spreading, and without the closure approximations necessary in the hierarchy method.[10] The

ISBN 0-12-434160-8

latter approximations clearly ignore the random fluctuations in certain operators depending on the level of the hierarchy in which it is made. What is involved is that an ensemble average of the form $\langle RL^{-1}R \ldots RL^{-1}Ry \rangle$, (where L is the deterministic part of the stochastic operator and R is the random part) is separated into $\langle RL^{-1}R \ldots RL^{-1}R \rangle \langle y \rangle$. The procedure is justified if R is sufficiently small. Essentially it is equivalent then to dropping terms after some power of ε depending again on the level. Thus writing $\langle RL^{-1}Ry \rangle = \langle RL^{-1}R \rangle \langle y \rangle$ is equivalent to dropping terms of ε^2 in the so-called method of "smoothing perturbation".[19] A variety of other approximate techniques for stochastic wave propagation are similar to the smoothing perturbation method. Thus diagram methods,[20,14,15,21] parabolic equation approximation, etc., all involve the assumption of a near negligible randomness. The iterative procedure on the other hand is independent of the presence of a small parameter, and it has been shown[8] that the results of perturbation theory are obtained where perturbation theory is applicable, i.e., where the random fluctuations are actually small. As Lerche and Parker[19], and earlier, Kraichnan[22], Frisch[15], Herring[23], and others, have pointed out, the first order smoothing gives incorrect results in general. As in the hierarchy method, it is simply incorrect to approximate the equations to be solved. Extensive computer results in preparation for publication[24] will show comparisons and the superior validity of the iterative procedure. As Lerche and Parker stated, the only way to be sure of the result is to solve the exact equations with a systematic mathematical approximation scheme. This has been our approach and it checks in the deterministic limit; it checks in the perturbation case and computer results show it corresponds extremely closely (as closely as desired, depending of course on computing sufficient terms) with the correct analytical solution (where that solution has been possible to obtain) while the other methods show larger and larger deviation as time progresses.

A general expression was obtained for the stochastic Green's

functions which yield the mean solution or the two point correlation of the solution. Relationships have been found between the various stochastic Green's functions for different statistical measures and also between stochastic Green's functions, random Green's functions, and ordinary Green's functions[8]. The stochastic Green's function is determined as a kernel of an integral expression yielding directly the desired statistical measure of the solution process - in our case, either the expectation of the solution or the correlation or covariance of the solution. The long-term objective is to develop methods of calculating the stochastic Green's function for the desired statistical measure (e.g., correlation or covariance) directly from knowledge of the given statistics (statistical measures) of the coefficients without the intermediate step of solving the equation.[7,8]

The Nonlinear Case. Recently, this work was significantly extended to nonlinear stochastic differential equations[25-27] making the results substantially more valuable in real physical problems. This paper reviews the status of newer work and a symmetrized form for the terms of the iteration which appears particularly useful. The equation considered is

$$\mathcal{L}y + N(y, \dot{y}, \ldots) = x(t)$$

where $x(t)$ is a stochastic process on a suitable index space T and probability space (Ω, F, μ), $\mathcal{L}$ is a linear stochastic differential operator of nth order given by $\mathcal{L} = \sum_{\nu=0}^{n} a_\nu(t,\omega) d^\nu/dt^\nu$, the a_ν for $\nu = 0,1,\ldots,n-1$ are possibly stochastic processes on $T \times \Omega$ statistically independent of $x(t)$, and N a nonlinear stochastic term of the form $N = \sum_{\mu=0}^{m_m} b_\mu(t,\omega)(y^{(\mu)})^{m_\mu}$ where $y^{(\mu)}$ is the μth derivative and b_μ are stochastic processes for $\mu = 0,1,\ldots,m$ on $T \times \Omega$.

The coefficients a_ν and b_μ are possibly stochastic processes defined on the same probability space as $x(t)$ purely for

convenience in notation. As pointed out in an earlier paper, there is no difficulty in allowing a_ν to be defined on Ω_a or even Ω_{a_ν}, $x(t)$ on Ω_x, etc. Hence no restriction is implied here.

As assumed in the earlier papers, $a_n > 0$ (we can let it conveniently be 1), and randomness may occur in all other terms $a_0,\ldots,a_{n-1}$, although it would normally be only one or two which would be random. The coefficient processes a_ν, b_μ are assumed statistically independent from the input process $x(t)$ for all allowed μ,ν. Further, the processes a_ν, b_μ are of class C^n on T for $\omega \in (\Omega, F, \mu)$ almost surely for allowed μ,ν We assume $\mathcal{L} = L + R$ where $L = \langle\mathcal{L}\rangle$, i.e., $L = \sum_{\nu=0}^{n} \langle a_\nu(t,\omega)\rangle d^\nu/dt^\nu$ and $R = \sum_{\nu=0}^{n-1} \alpha_\nu(t,\omega) d^\nu/dt^\nu$ where * $\langle a_\nu(t,\omega)\rangle$ exists and is continuous on T. We have

$$Ly + Ry + N(y,\dot{y},\ldots) = x(t)$$

$$Ly = x - Ry - N(y,\dot{y},\ldots)$$

Since L is assumed invertible

$$y = L^{-1}x - L^{-1}Ry - L^{-1}N(y,\dot{y},\ldots)$$

or in terms of the Green's function $\ell(t,\tau)$ for L,

$$y = \int_0^t \ell(t,\tau)x(\tau)d\tau - \int_0^t \ell(t,\tau)R[y(\tau)]d\tau - \int_0^t \ell(t,\tau)N(y(\tau),\dot{y}(\tau),\ldots)d\tau$$

Let $y = \sum_{i=0}^{\infty} (-1)^i y_i$. We get

$$y = \int_0^t \ell(t,\tau)x(\tau)d\tau - \int_0^t \ell(t,\tau)R[y_0(\tau)-y_1(\tau)+y_2(\tau)\ldots]d\tau - \int_0^t \ell(t,\tau)N(y_0(\tau) - y_1(\tau) + y_2(\tau)\ldots)d\tau$$

In order to proceed with the iteration we must assume the form of

---

* $a_\nu = \langle a_\nu\rangle + \alpha_\nu(t,\omega) \qquad \langle\alpha_\nu\rangle = 0$

N. As an example let $N = by^2$. We obtain

$$y = L^{-1}x - L^{-1}Ry_0 + L^{-1}Ry_1 - L^{-1}Ry_2 \cdot \cdot \cdot$$
$$-L^{-1}b(y_0^2 + y_1^2 + y_2^2 + \cdot \cdot \cdot$$
$$-2y_0y_1 + 2y_0y_2 \cdot \cdot \cdot$$
$$-2y_1y_2 + 2y_1y_3 \cdot \cdot \cdot)$$

The nonlinear term brings in all the cross products, e.g., terms involving y_1 appear in products involving y_2, y_3, etc. As in the linear case, we identify $y_0 = L^{-1}x = F(t)$* and identify as each y_i for $i > 0$ only the terms preceding y_i. We have now

$$y_0 = L^{-1}x$$
$$y_1 = L^{-1}Ry_0 + L^{-1}by_0^2$$
$$y_2 = L^{-1}Ry_1 - L^{-1}by_1^2 + L^{-1}by_0y_1$$
$$y_3 = L^{-1}Ry_2 + L^{-1}by_2^2 + L^{-1}by_0y_2 - L^{-1}by_1y_2$$
$$\vdots$$

Each y_i is calculable in terms of preceding terms until we get to y_0. At that point ensemble averages to be taken will separate because of the statistical independence of the x from the coefficients. We observe that if $N = 0$ we have the linear solution previously obtained in earlier papers. First and second order statistics are obtained by ensemble averaging the series for y to get $\langle y \rangle$ and a product of y's at different times to get correlations.

The $L^{-1}Ry$ term involves derivatives in R; it is replaced by the author in the previous linear analyses[1-3] using an adjoint operator $R^{\dagger}$ given by

$$R^{\dagger}\, \ell(t,\tau) = \sum_{k=0}^{n-1} (d^k/d\tau^k)[\alpha_k(\tau)\ell(t,\tau)]$$

* The homogeneous solution has been deleted here. It can be included in $F(t)$ as shown later.

Now y doesn't involve derivatives for the iteration, a procedure which can be viewed as use of a stochastic Green's formula[28].

$$\int_0^t \ell(t,\tau)R[y(\tau,\omega)]d\tau = \int_0^t R^+[\ell(t,\tau)]y(\tau,\omega)d\tau$$

Analogous to the ordinary Green's formula, we have a bilinear concomitant term - now a *stochastic bilinear concomitant* (s.b.c.). This s.b.c. term is zero if the initial conditions are zero but does not vanish in the case of random initial conditions. Hence the case of random initial conditions, which is essentially trivial in the case of a deterministic operator, becomes more difficult in the case of a stochastic operator since additional terms arise from the solution of the homogeneous equation and the value of the s.b.c. at $\tau = 0$, where the s.b.c. is given by

$$\sigma(y(t,\omega);\ell(t,\tau))\Big|_{\tau=0}^{\tau=t}$$

$$= \sum_{k=0}^{n-1} \sum_{\nu=0}^{k-1} (-1)^\nu [\ell(t,\tau)\alpha_k(\tau,\omega)]^{(\nu)} y^{(k-1-\nu)}(\tau,\omega)\Big|_{\tau=0}^{\tau=t}$$

It does vanish at the upper limit because $(d^r/d\tau^r)\ell(t,\tau)|_{\tau=t} = 0$ for $0 \leq r < n - 1$, a known property of the Green's function $\ell(t,\tau)$ most easily seen by differentiation applied to $\ell(t,\tau)$ (written in terms of determinants involving Φ, the solutions of $Ly = 0$) and noting that the determinants vanish if $r < n - 1$ if $\tau = t$.

Initial Conditions: Let ϕ_i, $i = 0,1,\ldots,n - 1$, be linearly independent solutions of $Ly = 0$ which satisfy the initial conditions $(d^j/dt^j)\phi_i(t) = \delta_i^j$, $0 \leq i$, $j \leq n - 1$, where δ_i^j is the Kronecker delta. We have $\phi = \sum_{\mu=0}^{n-1} c_\mu(\omega)\phi_\mu(t)$.

The solution to the stochastic operator equation in the previous work was written as a sum $y = y_0 + \sum_{i=1}^{\infty} (-1)^i y_i = y_0 + \sum_{i=1}^{\infty} (-1)^i L^{-1}Ry_{i-1}$ where $y_0 = F(t) = L^{-1}x + \phi$ represents the unique solution of $Ly_0 = x$ satisfying the initial conditions.

The formal sum may be written equivalently as

$$y = \sum_{i=0}^{\infty} (-1)^i (L^{-1}R)^i y_0$$

In this form it is clear that if convergence can be assured, y is the solution of $\mathcal{L}y = (L + R)y = x$ satisfying the given conditions.

In the previously assumed case of *zero initial conditions*, the stochastic bilinear concomitant (s.b.c) vanished as stated there. Hence $L^{-1}R$ could be written as the purely integral operator. However in general [7,28,29,30]

$$\begin{aligned} L^{-1}Ry_{i-1} &= \int_0^t \ell(t,\tau) \sum_{\nu=0}^{m} \alpha_\nu(\tau,\omega)(d^\nu/d\tau^\nu) y_{i-1}(\tau,\omega)\,d\tau \\ &= \int_0^t \{ \sum_{\nu=0}^{m} (-1)^\nu (d^\nu/d\tau^\nu)[\ell(t,\tau)\alpha_\nu(\tau,\omega)] y_{i-1}(\tau,\omega)\,d\tau \\ &\quad + \sum_{\nu=1}^{m} \sum_{k=0}^{\nu-1} (-1)^k (d^k/d\tau^k)[\ell(t,\tau)\alpha_\nu(\tau,\omega)](d^{\nu-k-1}/d\tau^{\nu-k-1}) \\ &\qquad \cdot\, y_{i-1}(\tau,\omega)\Big|_{\tau=0}^{\tau=t} \\ &= Ky_{i-1} + (s.b.c.)(y_{i-1})\Big|_0^t \end{aligned}$$

where K is the adjoint operator to $L^{-1}R$ and is defined if $\ell \in C^n$, $\alpha_\nu \in C^\nu$, $x \in C^0$. The s.b.c. vanishes for $0 \leq m < n$, $i \geq 2$. For $0 \leq r < n - 1$, it vanishes because * $(d^r/d\tau^r)\ell(t,\tau)|_{t=\tau} = 0$ hence, s.b.c. $(y_{i-1})|^{\tau=t} = 0$.

The s.b.c. vanishes at the lower limit $\tau = 0$ because, for $i \geq 2$, $0 \leq r < n - 1$, we have

$$\begin{aligned} &(d^r/d\tau^r)y_{i-1}(\tau)|_{\tau=0} \\ &\quad = (d^r/d\tau^r)L^{-1}Ry_{i-2}(\tau)|_{\tau=0} \end{aligned}$$

---

* $(d^{k-i}/dt^{k-i})[(d^i/dt^i)\ell(t,\tau)|_{t=\tau}] = 0$ since $(d^i/dt^i)\ell(t,\tau) = 0$ for $i < n - 1$.

$$= (d^r/d\tau^r) \int_0^\tau \ell(\tau,\tau')R(\tau')y_{i-2}(\tau')d\tau'|_{\tau=0}$$

$$= \int_0^\tau (d^r/d\tau^r)\ell(\tau,\tau')R(\tau')y_{i-2}(\tau')d\tau'|_{\tau=0}$$

$$+ \sum_{i=0}^{r-1} (d^i/d\tau^i)[(d^{r-i-1}/d\tau^{r-i-1})\ell(\tau,\tau')R(\tau')y_{i-2}(\tau')|_{\tau'=\tau}]|_{\tau=0}$$

$$= 0$$

because of vanishing of the derivative of $\ell(t,\tau)$ in the last term * and since $\ell(\tau,\tau')$ is smooth to $(n - 1)st$ order at zero and the range of integration vanishes. Thus s.b.c. $(y_{i-1})|_{\tau=0}$ vanishes for $i \geq 2$.

For *zero initial conditions*, $\phi = 0$ so that as above

$$(d^r/d\tau^r)y_0|_{\tau=0} = (d^r/d\tau^r) \int_0^\tau \ell(\tau,\tau')x(\tau')d\tau'|_{\tau=0} = 0$$

for $0 \leq r < n - 1$. Hence for zero initial conditions, one is able to replace $L^{-1}R$ by K throughout. This of course, coincides with the given solution in the previous papers.

In the case of *random initial conditions*, the s.b.c. does not vanish at $\tau = 0$ and hence must be carried along as part of the solution. Consider the sum

$$(1) \quad y = \sum_{i=0}^{\infty} (-1)^i (L^{-1}R)^i (L^{-1}x + \phi) = y_0 + \sum_{i=1}^{\infty} (-1)^i (L^{-1}R)^i y_0$$

representing the iterative solution. Since $L^{-1}R = K +$ (s.b.c.) and since s.b.c. $(y_{i-1}) = 0$ for $i \geq 2$, $y_n = L^{-1}Ry_{n-1} = (K + \text{s.b.c.})y_{n-1} = K^{n-1}(L^{-1}R)y_0$ and

$$y = y_0 + \sum_{i=1}^{\infty} (-1)^i y_i$$

or directly from (1)

$$y = y_0 + \sum_{i=1}^{\infty} (-1)^i K^{i-1} L^{-1} R y_0$$

$$= L^{-1}x + \phi + \sum_{i=1}^{\infty} (-1)^i K^{i-1} L^{-1} R (L^{-1}x + \phi)$$

(2) $$= \sum_{i=0}^{\infty} (-1)^i K^i L^{-1} x + \phi + \sum_{i=1}^{\infty} (-1)^i K^{i-1} (L^{-1} R)\phi$$

(3) $$= \sum_{i=0}^{\infty} (-1)^i K^i L^{-1} (x - R\phi) + \Phi$$

Consequently, we can view the solution as the sum of the zero initial condition solution and the homogeneous solution satisfying the initial condition (see Equation 2). Thus, the solution for random initial conditions involves both ϕ and the s.b.c. which vanished in the earlier solution. Since $y^{(i)}(0) = \phi^{(i)}(0)$, the s.b.c. can be written either as

$$\sum_{k=1}^{n-1} \sum_{\nu=0}^{k-1} (-1)^{\nu} [\ell(t,\tau)\alpha_k(\tau,\omega)]^{(\nu)} y^{(k-1-\nu)}(\tau,\omega)\Big|_{\tau=0}$$

or

(4) $$\sum_{k=1}^{n-1} \sum_{\nu=0}^{k-1} (-1)^{\nu} [\ell(t,\tau)\alpha_k(\tau,\omega)]^{(\nu)} \phi^{(k-1-\nu)}(\tau,\omega)\Big|_{\tau=0}$$

Another way to view this result is as follows:

$$y = L^{-1}x + \phi - L^{-1}Ry$$

$$= L^{-1}x + \phi - Ky - s.b.c.(y)\Big|_0^t$$

$$= L^{-1}x + \phi - Ky$$

$$+ \sum_{\nu=1}^{m} \sum_{k=0}^{\nu-1} (-1)^k (\partial^k/\partial\tau^k)[\ell(t,\tau)\alpha_\nu(\tau,\omega)]\Big|_0^t c_{\nu-k-1}(\omega)$$

since $y^{(k)}(\tau)\big|_0 = \phi^{(k)}(\tau)\big|_0 = \sum_{\mu=0}^{n-1} c_\mu(\omega)\phi_\mu^{(k)}(\tau)\big|_0$ $= \sum_{\mu=0}^{n-1} c_\mu(\omega)\delta_\mu^k = c_k(\omega)$. Let

$$\psi = \sum_{\nu=1}^{m} \sum_{k=0}^{\nu-1} (-1)^k (\partial^k/\partial\tau^k)[\ell(t,\tau)\alpha_\nu(\tau,\omega)]\Big|^{\tau=0} c_{\nu-k-1}(\omega).$$

Then

$$y = \sum_{i=0}^{\infty} (-1)^i K^i [L^{-1}x + \phi + \psi]$$

This result is, of course, the one obtained previously. Thus, to the previous results for zero initial conditions, one must add all iterates of ϕ and ψ by K.

The case in which, for an nth order differential equation, there are no random coefficients of the terms of order higher than the first, is particularly simple since then

$$\psi = \ell(t,0)\alpha_1(0,\omega)c_0.$$

Defining the Green's function

$$\ell(t,\tau) = \frac{(-1)^{n-1}}{W(\tau)} \begin{vmatrix} \phi_0(\tau) & \phi_1(\tau) & \cdots\phi_{n-1}(\tau) \\ \phi_0'(\tau) & \phi_1'(\tau) & \cdots\phi_{n-1}'(\tau) \\ \vdots & & \\ \phi_0^{(n-2)}(\tau) & \phi_1^{(n-2)}(\tau)\cdots & \phi_{n-1}^{(n-2)}(\tau) \\ \phi_0(t) & \phi_1(t) & \phi_{n-1}(t) \end{vmatrix}$$

we have

$$\psi = \phi_{n-1}(t)[\alpha_1(0,\omega)c_0(\omega)]$$

and

$$y = \sum_{i=0}^{\infty} (-1)^i K^i [L^{-1}x + \phi + \phi_{n-1}(t)\alpha_1(0,\omega)c_0(\omega)]$$

The kernel $k(t,\tau)$ of K is $-d\alpha_1(\tau)/d\tau = (d/d\tau)[\alpha_1(\tau)\ell(t,\tau)] + \alpha_0(\tau)\ell(t,\tau)$.

The more complicated case where R has no terms higher than $m = 2$ has

$$k(t,\tau) = (d^2/d\tau^2)[\ell(t,\tau)\alpha_2(\tau)] - (d/d\tau)[\ell(t,\tau)\alpha_1(\tau)] + \alpha_0(\tau)\ell(t,\tau)$$

$$\psi = \sum_{\nu=1}^{m} \sum_{k=0}^{\nu-1} (-1)^k (d^k/d\tau^k)[\ell(t,\tau)\alpha_k(\tau)]\,(d^{\nu-k-1}/d\tau^{\nu-k-1})y(\tau)|_0$$

$$=\{\ell(t,0)\alpha_1(0)\}c_0 + \{\ell(t,0)\alpha_2(0)\}c_1 - \{(d/d\tau)[\ell(t,\tau)\alpha_2(\tau)]|_0\}c_0$$

$$= \phi_{n-1}(t)\{\alpha_1(0)c_0 + \alpha_2(0)c_1 - \alpha_2^{(t)}(0)c_0\} + \phi_{n-2}(t)\{\alpha_2(0)c_0\}$$

where

$$\ell(t,0) = \phi_{n-1}(t)$$

$$\ell^{(1)}(t,0) = \det \begin{vmatrix} 1 & \ldots\ 0 & 0 & 0 \\ \vdots & \vdots & \vdots & \vdots \\ 0 & \ldots\ 1 & 0 & 0 \\ 0 & \ldots & 0 & 1 \\ \phi_0(t) & & \phi_{n-2}(t) & \phi_{n-1}(t) \end{vmatrix} = -\phi_{n-2}(t)$$

(the subdeterminant in the upper left is of order $n-2$ and has diagonal elements equal to 1 and off diagonal elements equal to 0).

Example: Second Order Equation: Let us consider a specific equation given by

$$\mathcal{L}y = d^2y/dt^2 + a_1(t,\omega)\ dy/dt + a_0(t,\omega)y = x$$

where $a_0 = 1 + \alpha_0(t,\omega)$, $\langle a_0\rangle = 1$, $\langle\alpha_0\rangle = 0$, and $a_1(t,\omega) = \alpha_1(t,\omega)$, $\langle a_1\rangle = 0$, $\langle\alpha_1\rangle = 0$. $\mathcal{L} = L + R$ where $L = (d^2/dt^2) + 1$ and $R = \alpha_1(d/dt) + \alpha_0$. The solutions of $Ly = 0$ are $\phi_0(t) = \cos t$ and $\phi_1(t) = \sin t$ hence, the Green's function $\ell(t,\tau)$ in terms of ϕ_0,ϕ_1 and the Wronskian W is given by:

$$\ell(t,\tau) = -\frac{1}{W(\tau)}\begin{vmatrix}\cos\tau & \sin\tau \\ \cos t & \sin t\end{vmatrix} = \sin\tau\cos t - \cos\tau\sin t$$

$$= \begin{cases} 0 & t \le \tau \\ \sin(\tau-t) & t \ge \tau \end{cases}$$

$$k(t,\tau) = -(d/dt)[\ell(t,\tau)\alpha_1(\tau)] + \alpha_0(\tau)\ell(t,\tau)$$

$$= \ell(t,\tau)\{\alpha_0(\tau) - (d/dt)\alpha_1(\tau)\} - (d/dt)\ell(t,\tau)\{\alpha_1(\tau)\}$$

$$= \sin(\tau-t)\{\alpha_0(\tau) - (d/dt)\alpha_1(\tau)\} - \cos(\tau-t)\alpha_1(\tau)$$

$$\psi = \sin(t-\tau)[\alpha_1(0)c_0(\omega)]$$

$$y = \sum_{i=0}^{\infty}(-1)^i K^i[F + \phi + \ell(t,0)\alpha_1(0)c_0(\omega)]$$

$$= \sum_{i=0}^{\infty}(-1)^i K^i[F + c_0\cos t + (c_1 - \alpha_1(0)c_0)\sin t]$$

We remark that physical problems will generally be simpler than the general equations considered for the nth order case.

Further, since, usually, only one or two coefficients would be random at most, the s.b.c. becomes quite simple.

Complete expressions obtained by Adomian and Lynch for the mean, correlation, and covariance of the solution of the linear stochastic operator equation, with random initial conditions included as well as stochastic coefficients and stochastic forcing function, appear elsewhere.[30]

When randomness enters the equation *only* through the initial conditions ($\mathcal{L} = L$, a deterministic operator) then the problem is relatively trivial since the solution $y(t)$ develops deterministically in time. We simply regard the initial values $y(0)$, $y^1(0)$, etc., as random variables with a given distribution and the joint distribution of $y(t)$, $y'(t)$, etc., is found from the joint distribution of $y(0)$, $y'(0),\ldots$ by standard methods of change of variables. The case of random initial conditions accompanying a stochastic operator has been discussed in a separate paper.

Obtaining First and Second Order Statistics: To obtain the mean or expected solution $\langle y\rangle$, the correlation $R_y(t_1,t_2)$ or the covariance $K_y(t_1,t_2)$, the solution process y must be averaged over the probability space Ω (or probability spaces $\Omega_\alpha,\Omega_\beta,\Omega_x$, if different measure spaces are allowed for α,β,x, or even over Ω_{α_ν}, etc., if different for each coefficient) to get $\langle y\rangle$. Similarly $y(t_1)\overset{*}{y}(t_2)$ is ensemble averaged to get $R_y(t_1,t_2)$. Thus in the linear case when $N = 0$,

$$\langle y\rangle = \langle F(t,\omega)\rangle - \int_0^t \langle\Gamma(t,\tau,\omega)\rangle\langle F(\tau)\rangle d\tau$$

where Γ is the resolvent kernel in the form used by Sibul.[6]

$$\Gamma(t,\tau,\omega) = \sum_{m=1}^{\infty} (-1)^{m-1} K_m(t,\tau,\omega)$$

$$K_m(t,\tau,\omega) = \int_0^t K(t,\tau_1,\omega)K_{m-1}(\tau_1,\tau,\omega)d\tau_1$$

$$K_1 = K(t,\tau,\omega)$$

In the general (nonlinear) case,[27]

$$\langle y(t)\rangle = \langle F(t)\rangle - \int_0^t \langle\Gamma(t,\tau)\rangle\langle F(\tau)\rangle d\tau - \langle\int_0^t \ell(t,\tau)N(y(\tau),\ldots)d\tau\rangle$$

$$\begin{aligned}R_y(t_1,t_2) = {} & \langle F(t_1)\overset{*}{F}(t_2)\rangle - \int_0^{t_1} \langle\Gamma(t_1,\tau_1)\rangle\langle F(\tau_1)\overset{*}{F}(t_2)\rangle d\tau_1 \\ & - \int_0^{t_2} \langle\overset{*}{\Gamma}(t_2,\tau_2)\rangle\langle F(t_1)\overset{*}{F}(\tau_2)\rangle d\tau_2 \\ & - \int_0^{t_1}\int_0^{t_2} \langle\Gamma(t_1,\tau_1)\overset{*}{\Gamma}(t_2,\tau_2)\rangle\langle F(\tau_1)\overset{*}{F}(\tau_2)\rangle d\tau_1 d\tau_2 \\ & - \langle\int_0^{t_1} \overset{*}{F}(t_2)\ell(t_1,\tau_1)N(y(\tau_1),\ldots)d\tau_1\rangle \\ & - \langle\int_0^{t_2} F(t_1)\overset{*}{\ell}(t_2,\tau_2)\overset{*}{N}(y(\tau_2),\ldots)d\tau_2\rangle \\ & + \langle\int_0^{t_1} \Gamma(t_1,\tau_1)F(\tau_1)d\tau_1 \int_0^{t_2} \overset{*}{\ell}(t_2,\tau_2)\overset{*}{N}(y(\tau_2),\ldots)d\tau_2\rangle \\ & + \langle\int_0^{t_1} \ell(t_1,\tau_1)N(y(\tau_1),\ldots)d\tau_1 \int_0^{t_2} \overset{*}{\Gamma}(t_2,\tau_2)\overset{*}{F}(\tau_2)d\tau_2\rangle \\ & + \langle\int_0^{t_1} \ell(t_1,\tau_1)N(y(\tau_1),\ldots)d\tau_1 \int_0^{t_2} \overset{*}{\ell}(t_2,\tau_2)\overset{*}{N}(y(\tau_2),\ldots)d\tau_2\rangle\end{aligned}$$

i.e., five more terms than in the linear case as we see by setting $N = 0$. Putting in the specific form for N, say for example, $N = \beta y^2(t)$ and iterating, we get the additional terms (over the linear case) involving βy_0^2, etc., as indicated arising from $\beta(y_0 - y_1 + y_2 - \ldots)^2$, but the separation of ensemble averages still occurs as in the linear case.

Form of $N(y,\dot{y},\ldots)$: A previous paper[25] showed that any $N(y(t))$ which can be expanded in a Taylor series can be treated by this method*. Additional assumptions (uniform convergence) will arise if $N = N(y(t),\dot{y}(t))$. For example, if we let

* Derivatives to appropriate order must exist a.e. (such questions will be discussed elsewhere).

$N(y) = b\ exp\{y\} = b(1+y+y^2/2!+\ldots)$** we obtain immediately

$$y_0 = L^{-1}x = F(t)$$
$$y_1 = L^{-1}Ry_0 + L^{-1}b + L^{-1}by_0 + L^{-1}by_0^2/2$$
$$y_2 = L^{-1}Ry_1 + L^{-1}by_1 - L^{-1}by_1^2/2 + L^{-1}by_0y_1$$

etc. The case $N = by^2$ yields

$$y_0 = L^{-1}x$$
$$y_1 = L^{-1}Ry_0 + L^{-1}by_0^2$$
$$y_2 = L^{-1}Ry_1 + L^{-1}by_1^2 + 2L^{-1}by_0y_1$$
$$y_3 = L^{-1}Ry_2 + L^{-1}by_2^2 + 2L^{-1}by_0y_2 - 2L^{-1}by_1y_2$$

etc. The case $N = \beta y^2 + \gamma y'^3$ yields

$$y_0 = L^{-1}x$$
$$y_1 = L^{-1}Ry_0 + L^{-1}\beta y_0^2 + L^{-1}\gamma y_0'^3$$
$$y_2 = L^{-1}Ry_1 - L^{-1}\beta y_1^2 + 2L^{-1}\beta y_0y_1 - L^{-1}\gamma(3y_0'y_1'^2 - 3y_0'^2y_1' - y_1'^3)$$

etc.

Convergence: Let us now examine the crucial convergence question.

$N = 0$ Case: Writing $F(t)$ for $L^{-1}x$ and any solution to the homogeneous equation $Ly = 0$, we get

$$y(t,\omega) = F(t,\omega) - L^{-1}Ry(t,\omega) \tag{2}$$

$$= F(t,\omega) - \int_0^t \ell(t,\tau) \sum_{\nu=0}^{n-1} \alpha_\nu(\tau,\omega)(d^\nu/d\tau^\nu)y(\tau,\omega)d\tau \tag{3}$$

where $\ell(t,\tau)$ is the Green's function for L. If the stochastic bilinear concomitant (s.b.c.)[29] vanishes[7,30], the equation becomes

$$y(t,\omega) = F(t,\omega) - \int_0^t R^{+}[\ell(t,\tau)]y(\tau,\omega)d\tau \tag{4}$$

where

$$R^{+}[\ell(t,\tau)] = \sum_{\nu=0}^{n-1} (-1)^k\partial^k/\partial\tau^k[\alpha_k(\tau)\ell(t,\tau)] \tag{5}$$

** Expansions in different senses, e.g., in m.s., a.e., etc. will be discussed in a forthcoming paper by Adomian and Malakian.

by the use of a stochastic Green's formula.[17]

The iterative solution has been given by Adomian[7] and Sibul[6] in the form:

(6) $$y(t,\omega) = F(t,\omega) - \int_0^t \Gamma(t,\tau;\omega)F(\tau,\omega)d\tau$$

where the resolvent kernel Γ is given by:

(7) $$\Gamma(t,\tau;\omega) = \sum_{m=1}^{\infty} (-1)^{m-1}K_m(t,\tau;\omega)$$

where

(8) $$K_m(t,\tau;\omega) = \int_0^t K(t,\tau_1)K_{m-1}(\tau_1,\tau)d\tau_1$$

$$K_1 = K = R^{\perp}[\ell(t,\tau)]$$

We assume the input process $x(t,\omega)$ is bounded almost surely on T and $\ell(t,\tau)$ is continuous on T; thus $|x(t,\omega)| < M_1$, a constant, or equivalently, $|F(t,\omega)| < M_1'$, a constant. Further, the α_ν, for $\nu = 0,\ldots,n-1$, are bounded almost surely; in fact, the kth derivatives of the α_ν are bounded for $k = 0,1,\ldots,n-1$, i.e., $|(\partial^k/\partial t^k)\alpha_\nu(t)| < M_2$, a constant, for $t \in T$, $\omega \in (\Omega,F,\mu)$. From (8),

$$|K_m(t,\tau)| \leq \int_0^t |K(t,\tau_1)||K_{m-1}(\tau_1,\tau)|d\tau_1$$

Since $(\partial^k/\partial t^k)\ell(t,\tau)$ is jointly continuous in t and τ on $T \times T$ for $0 \leq k \leq n-1$ and the derivatives of the α_ν are bounded a.s., we can assume a bound directly on the K_m. We observe that if $|K_1| \leq M$, then $|K_2| \leq \int_0^t K_1(t,\tau_1)K_1(\tau_1,\tau)d\tau_1 \leq M^2|t-\tau|$, and $|K_3| \leq M^3|t-\tau|^2/2!$, etc. Hence[2],

(8) $$|K_m(t,\tau)| \leq M^m|t-\tau|^{m-1}/(m-1)!$$

From (7),

$$|\Gamma(t,\tau;\omega)| = \sum_{m=1}^{\infty} (-1)^{m-1}|K_m(t,\tau;\omega)|$$

$$\leq \sum_{m=1}^{\infty} M^m |t - \tau|^{m-1}/(m - 1)!$$

$$\leq M \sum_{m=1}^{\infty} M^{m-1} |t - \tau|^{m-1}/(m - 1)!$$

$$\leq M \sum_{m=0}^{\infty} M^m |t - \tau|^m/m! = Me^{M|t-\tau|}$$

which exists for finite interval of observation $|t - \tau|$. The convergence of the series was first seen by Adomian and Sibul. It is guaranteed by the factorial in the denominator of series terms due to the multiple integration $\int \ldots \int dt$ with the bounded quantities taken outside the integrals. That similar considerations apply to the nonlinear case here will be seen best from the symmetrized form which follows, after some further remarks about the nonlinear case. For non-zero N, we assume the stochastic coefficients $b_\mu(t,\omega)$ are bounded a.s., on T for $\omega \in (\Omega, F, \mu)$. The iteration now leads to extra terms arising from $L^{-1}N$. However, y_0 is bounded, as before, by hypothesis. y_1 differs from the y_1 for the linear case by the addition of the term $L^{-1}N(y_0)$. y_2 differs from the y_2 for the linear case by addition of $L^{-1}N(y_0, y_1)$, etc. Thus

$$\begin{aligned} y(t,\omega) &= F(t,\omega) - \int_0^t \Gamma(t,\tau;\omega)F(\tau,\omega)d\tau - \int_0^t \ell(t,\tau)N(y,\dot{y},\ldots)d\tau \\ &= F(t,\omega) - \int_0^t \Gamma(t,\tau;\omega)F(\tau,\omega)d\tau - \int_0^t \ell(t,\tau) \sum_{\mu=0}^{m} b_\mu(\tau,\omega)(y^{(\mu)})^{m_\mu}\, d\tau \end{aligned} \tag{10}$$

Combining the first two terms,

$$y(t,\omega) = G(t) - \int_0^t \ell(t,\tau)[b_0(y^{(0)})^{m_0} + \ldots b_m(y^{(m)})^{m_m}]d\tau^*$$

Example: Suppose $N(y,\dot{y},\ldots) = N(y) = by^2$. The last term of (10) is $L^{-1}by_2$. The iteration for the linear case yields[7]

$$y_0 = F(t)$$

---

* Similar arguments apply for convergence of statistical measures with absolute values replaced by norms. This will be discussed in a subsequent paper by Adomian and Malakian.

$$y_1 = L^{-1}Ry_0$$
$$y_2 = L^{-1}Ry_1$$
$$y_3 = L^{-1}Ry_2$$

etc. The general case adds terms involving $L^{-1}by^2$. Thus

$$y_0 = F(t)$$

(11)
$$y_1 = L^{-1}Ry_0 + L^{-1}by_0^2$$
$$y_2 = L^{-1}Ry_1 - L^{-1}by_1^2 + 2L^{-1}by_0y_1$$
$$y_3 = L^{-1}Ry_2 + L^{-1}by_2^2 + 2L^{-1}by_0y_2 - 2L^{-1}by_1y_0$$

etc. Since y_0 is bounded, y_1 is bounded if ℓ and b are well behaved as assumed. Hence y_2 is bounded, etc. Each term of y, to whatever level of approximation we wish to carry it, depends only on *preceding* terms and hence finally on y_0 which is bounded by assumption, and, of course, on $\ell(t,\tau)$ and on the b_μ. The $\ell(t,\tau)$ is deterministic and bounded on T. R and b_μ are bounded a.s., on T for $\omega \in (\Omega,F,\mu)$. Consequently, the iterative solution exists.

Let us consider this further for $N = N(y)$. Abstracting these results, Equation (10) is of the form

(12)
$$y(t,\omega) = G(t) - \int_0^t g(t,\tau,y(\tau))d\tau$$

i.e., a nonlinear Volterra equation of the second kind. Let $g(t,\tau,y(\tau)) = \ell(t,\tau)N(\tau,y(\tau))$

Under the assumptions [7,25,25]

i) $G(t)$ is continuous and bounded a.s. on $[0,\infty]$

ii) $\ell(t,\tau)$ is continuous on $0 \leq \tau \leq t < \infty$

iii) There exists a constant $M > 0$ such that

$$\int_0^t |\ell(t,\tau)|d\tau < M \quad \text{for} \quad t \geq 0$$

The iterative series can be carried out as before; each term depends only on those before, and ultimately on y_0, hence the solution exists.

Symmetric Form Solution for the Nonlinear Case: A very promising modification of the iterative procedure by Adomian and Sibul[31] has been made for the quadratic nonlinear term $N(y) = by^2$ which leads to a useful symmetric form solution for the general term of the iterative series. With the modified procedure, the terms become rearranged. The general term becomes

$$y_{n+1} = L^{-1}K_n(y_0) \; *$$

where

$$K_n(y_0) = Ry_n + b(y_0 y_n + y_1 y_{n-1} + \cdots + y_n y_0).$$

The series for y is given by

$$y = y_0 - L^{-1}\Gamma(y_0)$$

where

$$\Gamma(y_0) = \sum_{i=0}^{\infty} (-1)^i K_i(y_0).$$

Consequently, we can write the general term in the above symmetric form easily or program on a computer. Statistics are determined as before leading to a stochastic Green's function for the desired statistics for this (nonlinear) case. Since it appears reasonable that such symmetric forms can be obtained for $N(y) = by^3$, etc., further work is in process on symmetrized forms for cubic, quartic, or general polynomial $N(y)$, as well as for other classes of nonlinearity such as product nonlinearities.

That convergence still holds in the nonlinear case here is an important result which can be seen best from the form of $K_n(y_0)$. If one examines the bracketed term $y_0 y_n + \cdots y_n y_0$ and replaces each y_i by y_{i-1} until y_0 is reached, each term yields n integrations or an $n!$ in the denominator as in the linear case. We now have n such terms in the general term yielding $1/(n-1)!$ and convergence follows. Further discussion appears in the forthcoming paper.[31]

---

* Initial conditions must be included in y_0 as previously discussed.

It is of interest now to see if the class of nonlinearities can be further extended. In at least one important case, it appears to be possible and is being studied.

ACKNOWLEDGEMENT: This work has been supported by a Sloan Foundation research grant. Appreciation is expressed also to M. K. Malakian for valuable editing.

REFERENCES

[1] Samuels, J. C. and Eringer, A. C., "*On Stochastic Linear Systems*," J. Math. and Phys., Vol. 38, (1959), pp. 83-103.

[2] Adomian, G., "*Linear Stochastic Operators*", Hughes Aircraft Company Research Study 278, August 9, 1961.

[3] Adomian, G., "*Linear Stochastic Operators*", Reviews of Modern Physics, Vol. 35, No. 1, pp. 185-207, Jan. 1963.

[4] Adomian, G., "*Stochastic Green's Function*", Stochastic Processes in Mathematical Physics and Engineering, Ann. Sympos. in Appl. Math., Vol. XVI, edited by R. Bellman, American Mathematical Society, 1964.

[5] Adomian, G., "*Theory of Random Systems*", Transactions of the Fourth Prague Conference on Information Theory, Statistical Decision Functions, and Random Processes, pp. 205-222, 1965, publ. 1968.

[6] Sibul, L. H., "*Application of Linear Stochastic Operator Theory*", Ph.D. Dissertation, Pennsylvania State University, Dec. 1968.

[7] Adomian, G., "*Random Operator Equations in Mathematical Physics, Part I*", J. of Math. Phys., Vol. 11, No. 3, pp. 1069-1084, March 1970.

[8] Adomian, G., "*Random Operator Equations in Mathematical Physics, Part II*", J. of Math. Phys., Vol. 12, No. 9, pp. 1944-1948, Sept. 1971.

[9] Adomian, G., "*Random Operator Equations in Mathematical Physics, Part III*", J. of Math. Phys., Vol. 12, No. 9,

pp. 1948-1955, Sept. 1971.

[10] Adomian, G., "*The Closure Approximation in the Hierarchy Equations*", J. of Statistical Physics, Vol. 3, No. 2, pp. 127-133, July 1971.

[11] Adomian, G., "*On Random Parameter Systems*", Izvestia Matematika, Academy of Sciences, Erevan, U.S.S.R., Vol. VII, No. 1, pp. 14-21, Jan. 1972.

[12] Keller, J. B., " *Stochastic Equations and Wave Propagation in Random Media*", Proc. Sympos. in Appl. Math., Vol. XVI, pp. 145-170, Amer. Math. Soc., 1964.

[13] Bharucha-Reid, A. T., "*On the Theory of Random Equations*", Proc. Sympos. in Appl. Math., Vol. XVI, pp. 40-69, Amer. Math. Soc. (1964).

[14] Van Kampen, N. G., "*Stochastic Differential Equations*", Physics Reports, Vol. 24C, No. 3, pp. 172-228, North Holland Publishing Company, March 1976.

[15] Frisch, U., "*Wave Propagation in Random Media*", Probabilistic Methods in Applied Mathematics I, Ed. Bharucha-Reid, pp. 75-191 (1968).

[16] Hoffman, W. C., "*Wave Propagation in a Random Medium*", Proc. Sympos. in Appl. Math., Vol. XVI, pp. 117-144, Amer. Math. Soc. (1964).

[17] Chen, K. K. and Soong, T. T., "*Covariance Properties of Waves Propagating in a Random Medium*", J. Acoust. Soc. Amer. Vol. 49, pp. 1639-1642 (1971).

[18] Papanicolaou, G. C., "*Wave Propagation in a One-Dimensional Random Medium*", SIAM J. Math., Vol. 21, pp. 13-18 (1971).

[19] Lerche, I. and Parker, E. N., "*Random Function Theory Revisited: Exact Solutions Versus the First Order Smoothing Conjecture*", J. Math. Phys., Vol. 16, No. 9, pp. 1838-1839, Sept. 1975, and Vol. 14, p. 1949, 1973.

[20] Molyneux, J. E., "*Wave Propagation in Certain One-Dimensional Random Media*", J. Math. Phys., Vol. 13, pp. 58-69 (1972).

[21] LoDato, V. A., "*The Renormalized Projection Operator Technique for Linear Stochastic Differential Equations*", J. Math. Phys., Vol. 14, No. 3, pp. 340-345, March 1973.

[22] Kraichnan, R. H., "*Dynamics of Nonlinear Stochastic Systems*", J. Math. Phys., Vol. 2, No. 1, pp. 124-148, 1961.

[23] Herring, V. R., "*Statistical Theory of Thermal Convection at Large Prandtl Number*", Phys. Fluids, Vol. 12, No. 1, p. 39, Jan. 1969.

[24] Elrod, M., "*Numerical Methods for Stochastic Differential Equations*", Ph.D. Dissertation, University of Georgia, 1973.

[25] Adomian, G., "*Nonlinear Stochastic Differential Equations*", J. of Math. Anal. and Applic., Vol. 55, No. 2, pp. 441-452, Aug. 1976.

[26] Adomian, G., "*The Solution of Linear and Nonlinear Stochastic Systems*", Norbert Wiener Memorial Volume, Published by the World Organization of General Systems and Cybernetics, edited by J. Rose, Blackburn College, London, 1976, to appear.

[27] Adomian, G., "*Obtaining First and Second Order Statistics in Stochastic Differential Equations for the Nonlinear Case*", Math. Bull. of Acad. of Sci. -- Isvestia Matematika Akademi Nauk, Armenskoi CCP, U.S.S.R., Vol. X, No. 6, 1975.

[28] Adomian, G. and Sibul, L. H., "*Stochastic Green's Formula and Application to Stochastic Differential Equations*", J. Math. Anal. and Applic., Vol. 60, pp. 1-4, Sept. 1977.

[29] Miller, K. S., "*Theory of Differential Equations*", Norton Co., New York.

[30] Adomian, G. and Lynch, T. E., "*On the Stochastic Operator Equation with Random Initial Condtions*", J. Math. Anal. and Applic., to appear.

[31] Adomian, G. and Sibul, L. H., "*Symmetrized Solutions for Nonlinear Stochastic Differential Equation*", J. of Nonlinear Anal. and Applic., to appear.

POSITIVE OPERATORS AND STURMIAN THEORY OF NONSELFADJOINT SECOND-ORDER SYSTEMS

Shair Ahmad
Oklahoma State University

and

Alan C. Lazer
University of Cincinnati

I. INTRODUCTION

The purpose of this paper is to extend Sturmian theory, which originated in [13], to systems of the form

$$x''(t) + P(t)x(t) = 0 \tag{S}$$

where $P(t)$ is an $n \times n$ matrix with non-negative elements. Actually all of our results will be true if it is only assumed that the off-diagonal elements of $P(t)$ are non-negative. Indeed, in this case, by making a simple change of both the dependent and independent variables in (S), one can transform (S) into another system of the same form in which all elements of the new matrix are non-negative. Moreover, this transformation preserves zeros of solutions. We refer the reader to [2, p. 18] for details.

Although the methods used in this paper are similar to those used in [3], all but one of the theorems concerning (S), given here, appear to be new. Subsequent to the announcement of the results which will appear in [3] (See [1]), K. Schmitt and H. Smith [12] considered systems of the form (S) where $P(t)$ satisfies the condition: $P(t)K \subseteq K$ for all t, where K is some

ISBN 0-12-434160-8

fixed cone in n-space. The new theorems appearing in this paper can also be extended to such systems provided that the cone K has certain properties. We have also learned that S. Cheng [5] has recently obtained some of the results of [3] under the stronger assumption that all of the elements of the matrix $P(t)$ are strictly positive.

There is an extensive literature dealing with zeros of solutions of differential systems which include the type (S) if $P(t)$ is symmetric. (See for example [6], [10], and [11].) However, the theorems given here are new even if P is symmetric.

Our main tool in studying (S) will be an extremal characterization of positive characteristic values of a class of linear positive operators defined on ordered Banach spaces. Although this extremal characterization is simple, it does not seem to be given in any of the classical references dealing with linear positive operators. (See [7], [8] and [9].)

II. SOME KNOWN RESULTS ABOUT POSITIVE LINEAR OPERATORS

Let E be a real Banach space. A closed set $K \subseteq E$ is a cone if the following conditions are satisfied:

(A) if $x \in K$ and $y \in K$ then $x + y \in K$;

(B) if $x \in K$ and $t \geqslant 0$ then $tx \in K$;

(C) if $x \in K$ and $x \neq 0$ then $-x \notin K$.

Given a Banach space with a cone K and $x,y \in E$ we write $x \leqslant y$ or $y \geqslant x$ if $y - x \in K$. Thus, given $x \in E$, $x \in K$ iff $x \geqslant 0$.

The proof of the following result follows readily from the above definitions. (See, for example, [7, p. 241]).

Lemma 1. Let $x \in K$ and $u \in K$ with $u \neq 0$. Then there exists a number γ_0 such that

$$x - tu \in K \quad \text{if} \quad t \leqslant \gamma_0,$$

$$x - tu \notin K \quad \text{if} \quad t > \gamma_0.$$

If E is a Banach space with a Cone K and $A\colon E \to E$ is linear then A is positive (with respect to K) if $A(K) \subseteq K$. If A is positive then $x \in K$ is called a positive characteristic vector of A if $x \in K$, $x \neq 0$, and $x = \lambda Ax$ for some real number λ, which is necessarily positive. The number λ is called a positive characteristic value of A corresponding to x.

The following result gives a condition for the existence of a positive characteristic vector of A. (See [7, p. 257] or [8, p. 67]).

Theorem 1. Let A be linear, positive, and completely continuous. Suppose there exists $u \in E$ such that $-u \notin K$, $u = v - w$ with $v, w \in K$, and there exists a number $c > 0$ and an integer p such that $cA^p u \geqslant u$. Then A has a characteristic vector $x_0 \in K$:

$$x_0 = \lambda_0 A x_0$$

where the positive characteristic value λ_0 satisfies $\lambda_0 \leqslant (c)^{1/p}$.

III. STRICTLY POSITIVE LINEAR OPERATORS

A cone $K \subset E$ is called solid if it contains interior points. If $A\colon E \to E$ is linear, A will be called strictly positive (with respect to the solid cone K) if $x \in K$ and $x \neq 0$ implies that Ax is in the interior of K. As a consequence of Theorem 1 we have

Theorem 2. Let A be linear, completely continuous, and strictly positive with respect to the solid cone $K \subset E$. There exists $x_0 \in K$ with $x_0 \neq 0$ and $\lambda_0 > 0$ such that

$$x_0 = \lambda_0 A x_0 .$$

If $x_1 \in K$, $x_1 \neq 0$, and

$$x_1 = \lambda_1 A x_1,$$

then $\lambda_0 = \lambda_1$ and $x_0 = \lambda_0 x_1$ for some $\gamma_0 > 0$.

Proof: Let $v \in K$ and $v \neq 0$. Since Av is an interior point of K there exists $s > 0$ such that $Av - sv \in K$ or $(1/s)Av \geqslant v$. Therefore, the hypotheses of Theorem 1 are satisfied with $c = 1/s$, $p = 1$, and $w = 0$. This proves the first assertion of the theorem. Suppose $x_j = \lambda_j A x_j$ with $x_j \in K$, $x_j \neq 0$ for $j = 0, 1$ and $\lambda_1 \geqslant \lambda_0$ where $\lambda_0 > 0$, by necessity. By Lemma 1, there exists γ_0 such that $x_1 - tx_0 \in K$ iff $t \leqslant \gamma_0$. Clearly $\gamma_0 > 0$ and $x_1 - \gamma_0 x_0 \in \partial K$. We have

$$x_1 - \gamma_0 x_0 = A(\lambda_1 - \lambda_0)x_1 + A\lambda_0(x_1 - \gamma_0 x_0). \tag{1}$$

Now, $A\lambda_0(x_1 - \gamma_0 x_0) \in$ Interior K unless $x_1 = \gamma_0 x_0$, and $A(\lambda_1 - \lambda_0)x_1 \in$ Interior K unless $\lambda_1 = \lambda_0$. Since $K + K \subseteq K$, and for $y \in E$ the translation T_y: $E \to E$ defined by $T_y(x) = x + y$ is a homeomorphism, $K +$ Interior $K \subseteq$ Interior K. Since $x_1 - \gamma_0 x_0 \in \partial K$, it follow from (1) that $x_1 = \gamma_0 x_0$ and $\lambda_1 = \lambda_0$. This proves the result.

Given an operator A which satisfies the hypotheses of Theorem 2, we denote by $\lambda_0(A)$ the unique positive number such that there exists $x_0 \in K$ with $x_0 \neq 0$ and $x_0 = \lambda_0(A)Ax_0$.

The following result gives a useful extremal characterization of $\lambda_0(A)$. This characterization will be the unifying principle in our treatment of some diverse results in Sturmian theory for second-order linear differential systems.

Theorem 3. Let A and K satisfy the hypotheses of Theorem 2. Let $\Lambda(A)$ be the set of numbers $\lambda > 0$ such that $\lambda \in \Lambda(A)$ if and only if $x \leqslant \lambda Ax$ for some $x \in K$ with $x \neq 0$. Then $\lambda_0(A) = \inf\{\lambda \mid \lambda \in \Lambda(A)\}$.

Proof. Let $\lambda \in \Lambda(A)$ and suppose $v \in K$, $v \neq 0$, and $v \leqslant \lambda Av$. It follows that the hypotheses of Theorem 1 are satisfied with $c = \lambda$, $u = v$, $w = 0$, and $p = 1$. Therefore there exists $\lambda_1 > 0$ and $x_1 \in K$ with $x_1 \neq 0$ such that $x_1 = \lambda_1 Ax_1$ and $\lambda_1 \leqslant \lambda$. Since, according to Theorem 2, $\lambda_1 = \lambda_0(A)$, this proves the result.

Remark. Let E be a Banach space with a cone K, which

is not necessarily solid, and $A\colon E \to E$ a completely continuous linear operator, which is positive with respect to K. If A has a positive characteristic value and $\lambda_0(A)$ denotes the smallest positive characteristic value, then the above proof shows that $\lambda_0(A)$ has the same extremal characterization.

Although this simple result does not seem to be stated explicitly in the literature, Bellman [4, p. 288] has given an analogous extremal characterization of the Frobenius-Perron eigenvalue of a strictly positive matrix.

In the applications the following elementary consequence of Theorem 3 will be useful.

Corollary 3.1. Let A and K satisfy the same hypotheses as in Theorem 3. If $\lambda > 0$ and there exists $x \in K$ with $x \leqslant \lambda Ax$ but $x \neq \lambda Ax$, then $\lambda_0(A) < \lambda$.

Proof. Since $\lambda Ax - x \in K$, $\lambda Ax - x \neq 0$, it follows that $A(\lambda Ax - x) \in$ Interior K. Consequently, if $y = Ax \neq 0$, then $\lambda Ay - y \in$ Interior K, so there exists λ_1 with $0 < \lambda_1 < \lambda$ such that $\lambda_1 Ay - y \in K$ or equivalently $y \leqslant \lambda_1 Ay$. According to Theorem 3, $\lambda_0(A) \leqslant \lambda_1 < \lambda$ and the corollary is proved.

IV. TWO BASIC EXAMPLES

We identify Euclidean n-space R^n with the set of all real $n \times 1$ matrices. If $x = \text{col}\,(x_1, \ldots, x_n) \in R^n$ we write $x \geqslant 0$ or $0 \leqslant x$ if $x_k \geqslant 0$ for $1 \leqslant k \leqslant n$, and we write $x > 0$ or $0 < x$ if $x_k > 0$ for $1 \leqslant k \leqslant n$. If $x,y \in R^n$ we write $x \leqslant y$ or $y \geqslant x$ if $y - x \geqslant 0$ and we write $x < y$ or $y > x$ if $y - x > 0$. If $a < b$ we let $C^1([a,b], R^n)$ denote the Banach space of functions $u\colon [a,b] \to R^n$ such that the components of u are continuously differentiable on $[a,b]$ with norm $|\;|$ given by

$$|u| = \max_{t\in[a,b]} \max_{1\leq k\leq n} \max \{|u_k(t)|, |u_k'(t)|\} \text{ where } u = \text{col}(u_1,\ldots,u_n).$$

We define subsets $K_0^n\,[a,b]$ and $K_1^n\,[a,b]$ of $C^1([a,b],R^n)$ as follows:

(i) $u \in K_0^n[a,b]$ iff $0 \leqslant u(t)$ for all $t \in [a,b]$ and $u(a) = u(b) = 0$;

(ii) $u \in K_1^n[a,b]$ iff $0 \leqslant u(t)$ for all $t \in [a,b]$ and $u(b) = 0$.

The proof of the next result follows easily from the definitions.

Lemma 4. The sets $K_0^n[a,b]$ and $K_1^n[a,b]$ are solid cones in $C^1([a,b], R^n)$. A point $u \in K_0^n[a,b]$ is an interior point of $K_0^n[a,b]$ iff $u'(a) > 0$, $u'(b) < 0$ and $u(t) > 0$ for $a < t < b$. A point $u \in K_1^n[a,b]$ is an interior point of $K_1^n[a,b]$ iff $u(t) > 0$ for $a \leqslant t < b$ and $u'(b) < 0$.

Let $Q(t)$ be an $n \times n$ matrix defined on $[a,b]$ such that if $Q(t) = (q_{ij}(t))$ then $q_{ij}(t)$ is continuous and $q_{ij}(t) > 0$ for $1 \leqslant i,j \leqslant n$. Let $G_0\,(s,t,a,b)$ be defined for $(s,t) \in [a,b] \times [a,b]$ by

$$G_0(s,t,a,b) = \begin{cases} \dfrac{(s-a)\,(b-t)}{b-a}, & a \leqslant s \leqslant t \leqslant b \\[2ex] \dfrac{(t-a)\,(b-s)}{b-a}, & a \leqslant t \leqslant s \leqslant b. \end{cases} \tag{2}$$

If $u\colon\ [a,b] \to R^n$ is continuous we define

$$(Au)(t) = \int_a^b G_0(s,t,s,a)\ Q(s)u(s)ds. \tag{3}$$

A simple computation shows that $Au \in C^2([a,b], R^n)$ and

$$(Au)''(t) = -Q(t)u(t),\ (Au)(a) = (Au)(b) = 0. \tag{4}$$

Lemma 5. If $u\colon\ [a,b] \to R^n$ is continuous, if $u(t) \geqslant 0$ for all $t \in [a,b]$ and if $u(t) \not\equiv 0$ then

$$A(u) \in \text{Interior}\ K_0^n[a,b].$$

If A is considered as a linear operator on $C^1([a,b], R^n)$ then A is completely continuous and strictly positive with respect to $K_0^n[a,b]$.

Proof. If $u(t) \geqslant 0$ for $t \in [a,b]$ and $u(t) \neq 0$ then

$Q(s)u(s) \geqslant 0$ for $s \in [a,b]$ and there exists $s_1 \in (a,b)$ such that $Q(s_1)u(s_1) > 0$. From (2) we see that $G_0(s,t,a,b) > 0$ for $a < s < b$ and $a < t < b$. Hence for $a < t < b$,

$$(Au)(t) = \int_a^b G_0(s,t,a,b)\ Q\ (s)u(s)ds > 0.$$

Simple computations show that

(5) $$(Au)'(a) = \int_a^b \left(\frac{b - s}{b - a}\right) Q(s)u(s)ds > 0$$

and

(6) $$(Au)'(b) = -\int_a^b \left(\frac{s - a}{b - a}\right) Q(s)u(s)ds < 0.$$

Since $Au \in C^2([a,b], R^n)$, it follows from Lemma 4 that $Au \in$ Interior $K_0^n[a,b]$. To see that $A\colon C^1([a,b], R^n) \to C^1([a,b], R^n)$ is completely continuous, let $\{u_m\}_1^\infty$ be a sequence in $C^1([a,b], R^n)$ which is bounded with respect to the norm defined above. Since the second derivatives $(Au_m)''(t) = -Q(t)u_m(t)$ are uniformly bounded on $[a,b]$ the sequence $\{(Au_m)'\}_1^\infty$ is equicontinuous on $[a,b]$. Moreover, according to (5), the sequence $\{(Au_m)'(a)\}_1^\infty$ is bounded. Hence, by Ascoli's lemma some subsequence of $\{(Au_m)'\}_1^\infty$ converges uniformly on $[a,b]$. Since $(Au_m)(a) = 0$ for all m, the subsequence converges with respect to the norm in $C^1([a,b], R^n)$ and the Lemma is proved.

To give our second basic example we define $G_1(s,t,a,b)$ for $(s,t) \in [a,b] \times [a,b]$ by

(7) $$G_1(s,t,a,b) = \begin{cases} b - t, & a \leqslant s \leqslant t \leqslant b, \\ b - s, & a \leqslant t \leqslant s \leqslant b. \end{cases}$$

If $u\colon [a,b] \to R^n$ is continuous, we define for $a \leqslant t \leqslant b$.

(8) $$(Bu)(t) = \int_a^b G_1(s,t,a,b)Q(s)u(s)ds.$$

A trivial computation shows that $Bu \in C^2([a,b], R^n)$ and

(9) $$(Bu)''(t) = -Q(t)u(t), \quad u'(a) = u(b) = 0.$$

Moreover,

(10) $$(Bu)'(b) = - \int_a^b Q(s)u(s)ds.$$

Using (7), (8), (9), (10), Lemma 4 and the same reasoning as in the proof of Lemma 5 we establish

Lemma 6. If $u: [a,b] \to R^n$ is continuous, if $u(t) \geqslant 0$ for all $t \in [a,b]$, and if $u(t) \not\equiv 0$, then $B(u) \in$ Interior $K_1^n[a,b]$. If B is considered as a linear operator on $C^1([a,b], R^n)$ then B is completely continuous and strictly positive with respect to $K_1^n[a,b]$.

From Lemma 4, Lemma 5, and Theorem 2 we arrive at the main result of this section.

Theorem 4. Let $Q(t)$ be an $n \times n$ matrix whose elements are positive and continuous on the interval $[a,b]$. There exist numbers $\lambda_0(a,b) > 0$ and $\mu_0(a,b) > 0$ and $x(t), y(t) \in C^1([a,b], R^n)$ such that $x(t) = \lambda_0(a,b)\ (Ax)(t)$, $y(t) = \mu_0(a,b)(By)(t)$, and $x(t) > 0$, $y(t) > 0$ for $t \in (a,b)$. Consequently,

(11) $$x''(t) + \lambda_0(a,b)Q(t)x(t) = 0, \quad x(a) = x(b) = 0,$$

(12) $$y''(t) + \mu_0(a,b)Q(t)y(t) = 0, \quad y'(a) = y(b) = 0.$$

V. MONOTONICITY AND CONTINUITY PROPERTIES OF $\lambda_0(a,b)$ AND $\mu_0(a,b)$

In this section we again let Q satisfy the condition of Theorem 4 but we consider the effect of varying the endpoints of the interval. By means of Theorem 4 we can consider $\lambda_0(c,d)$ and $\mu_0(c,d)$ to be defined for $a \leqslant c < d \leqslant b$ with the same Q.

Lemma 7. $\mu_0(a,b) < \lambda_0(a,b)$.

Proof. From (2) and (7) we see at once that $G_0(s,t,a,b) < G_1(s,t,a,b)$ if $a < s < b$ and $a < t < b$. Consequently if $x \in K_0^n[a,b]$ is such that $x(t) \not\equiv 0$ and $x(t) = \lambda_0(a,b)(Ax)(t)$, then for $a < t < b$, $x(t) = \lambda_0(a,b) \int_a^b G_0(s,t,a,b)\ Q(s)x(s)ds < \lambda_0(a,b) \int_a^b G_1(s,t,a,b)Q(s)x(s)ds = \lambda_0(a,b)(Bx)(t)$. We have $x \in K_0^n[a,b] \subset K_1^n[a,b]$, and the above shows, that with respect to the ordering defined by $K_1^n[a,b]$, $x \leqslant \lambda_0(a,b)\ Bx$, $x \neq \lambda_0(a,b)Bx$.

Recalling that $\mu_0(a,b)$ is the unique positive characteristic value of B, it follows from Corollary 3.1 that

$$\mu_0(a,b) < \lambda_0(a,b).$$

Lemma 8. If $[c,d] \subset [a,b]$ and $[c,d] \neq [a,b]$, then $\lambda_0(a,b) < \lambda_0(c,d)$. If $a < b' < b$, then $\mu_0(a,b) < \mu_0(a,b')$.

Proof. Using the definition (2), a trivial calculation shows that if $a \leqslant c < d \leqslant b$ but $[c,d] \neq [a,b]$ then whenever $s,t \in [c,d] \cap (a,b)$

(13) $$G_0(s,t,c,d) < G_0(s,t,a,b).$$

Let $x \in K_0^n[c,d]$ satisfy

(14) $x(t) = \lambda_0(c,d) \int_c^d G\ (s,t,c,d)Q(s)x(s)ds$, with $x(t) > 0$ on $[c,d]$.

$$\text{Define } \hat{x}(t) = \begin{cases} 0, & a \leqslant t \leqslant c \\ x(t), & c \leqslant t \leqslant d \\ 0, & d \leqslant t \leqslant b. \end{cases}$$

From (2), (13) and (14) we see that for $a < t < b$, $\hat{x}(t) < \lambda_0(c,d) \int_a^b G_0(s,t,a,b)Q(s)\ \hat{x}(s)ds = \lambda_0(c,d)(A\hat{x})(t)$. If $w = A\hat{x}$ then according to Lemma 5 and the above, $A(\lambda_0(c,d)w - \hat{x}) \in$ Interior $K_0^n[a,b]$. Therefore, with respect to the ordering defined by $K_0^n[a,b]$, $w \leqslant \lambda_0(c,d)Aw$ and $w \neq \lambda_0(c,d)Aw$. Hence, from Corollary 3.1 $\lambda_0(a,b) < \lambda_0(c,d)$.

The final assertion of the lemma is proved in a similar manner.

Lemma 9. For $a \leqslant t \leqslant b$, let $\|Q(t)\| = \max_{1 \leq i \leq n} \Sigma_{j=1}^n q_{ij}(t)$. If $a \leqslant c \leqslant d \leqslant b$, then

(15) $$\lambda_0(c,d) \geqslant 4/(d - c) \int_c^d \|Q(t)\|\ dt$$

and

(16) $$\mu_0(c,d) \geqslant 1/(d - c) \int_c^d \|Q(t)\|\ dt.$$

Proof: Let $x \in K_0^n(a,b)$, with $x \not\equiv 0$, satisfy

$$x(t) = \lambda_0(c,d) \int_c^d G_0(s,t,c,d)\, Q(s)x(s)ds.$$

Let $1 \leqslant k \leqslant n$ and $\bar{t} \in (a,b)$ be such that $x_k(\bar{t})$

$$= \max_{t\in[a,b]} \max_{1\leq j\leq n} x_j(t).$$

From (2) we see that if $c \leqslant s \leqslant d$ and $c \leqslant t \leqslant d$, then

$$G_0(s,t,c,d) \leqslant G_0(t,t,c,d) = \frac{(t - c)(d - t)}{d - c}, \text{ thus}$$

$$G_0(s,t,c,d) \leqslant (d - c)/4.$$

Therefore,

$$x_k(\bar{t}) = \lambda_0(c,d) \int_c^d G_0(s,\bar{t},c,d) \sum_{j=1}^n q_{kj}(s)x_j(s)ds \leqslant$$

$$\lambda_0(c,d)\left(\frac{d - c}{4}\right)x_k(\bar{t}) \int_c^d \sum_{j=1}^n q_{kj}(s)ds \leqslant \lambda_0(c,d)\left(\frac{d - c}{4}\right) x_k(\bar{t}) \int_c^d$$

$\|Q(s)\|\, ds$ and (15) follows.

Lemma 10. For fixed c, $\lambda_0(c,d)$ and $\mu_0(c,d)$ depend continuously on d for $d > c$.

Proof: From Lemma 8 it follows that if $c < d < d'$ then $\lambda_0(c,d) > \lambda_0(c,d')$. Therefore if $c < \bar{d}$, then both the right-hand and left-hand limits $\lambda_0(c, \bar{d} + 0)$ and $\lambda_0(c,\bar{d} - 0)$ exist. To establish the first claim we must show that $\lambda_0(c,\bar{d} + 0) = \lambda_0(c,\bar{d} - 0) = \lambda_0(c,d)$. To this end, let $\{d_m\}_1^\infty$ be a sequence of numbers such that $\bar{d} < d_{m+1} < d_m$ for all m and $\lim_{m\to\infty} d_m = \bar{d}$. Let $\bar{\lambda} = \lim_{m\to\infty} \lambda_0(c,d_m)$. For each $m \geqslant 1$, let $x_m \in K_0^n[c,d_m]$, with $x_m \not\equiv 0$, satisfy $x_m''(t) + \lambda_0(c,d_m)\, Q(t)x_m(t) = 0$, $c \leqslant t \leqslant d_m$, $x_m(c) = x_m(d_m) = 0$ (see Theorem 4). Since $x_m'(c) \neq 0$, we may assume, by multiplying x_m by a suitable positive constant, that $|x_m(c)| = 1$ where $|\ |$ is the Euclidean norm. Let $\{x_{mk}(c)\}_{k=1}^\infty$ be a convergent subsequence and $v = \lim_{k\to\infty} x_{mk}(c)$ so that $|v| = 1$. If $u(t)$ is the solution of the initial-value problem

$$u''(t) + \bar{\lambda}Q(t)u(t) = 0, \quad u(c) = 0, \quad u'(c) = v,$$

then, by standard results concerning continuity of solutions of differential equations with respect to initial conditions and parameters, it follows that $\lim_{k\to\infty} x_{mk}(t) = u(t)$ uniformly on $[c,b]$. Therefore, since $x_{mk}(t) \geqslant 0$ for $c \leqslant t \leqslant d_{mk}$, $u(t) \geqslant 0$ for $c \leqslant t \leqslant \bar{d}$. Moreover, $u(\bar{d}) = \lim_{k\to\infty} x(d_{mk}) = 0$. Hence $u(t) = \bar{\lambda} \int_c^d G(s,t,c,\bar{d})Q(s)u(s)ds$. Since $u \in K_0^n[c,\bar{d}]$ and $u'(c) \neq 0$, it follows from the uniqueness part of Theorem 2 that $\bar{\lambda} = \lambda_0(c,\bar{d})$. This proves that $\lambda_0(c,\bar{d}+0) = \lambda_0(c,\bar{d})$ and the proof that $\lambda_0(c,\bar{d}-0) = \lambda_0(c,\bar{d})$ is similar.

The proof of the assertion regarding $\mu_0(c,d)$ follows in a parallel manner.

Lastly, we consider the effect of varying the matrix Q.

Lemma 11. Let $Q(t) = (q_{ij}(t))$ and $\hat{Q}(t) = (\hat{q}_{ij}(t))$ be $n \times n$ matrix functions defined on the interval $[a,b]$ such that the elements of both $Q(t)$ and $\hat{Q}(t)$ are positive and continuous on $[a,b]$. Let A and $\hat{A}$ be the linear operators defined on $C^1([a,b], R^n)$ by

$$(Ax)(t) = \int_a^b G_0(s,t,a,b)\, Q(s)x(s)ds$$
$$(\hat{A}x)(t) = \int_a^b G_0(s,t,a,b)\, \hat{Q}(s)x(s)ds.$$

If $\hat{q}_{ij}(t) > q_{ij}(t)$ for $1 \leqslant i,j \leqslant n$ and $t \in [a,b]$ and if $\lambda_0(a,b)$ and $\hat{\lambda}_0(a,b)$ denote the unique positive characteristic values of A and $\hat{A}$ respectively with respect to $K_0^n[a,b]$ then $\hat{\lambda}_0(a,b) < \lambda_0(a,b)$. Similarly, if B and $\hat{B}$ are the operators defined on $C^1([a,b], R^n)$ by

$$(Bx)(t) = \int_a^b G_1(s,t,a,b)\, Q(s)\, x(s)ds$$

and

$$(\hat{B}x)(t) = \int_a^b G_1(s,t,a,b)\, \hat{Q}(s)\, x(s)ds,$$

and if $\mu_0(a,b)$ and $\hat{\mu}_0(a,b)$ denote the unique positive characteristic values of B and $\hat{B}$, with respect to $K_1^n[a,b]$ then $\hat{\mu}_0(a,b) < \mu_0(a,b)$.

Proof: Let $x \in K_0^n[a,b]$, with $x \not\equiv 0$, satisfy

$$x(t) = \lambda_0(a,b)(Ax)(t) = \lambda_0(a,b) \int_a^b G_0(s,t,a,b)Q(s)x(s)ds.$$

Since $G_0(s,t,a,b) > 0$ for $a < s < b$, $a < t < b$, and $x \in$ Interior $K_0^n[a,b]$, it follows that for $a < t < b$,

$$x(t) < \lambda_0(a,b) \int_a^b G_0(s,t,a,b)\, \hat{Q}(s)x(s)ds = \lambda_0(a,b)(\hat{A}x)(t).$$

Since, with respect to the ordering defined by $K_0^n[a,b]$, $x \leqslant \lambda_0(a,b)\hat{A}x$ and $x \neq \lambda_0(a,b)\hat{A}x$, Corollary 3.1 implies that $\hat{\lambda}_0(a,b) < \lambda_0(a,b)$. The proof that $\hat{\mu}_0(a,b) < \mu_0(a,b)$ is similar.

VI. STURMIAN THEORY FOR $x'' + P(t)x = 0$

Our first application of the preceeding development is an extension of the Sturm separation theorem.

Theorem 5. Let $P(t) = (p_{ij}(t))$ be a continuous $n \times n$ matrix function whose elements are non-negative on $[a,b)$, where b may equal $+\infty$]. Suppose that there exists no non-trivial solution $x(t)$ of

$$x''(t) + P(t)\,x(t) = 0 \tag{17}$$

such that $x(a) = x(c) = 0$ whenever $a < c < b$. Then there exists no nontrivial solution $x(t)$ of (17) such that $x(t_1) = x(t_2) = 0$ if $a < t_1 < t_2 < b$.

Proof: Suppose on the contrary that there exists a nontrivial solution $x(t)$ of (17) with $x(t_1) = x(t_2) = 0$ and $a < t_1 < t_2 < b$. We have $x(t) = \int_{t_1}^{t_2} G_0(s,t,t_1,t_2)\, P(s)x(s)ds$ for $t_1 \leqslant t \leqslant t_2$. If $x(t) = \text{col}\,(x_1(t), \ldots, x_n(t))$ and $w(t) = \text{col}\,(|x_1(t)|, \ldots, |x_n(t)|)$, then since $G_0(s,t,t_1,t_2)$ and the elements of P are non-negative we see that for $t_1 \leqslant t \leqslant t_2$,

$$w(t) \leqslant \int_{t_1}^{t_2} G_0(s,t,t_1,t_2)\, P(s)w(s)ds. \tag{18}$$

For each integer $m \geqslant 1$ let $P_m(t)$ be the $n \times n$ matrix defined by $P_m(t) = (P_{ij}(t) + \frac{1}{m})$. Since the elements of $P_m(t)$ are strictly positive it follows that if $[c,d] \subset [a,b]$, then there exists a unique positive number $\lambda_{om}(c,d)$ such that the equation

$$u(t) = \lambda_{om}(c,d) \int_c^d G_0(s,t,c,d)\, P_m(s)u(s)ds$$

has a nontrivial solution $u \in K_0^n[c,d]$. Let m be fixed. As the elements of $P_m(t)$ are greater than the corresponding ele- elements of $P(t)$, we infer from (17) that for $a \leqslant t \leqslant b$,

$$w(t) \leqslant \int_{t_1}^{t_2} G_0(s,t,t_1,t_2)\, P_m(s)w(s)ds.$$

If

$$v(t) = \int_{t_1}^{t_2} G_0(s,t,t_1,t_2)\, P_m(s)w(s)ds$$

for $a \leqslant t \leqslant b$, then since $w(t) \leqslant v(t)$,

(19) $$v(t) \leqslant \int_{t_1}^{t_2} G_0(s,t,t_1,t_2)\, P_m(s)v(s)ds.$$

Since $x \not\equiv 0$, $w \not\equiv 0$; hence $v \not\equiv 0$. Therefore, as $v \in K_0^n[t_1,t_2]$, we infer from Theorem 3 that $\lambda_{om}(t_1,t_2) \leqslant 1$. From the inclusion $[t_1,t_2] \subset [a,t_2]$ and Lemma 8 it follows that $\lambda_{om}(a,t_2) < 1$. By Lemma 9, $\lambda_{om}(a,d) \to +\infty$ as $d \to a$ and, according to Lemma 8, $\lambda_{om}(a,d)$ is a continuous, strictly-decreasing function of d. Hence, there exists a unique $c_m \in (a,t_2)$ such that $\lambda_{om}(a,c_m) = 1$ and this implies the existence of a nontrivial solution $u_m(t)$ of

(20) $$u_m''(t) + P_m(t)\, u_m(t) = 0$$

such that

(21) $$u_m(a) = u_m(c_m) = 0.$$

Letting m vary, without loss of generality we may assume that $|u_m'(a)| = 1$ for all m.

Since $m_2 > m_1$ implies that every element of $P_{m2}(t)$ is less than the corresponding element of $P_{m1}(t)$, it follows from Lemma 11 that $\lambda_{om_2}(a,c_{m1}) > \lambda_{om_1}(a,c_{m1}) = 1$. Hence, by Lemma 8, $c_{m1} < c_{m2}$. Let $c = \lim_{m\to\infty} c_m$; clearly $c \leqslant t_2$. Let $\{u_{m\ell}(a)\}$ be a subsequence of $\{u_m(a)\}$ which converges to v with $|v| = 1$. If $u(t)$ is the nontrivial solution of the initial value problem

$$u''(t) + P(t)u(t) = 0, \quad u(a) = 0, \quad u'(a) = v,$$

then, since $P_m(t)$ converges uniformly to $P(t)$, $\lim_{\ell\to\infty} u_{m\ell}(t) = u(t)$ uniformly on compact subintervals of $[a,c)$. In particular, $u(c) = \lim_{\ell\to\infty} u_{m\ell}(c_{m\ell}) = 0$. Since $c \leqslant t_2 < b$, this contradicts the hypotheses of Theorem 5 and the proof is complete.

The following result gives another condition which implies the assertion of Theorem 5.

Theorem 6. Let $P(t)$ satisfy the same hypotheses as in Theorem 5. Suppose that there exists no nontrivial solution $x(t)$ of (17) such that $x'(a) = x(c) = 0$ whenever $a < c < b$. Then there exists no nontrivial solution $x(t)$ of (17) such that $x(t_1) = x(t_2) = 0$ if $a < t_1 < t_2 < b$.

Proof: Assume on the contrary that there exist numbers t_1 and t_2 with $a < t_1 < t_2 < b$ and a nontrivial solution $x(t)$ of (17) with $x(t_1) = x(t_2) = 0$. Defining the matrices $P_m(t)$ and the numbers $\lambda_{0m}(c,d)$ as in the proof of Theorem 5, the exact same reasoning as in the proof of Theorem 5 implies that there exists a sequence of numbers $\{c_m\}_1^\infty \subset (a,t_2)$ with $c_m < c_{m+1}$ such that $\lambda_{0m}(a,c_m) = 1$. If $\mu_{0m}(c,d)$ denotes the unique positive number such that there exists $x \in K_1^n[c,d]$ with $x \not\equiv 0$ and

$$x(t) = \mu_{0m}(c,d) \int_c^d G_1(s,t,c,d)\, P_m(s)x(s)ds,$$

then, according to Lemma 7, $\mu_{0m}(a,c_m) < \lambda_{0m}(a,c_m) = 1$. Using Lemmas (8), (9), and (10) and the same reasoning as in Theorem 5, we infer the existence of a unique number $d_m \in (a,c_m) \subset (a,t_2)$ such that $\mu_{0m}(a,d_m) = 1$. Moreover, since every element of $P_m(t)$ is strictly greater than the corresponding element of $P_{m+1}(t)$, Lemma 11 implies that $\mu_{0(m+1)}(a,d_{m+1}) > \mu_{0m}(a,d_m) = 1$, and hence $d_m < d_{m+1}$. For each integer $m \geqslant 1$ this implies the existence of a nontrivial solution $u_m(t)$ of the boundary value problem $x'' + P_m(t)x = 0$, $x'(a) = x(d_m) = 0$. If $\lim_{m\to\infty} d_m = d \leqslant t_2$, then the reasoning used in the proof of Theorem 5 implies the existence of a nontrivial solution of

$$u''(t) + P(t)u(t) = 0, \quad u'(a) = u(d) = 0.$$

This contradicts the hypothesis of the theorem and the result is proved.

Theorem 7. Let $P(t) = (p_{ij}(t))$ be a continuous $n \times n$ matrix function defined on $[a,b]$ whose elements are non-negative and suppose there exists a nontrivial solution $x(t)$ of $x'' + P(t)x = 0$ with $x'(a) = x(b) = 0$. If $Q(t) = (q_{ij}(t))$ is a continuous $n \times n$ matrix function defined on $[a,b]$ such that $q_{ij}(t) \geqslant p_{ij}(t)$ for $1 \leqslant i,j \leqslant n$ and all $t \in [a,b]$, and $q_{ij}(\bar{t}) > p_{ij}(\bar{t})$ for $1 \leqslant i,j \leqslant n$ and some $\bar{t} \in [a,b]$, then there exists a nontrivial solution $y(t)$ of $y'' + Q(t)y = 0$ with $y'(a) = y(c) = 0$ where $a < c < b$.

Proof: We use the same type of perturbational argument given in the proofs of Theorems 5 and 6. First, since

$$x(t) = \int_a^b G_1(s,t,a,b)\, P(s)x(s)ds,$$

it follows that if

$$w(t) = \text{col}\,(|x_1(t)|, \ldots, |x_n(t)|)$$

then

$$w(t) \leqslant \int_a^b G_1(s,t,a,b)\, P(s)w(s)ds.$$

Next, if for each integer $m \geqslant 1$ we define the matrix $Q_m(t)$ by $Q_m(t) = (q_{ij}(t) + \frac{1}{m})$ then

(22) $$w(t) \leqslant \int_a^b G_1(s,t,a,b)\, Q_m(s)w(s)ds.$$

Fix m for the time being. If for $a \leqslant t \leqslant b$,

$$v(t) = \int_a^b G_1(s,t,a,b)Q_m(s)w(s)ds$$

then the preceeding inequality gives

(23) $$v(t) \leqslant \int_a^b G_1(s,t,a,b)Q_m(s)v(s)ds.$$

Since $x \not\equiv 0$, $w \not\equiv 0$; and hence $v \not\equiv 0$. Therefore, if for $[c,d] \subseteq [a,b]$ we let $\mu_{0m}(c,d)$ denote the unique positive number such that there exists a non-zero $x \in K_1^n[c,d]$ satisfying $x(t) = \mu_{0m}(c,d) \int_c^d G_1(s,t,c,d)Q_m(s)x(s)ds$, then (23) and

Theorem 3 imply that $\mu_{0m}(a,b) \leqslant 1$. Now letting m vary and using the same reasoning as in the proof of Theorem 5, for each $m = 1, 2, \ldots$ we infer the existence of a nontrivial solution $u_m(t)$ of $u_m'' + Q_m(t)u_m = 0$ with $u_m'(a) = u_m(c_m) = 0$ with $a < c_m < b$. Moreover, $u_m(t) \geqslant 0$ for $t \in [a,c_m]$, $|u_m(a)| = 1$ for all m, and $c_m \leqslant c_{m+1}$. Using the same type of limit argument as before, we ascertain the existence of a nontrivial solution $u(t)$ of

$$u'' + Q(t)\, u = 0, \quad u'(a) = u(c) = 0$$

with $a < c \leqslant b$. If $c < b$, we are done, so we need only show that $c = b$ is impossible. Suppose then that $c = b$. We have

(24) $$u(t) = \int_a^b G_1(s,t,a,b)Q(s)u(s)ds$$

for $a \leqslant t \leqslant b$. Since $G_1(s,t,a,b) > 0$ for $a \leqslant s < b$, $a \leqslant t < b$, $u(t) \geqslant 0$ for $a \leqslant t \leqslant b$, the elements of $Q(t)$ are non-negative, the elements of $Q(\bar{t})$ are strictly positive, and the components of u cannot vanish simultaneously on any subinterval, (24) implies that

(25) $$u(t) > 0 \quad \text{for} \quad a \leqslant t < b.$$

Moreover, by (10)

(26) $$u'(b) = -\int_a^b Q(s)u(s)ds < 0.$$

It follows from (24) and the hypotheses of the theorem that

(27) $$u(t) > \int_a^b G_1(s,t,a,b)\, P(s)u(s)ds, \quad a \leqslant t < b$$

and

(28) $$u'(b) < -\int_a^b P(s)u(s)ds.$$

Since $u(b) = x(b) = 0$, (25) and (26) imply that if $\alpha > 0$ is small, then

(29) $$u_k(t) - \alpha x_k(t) > 0 \quad \text{if} \quad 1 \leqslant k \leqslant n, \quad a \leqslant t < b,$$

and

(30) $$u_k'(b) - \alpha x_k'(b) < 0 \quad \text{if} \quad 1 \leqslant k \leqslant n.$$

Since x is a nontrivial solution of $x'' + P(t)x = 0$ we may assume without loss of generality that some component of $x(t)$ is positive somewhere on $[a,b)$. Therefore (29) and (30) cannot hold for all $\alpha > 0$. If $\bar{\alpha} > 0$ is the least upper bound of all numbers α for which (29) and (30) hold, then

(31) $$u(t) - \bar{\alpha}x(t) \geqslant 0 \quad \text{if} \quad a \leqslant t < b,$$

and either:

(i) $u_i(\bar{t}) - \bar{\alpha}x_i(\bar{t}) = 0$ for some i and some $\bar{t} \in [a,b)$, or

(ii) $u_i'(b) - \bar{\alpha}x_i'(b) = 0$ for some i.

Since

$$x(t) = \int_a^b G_1(s,t,a,b)P(s)x(s)ds, \quad x'(b) = -\int_a^b P(s)x(s)ds,$$

it follows from (27), (28) and (31) that

$$u(t) - \bar{\alpha}x(t) > \int_a^b G_1(s,t,a,b)P(s)[u(s) - \bar{\alpha}x(s)]ds \geqslant 0$$

if $a \leqslant t < b$, and

$$u'(b) - \bar{\alpha}x'(b) < -\int_a^b P(s)[u(s) - \bar{\alpha}x(s)]ds \leqslant 0.$$

Therefore both (i) and (ii) are impossible. This contradiction shows that $c < b$ and the proof is complete.

By a similar argument one can establish

Theorem 8. Assume that the matrices P and Q satisfy the same conditions as in Theorem 7. If there exists a nontrivial solution $x(t)$ of $x'' + P(t)x = 0$ with $x(a) = z(b) = 0$, then there exists a nontrivial solution $u(t)$ of $u'' + Q(t)u = 0$ with $u(a) = u(c) = 0$, where $a < c < b$.

VII. REFERENCES

[1] Ahmad, S., and Lazer, A. C., "*Separation and comparison*", Notices Amer. Math. Soc., 76T-B205, 23 (1976), A-647.

[2] Ahmad, S., and Lazer, A. C., "*On the components of extremal solutions of second order systems*", SIAM J. Math. Anal., 8

(1977), pp. 16-23.

[3] Ahmad, S., and Lazer, A. C., "*An n-dimensional extension of the Sturm separation and comparison theory to a class of nonselfadjoint systems*", SIAM J. Math. Anal., to appear.

[4] Bellman, R., *Matrix Analysis*, 2nd Edition, McGraw-Hill, New York, (1970).

[5] Cheng, S., "*On the n-dimensional harmonic oscillator*", preprint.

[6] Coppel, W. A., *Disconjugacy*, Lecture Notes in Mathematics, Vol. 220, Springer-Verlag, Berlin, 1971.

[7] Krasnoselski, M. A., *Topological Methods in the Theory of Nonlinear Integral Equations*, Pergamon, New York, 1964.

[8] Krasnoselski, M. A., *Positive solutions of operator equations*, Noordhoff, Groningen, 1964.

[9] Krein, M. G., and Rutman, M. A., "*Linear operators leaving invariant a cone in a Banach space*", Uspehi Math - Nauk (N.S.), 3 (1948), No. 1 (23, 3-95, Russian (English Translation: Amer. Math. Soc. Tran. No. 26).

[10] Morse, M., *Variational Analysis: Critical Extremals and Sturmian Extensions*, John Wiley, New York, 1973.

[11] Reid, W. T., *Ordinary Differential Equations*, John Wiley, New York, 1971.

[12] Schmitt, K., and Smith, H. L., *Positive solutions and conjugate points for systems of ordinary differential equations*, preprint.

[13] Sturm, J. C. F., "*Memoire sur les equations differentielles lineaires de second ordre*", Journal de Mathematiques Pures et Appliquees, 1 (1836), pp. 106-186.

NONLINEAR EQUATIONS IN ABSTRACT SPACES

NONLINEAR SUPERPOSITION FOR OPERATOR EQUATIONS

W. F. Ames
Center for Applied Mathematics
University of Georgia

I. INTRODUCTION

The great utility of the ad hoc techniques of the linear theory, such as separability, rests upon the principle of (linear) superposition. In accordance with that principle, elementary solutions of the equations are combined to yield more flexible ones, namely solutions which can satisfy initial and boundary conditions arising from modeled phenomena. In nonlinear problems this (linear) principle does not apply. But that is no reason to assert that linear equations have the monopoly of the superposition principle. It is certainly not essential that the superposition be additive in order to obtain solutions of an equation by composing known solutions. All that is required is a knowledge of how to compose a proper number of solutions to arrive at other solutions. If such a composition is not linear we call it a <u>nonlinear superposition principle</u>. There exist classes of nonlinear, <u>and even linear,</u> equations which possess nonlinear superposition principles. Of course there is no universal nonlinear superposition - such a situation would be "beyond our fondest dreams".

The Riccati equation (Davis [1], Reid [2]), $y' + Qy + Ry^2 = P$, lies on the "boundary" between the linear and nonlinear worlds because by means of the transformation $y = u'/Ru$ the <u>linear</u> second order equation $Ru'' - (R' - QR)u' - PR^2u = 0$ is obtained. And, more to the point here, this Riccati equation has the important cross ratio theorem: <u>The cross ratio of any four linearly</u>

ISBN 0-12-434160-8

independent solutions of the Riccati equation is a constant. Thus if y_1, y_2, y_3 are three linearly independent solutions we can calculate a fourth, y_4 (and hence infinitely many), from the cross ratio

$$R = \frac{(y_1 - y_3)(y_2 - y_4)}{(y_1 - y_4)(y_2 - y_3)} \quad (R = \text{constant}) \tag{1.1}$$

without quadrature! This composition law is one of the earliest known nonlinear superposition principles.

The question of the existence of nonlinear superpositions is a natural one. So it is not surprising that several mathematicians have considered the subject. Probably first among these was Abel [3] who considered the general, but closely related, question of linearization of nonlinear operators. His study lead to what is sometimes called the Abel-Schröder functional equation (see Aczel [4, p. 253]).

Rather a more direct approach was taken by Vessiot [5] who considered a generalization of the (linear) superposition principle in 1893. He studied the problem - for which class of nonlinear ordinary differential equations

$$\frac{dy}{dx} = f(x,y) \tag{1.2}$$

does there exist a function ϕ such that the general solution, y, of (1.2) can be expressed as

$$y = \phi(y_1, y_2, \ldots, y_m; \;\; k) \tag{1.3}$$

where y_i, $i = 1, 2, \ldots, m$ are independent particular solutions of (1.2), k is an arbitrary constant and ϕ is independent of the particular solutions used? (He was searching for a nonlinear superposition!) Using the theory of transitive groups in three variables Vessiot demonstrated that the only nonlinear differential equation of type (1.2) having such a property is the Riccati equation. As previously noted four solutions of that equation have the cross ratio property (1.1). While disappointing this result demonstrates the special position the Riccati equation occupies. During the same year Guldberg [6] obtained a generalization of Vessiot's result for systems of first-order equations.

Vessiot was severely reprimanded by S. Lie who, quite correctly, remarked that this work was but a special case of his own theory of "fundamental solutions" of differential equations (1893). In Lie's theory it is supposed that there are n special or fundamental solutions $y_1, y_2, \ldots, y_n$ such that the general solution can be expressed in the form

$$y = g(y_1, y_2, \ldots, y_n; \ a), \quad \text{a constant}$$

and there is no question of this relation forming a general law of composition. In fact we may not know which particular solutions are fundamental. Thus Lie's general theory does not seem as useful as Vessiot's. The heavy hand of Lie quashed a budding development.

In 1960 Temple [7] re-examined the problem and refined Vessiot's approach. He concluded that even though Guldberg's theory can be used to extend the class of equations soluble by composition-equation (1.3) - still, only a limited class can be so treated. At the risk of repetition let me repeat - the superposition principle as treated by the foregoing authors is understood to be a way of expressing the general solution of a nonlinear equation as a function of a certain finite number of particular solutions, the function having the same form whatever the particular solutions.

At approximately the same time Inselberg [8] and Jones and Ames [9] removed the general solution restriction and inquired how two or more solutions could be combined in order to form another solution. This notion of superposition is an extension of the idea considered by Vessiot and Temple. After these papers there followed a number of published results consolidating and applying the theory laid down. In particular there are papers by Oppenheim [10], Foerster et al. [11], Inselberg [12, 13, 14, 15], Weston [16], Peterson [17], Levin [18], Spijker [19] and Kečkić [20, 21]. For some of the ensuing results applications have been found (see Oppenheim [10], Foerster et al. [11], Inselberg [12, 13], Weston [16] and Peterson [17].

While most of the basic theory seems applicable to nonlinear partial differential operators the usual application has been to ordinary differential operators. Exceptions include the papers by Jones and Ames [9], Levin [18] and Kečkić [20, 21]. In particular it is known that some classes of linear and nonlinear equations possess a noncountable infinity of nonlinear superposition principles.

Lastly it should be remarked that the Bäcklund and other general transformations sometimes permit development of a "theorem of permutability" which is a nonlinear superposition. In many cases this seems to be related to the Riccati cross ratio theorem. Some developments and applications are found in Ames [22], Forsyth [23], Lamb [24] and Anderson-Ibragimov [25].

This paper serves the dual purposes of a general survey and a description of new results and applications. It consists of three parts: Constants of Superposition and Applications; Construction and Analysis of Some Superpositions; Bäcklund Transformations.

II. CONSTANTS OF SUPERPOSITION

Perhaps the simplest and best known superposition for a linear homogeneous equation is the fact that if $u(x)$ is a solution of $Lu(x) = 0$, then so is $\lambda u(x)$, λ any scalar. Unless each term of the nonlinear equation has the same degree this is not true for nonlinear equations. For nonlinear ordinary equations a reasonable question is: If $U(x)$ is a solution of

$$u^{(n)} = f(x,u,u',u'', \ldots, u^{(n-1)}) \tag{2.1}$$

do there exist constants λ and μ such that

$$u(x) = \lambda U(\mu x) = \lambda U(X) \tag{2.2}$$

is also a solution of (2.1)? If such constants exist, possibly related, they are called constants of superposition and (2.2) is a superposition and a logical extension of the simple classical idea. Clearly (2.2) is expressible as the two parameter group of dilatations

$$\begin{aligned} u &= \lambda U \\ x &= \mu^{-1} X \end{aligned} \tag{2.3}$$

This problem has been thoroughly studied by Klamkin [26] who also presents a brief historical sketch. In particular he is interested in characterizing those problems which permit conversion of boundary value to initial value problems with (2.3) and a nonlinear group. This idea was certainly also known by Prandtl and Blasius in the first decade of this century. In this section we describe the characterization of those second-order equations which possess property (2.2) and its reasonable extension to partial differential equations.

If $u'' = f(x,u,u')$ is to be invariant under the two parameter group (2.3) f would have to satisfy the function equation

$$f(\mu^{-1}X, \lambda U, \lambda\mu P) = \lambda\mu^2 f(X,U,P), \tag{2.4}$$

for all λ, μ, X, U and P*. This is solved by the same technique as that employed to establish Euler's theorem on homogeneous functions. Assuming f is C^1 in all of the arguments one differentiates partially with respect to λ and then with respect to μ. Then setting both parameters equal to one the following pair of equations are obtained

$$U\frac{\partial f}{\partial U} + P\frac{\partial f}{\partial P} = f \tag{2.5}$$

---

* A number of examples exist for which λ is a function of μ.

$$X \frac{\partial f}{\partial x} - P \frac{\partial f}{\partial P} = -2f \tag{2.6}$$

With $g \in C'$ arbitrary, a general solution of (2.6) is

$$f = X^{-2} g(XP, U). \tag{2.7}$$

Substituting this in (2.5) it follows that g must be homogeneous of first degree in XP/U Thus the equation possessing invariance under (2.3) is

$$u'' = ux^{-2} h(xu'/u) \tag{2.8}$$

where h is arbitrary. In an exactly similar way one can treat an nth order equation.

An extension of this idea to partial differential equations is immediate. Thus if $U(x,y)$ is a solution of $Lu = 0$ do there exist constants of superposition λ, μ, and ν such that

$$u(x,y) = \lambda U(\mu x, \nu y)$$

is also a solution, for† all λ, μ, ν? By exactly the same procedure as that described in the preceding paragraph one could, for example, characterize all those diffusion equations

$$u_y = f(x, y, u, u_x, u_{xx})$$

which possess invariance under the transformation

$$u = \lambda U, \quad x = \mu^{-1} X, \quad y = \nu^{-1} Y$$

In many examples the invariance requirement places restrictions on the parameters such as one is a function of the other. When boundary and initial data are introduced additional restrictions may occur.

III. APPLICATIONS OF CONSTANTS OF SUPERPOSITION

In this section two applications to ordinary equations will

---

† Restricted cases include various functional relations among the three parameters.

be discussed. The first of these concerns "exact shooting" which has been widely promoted by Klamkin [26] and the second concerns nonlinear eigenparameter problems.

A. Exact Shooting (Ames and Adams [27])

A substantial number of problems in fluid mechanics and diffusion possess nonlinearities of "power" type - that is the nonlinearities appear as powers of the dependent variables and/or their derivatives. A substantial subset of these possess invariant (similar) solutions and the "constants of superposition" property. Here we examine one example from non-Newtonian (power law) fluid mechanics. In the stream function ψ the boundary layer flow over a flat plate of a power law fluid, with $\psi_{yy} > 0$, is governed by the equation (see E.G. BIZELL and SLATTERY [28])

$$\psi_y\psi_{xy} - \psi_x\psi_{yy} = [(\psi_{yy})^n]_y, \quad 0 < n < 2, \tag{3.1}$$

together with boundary conditions leading to (3.3). The group invariants (similar variables) are found to be

$$\phi = y\, x^{-1/(n+1)}, \quad f(\phi) = \psi\, x^{-1/(n+1)}$$

where $f(\phi)$ satisfies the boundary value problem

$$n(n + 1)\, f''' + (f'')^{2-n} f = 0 \tag{3.2}$$

$$f(0) = f'(0) = 0, \quad \lim_{\phi\to\infty} f'(\phi) = 1 \tag{3.3}$$

For $n = 2$ equation (3.2) is linear and the boundary value problem does not possess a solution! However, this value of n is outside the range of interest here.

Let $F(\phi)$ be any solution of (3.2). Upon introducing

$$f = \lambda F, \quad \phi = \mu^{-1}\,\overline{\phi} \tag{3.4}$$

and requiring that (3.2) be invariant under (3.4), it follows that

$$f = \lambda F(\mu\phi)$$

is also a solution of (3.2) if

$$\lambda = \mu^{(2n-1)/(2-n)}, \quad n \neq 2.$$

This constants of superposition property permits the conversion of the original boundary value problem (3.2) and (3.3) to two initial value problems - an exact shooting procedure. The first initial value problem consists of F satisfying (3.2) subject to the initial conditions $F(0) = F'(0) = 0$, $F''(0) = 1$, followed by the second initial value problem wherein f satisfies (3.2) subject to

$$f(0) = f'(0) = 0, \quad f''(0) = \mu^{3/(2-n)}. \tag{3.5}$$

The value of μ to be used in (3.5), obtained from the third condition in (3.3), is found to be

$$\mu = [F'(\infty)]^{(n-2)/(n+1)}.$$

Monotonically convergent upper and lower bounds are obtainable for either initial value problem. With $0 < n < 1$, consider (3.2) with $f(0) = f'(0) = 0$, $f''(0) = \alpha > 0$. Upon integrating, using the initial conditions, there results for $g(\phi) = f''(\phi)$

$$g(\phi) = \left[\frac{1-n}{n(n+1)} \int_0^\phi (\phi - u)^2 g(u)du + \alpha^{n-1}\right]^{-1/(1-n)} = Tg(\phi).$$

For $0 < n < 1$ and $g(\phi) \geqslant 0$, $Tg > 0$ successive approximations are computed by means of the algorithm $g_1 = 0$, $g_m = Tg_{m-1}$, $m \geqslant 2$. Once $g = f''$ has been obtained f' can be calculated by quadrature. If $0 < g < h$ it is clear from Tg that $Tg > Th$, that is T is antitone. With $g_1 \equiv 0$, $g_2 = Tg_1 = \alpha$ so $g_1 < g_2$. By induction there follows the convergent two-sided bounds

$$0 \equiv g_1 < g_3 < g_5 < \ldots < g_{2k+1} < \ldots < g_{2m} < \ldots < g_6 < g_4 < g_2$$

for all positive integers k and m. Uniform convergence to a solution of the original problem is demonstrated by means of integral inequalities.

Other applications can be found in Ames-Adams [27], Klamkin [26] and Ames [22].

B. Eigenparameter Calculations (Ames-Adams [29])

Here we show how the versatile constants of superposition concept can be applied to the determination of critical values for certain classes of nonlinear differential equations. Success of the analysis depends upon the manner in which the eigenparameter is employed in the transformation. As in Subsection (A) we shall only discuss second order equations with homogeneous Dirichlet boundary conditions. More general boundary conditions and higher order equations are analyzed in Ames-Adams [29].

Let ω be the (positive) eigenparameter sought and consider the general problem

$$u'' = f(x,u,u',\omega) \tag{3.6}$$

with boundary conditions

$$u(0) = 0, \quad u(1) = 0 \tag{3.7}$$

The form of f is further restricted so that under the transformation

$$R = g(\omega)r, \quad U = h(\omega)u$$

the eigenparameter, ω, is eliminated from (3.6). For this f must have the property that

$$f[R/g(\omega),U/h(\omega),(g(\omega)/h(\omega))dU/dR,\omega] = (g^2(\omega)/h(\omega))F[R,U,dU/dR] \tag{3.8}$$

as well as a condition found subsequently. With (3.8) equation (3.6) becomes

$$\frac{d^2U}{dR^2} = F\left(R,U,\frac{dU}{dR}\right) \tag{3.9}$$

and the boundary conditions are

$$U(0) = 0, \quad U(g(\omega)) = 0 \tag{3.9a}$$

-- these now include the eigenparameter.

Second, for (3.9) to be invariant under the one-parameter group

$$R = \omega^{\alpha}\overline{R}, \quad U = \omega^{\beta}\overline{U}$$

$$F\left[\omega^{\alpha}R, \omega^{\beta}U, \omega^{\beta-\alpha}\frac{dU}{dR}\right] = \omega^{\beta-2\alpha}F\left[R, U, \frac{dU}{dR}\right],$$

a functional equation which can be solved as in Euler's homogeneity studies. The well known procedure is to differentiate with respect to ω and set $\omega = 1$, thereby generating the equation for F,

$$\alpha R\frac{\partial F}{\partial R} + \beta U\frac{\partial F}{\partial U} + (\beta - \alpha)\, p\frac{\partial F}{\partial p} = (\beta - 2\alpha)\, F,$$

where $p = dU/dR$. Integrating this, by means of the theory of characteristics, gives

$$F = R^{(\beta/\alpha)-2}G[UR^{-\beta/\alpha},\ pR^{-(\beta/\alpha)+1}], \quad \text{if } \alpha \neq 0, \tag{3.10}$$

where G is a differentiable arbitrary function of its arguments. This provides all but solutions called special.

Applying condition (3.8) to (3.10) we find that admissible functions f of (3.6) are

$$f[r,u,u',\omega] = h^{-1}g^{\beta/\alpha}r^{\beta/\alpha-2}G[hu/(gr)^{\beta/\alpha}, (h/g)u'/(gr)^{\beta/\alpha-1}] \tag{3.11}$$

where $u' = du/dr$ in (3.11).

So that we can search for positive eigenparameters ω, assume $f(\overline{R},\overline{U},\overline{U}') < 0$ for $\overline{U} \in [0,\overline{U}_{max}]$ and $\overline{U}' \in [-a,a]$ with a, U_{max} suitable, in the first initial value problem below, i.e. $\overline{U}$ is concave downward.

For conditions (3.9a) the first i.v.p. is

$\overline{U}(\overline{R})$ satisfies (3.9) with (3.10)

$$\overline{U}(0) = 0, \quad (d\overline{U}/d\overline{R})(0) = 1 \tag{3.12}$$

and the second i.v.p. is

$U(R)$ satisfies (3.9) with (3.10)

$$U(0) = 0, \quad (dU/dR)(0) = \omega^{\beta-\alpha} \tag{3.13}$$

where the last condition of (3.13) follows from the relation $(dU/dR) = \omega^{\beta-\alpha}(d\overline{U}/d\overline{R})$. One of the parameters α, β is arbitrary to this point. Here we choose $\beta - \alpha = 1$ to provide the third condition required to uniquely fix the value (cf. Moroney [30]) of the eigenparameter. From the first i.v.p., because of the requirement $U(g(\omega)) = 0$ in (3.9a), take the first zero of $\overline{U}(\overline{R})$, say $\overline{R}_1$, whereupon

$$\omega^{\alpha}\overline{R}_1 = g(\omega) \tag{3.14}$$

yields an equation for ω. Under the assumption that a smallest positive solution, ω_1, exists, that value is used to solve the second i.v.p., (3.13), generating $U = U(R)$. Since $R = g(\omega)r$, $U = h(\omega)u$ the solution $u = u(r)$ is directly obtainable by transformation and thus the first zero of $u(r)$ is found for $r \in 0, 1$. Correspondingly, (3.14) may possess a second, third, etc., root for ω.

IV. CONSTRUCTION OF A NONLINEAR SUPERPOSITION

Consider the first order semilinear equation in two independent variables (the procedure is the same in n variables).

$$Z_x + a(x,y)\, Z_y + b(x,y)\, f(Z) = C(x,y). \tag{4.1}$$

Two cases will be distinguished. First, suppose $C \equiv 0$ and that u and ν are two independent solutions of (4.1). We wish to construct a composition law, g, so that

$$Z = g(u,\nu) \tag{4.2}$$

is also a solution of (4.1). When (4.2) is substituted in (4.1) and the result rearranged it follows that (4.2) is a solution if

$$g_u[u_x + au_y + bf(u)] + g_\nu[\nu_x + a\nu_y + bf(\nu)]$$
$$+ b[f(g) - f(u)g_u - f_\nu g_\nu] = 0 \tag{4.3}$$

Since u and ν satisfy (4.1), with $C \equiv 0$, equation (4.3) is satisfied if

$$f(u)g_u + f(\nu)g_\nu = f(g). \tag{4.4}$$

A sub-family of the general solution is

$$\int \frac{dg}{f(g)} = \int \frac{d\nu}{f(\nu)} + H\left(\int \frac{du}{f(u)} - \int \frac{d\nu}{f(\nu)} \right) \tag{4.5}$$

where $H \in C^1$ is arbitrary. Equation (4.5) provides the general infinite family of nonlinear superpositions.

Some special cases are of interest.

i) If $f(Z) = Z$ then

$$g(u,\nu) = \nu H(u/\nu). \tag{4.6}$$

Equation (4.6) includes the classic linear superposition if H is chosen as $H(u/\nu) = A(u/\nu) + B$ where A and B are constants. The important point here is that a family of linear equations can have an infinity of nonlinear superpositions! This has been observed by Jones and Ames [9] and Kečkić [20].

ii) If $f(Z) = Z^2$ then

$$g(u,\nu) = \frac{1}{\nu^{-1} + H(\nu^{-1} - u^{-1})}, \tag{4.7}$$

again with arbitrary H.

As an example of the application of (4.7) to simple waves let us consider the equation

$$Z_t + ZZ_x = 0 \tag{4.8}$$

which possesses the Lagrange system

$$\frac{dt}{1} = \frac{dx}{Z} = \frac{dZ}{0}$$

Thus Z is constant along the characteristic $\frac{dx}{dt} = Z$. In addition (4.8) possesses the invariant solution

$$Z = \frac{x}{t-a}$$

for any constant a. We write

$$u = \frac{x}{t-a_1}, \quad v = \frac{x}{t-a_2}, \quad a_1 \neq a_2. \tag{4.9}$$

Upon calculating the x derivative of (4.8) there results, with $\omega = Z_x$,

$$\frac{d\omega}{dt} + \omega^2 = 0. \tag{4.10}$$

Equation (4.10) has a family of nonlinear superpositions (4.7). Thus (4.8) possesses a <u>family of nonlinear superpositions of the x derivatives</u>! Setting u_x and v_x of (4.9) into the composition we find

$$\omega = \frac{1}{t-a_2 + H(t-a_2 - t + a_1)} = \frac{1}{t-c}$$

where c is arbitrary. Thus the general invariant solution $Z = x/(t - c)$ is recovered.

The second case with $C \neq 0$ is of interest for in that case a superposition of three independent solutions u, v, ω is required. Using the same argument as that leading to (4.4) $Z = g(u, v, \omega)$ must satisfy the two equations

$$g_u + g_v + g_\omega = 1 \tag{4.11}$$

$$f(u)\, g_u + f(v)\, g_v + f(\omega)\, g_\omega = f(g) \tag{4.12}$$

Some special cases are of interest.

i) If $f(Z) = Z$ then the solution of (4.11) and (4.12) is

$$g(u, \nu, \omega) = u + (\nu - u)\, H\left(\frac{\omega - u}{\nu - u}\right) \tag{4.13}$$

with $H \in C^1$ arbitrary. Kečkić [20] first observed this result.

ii) If $f(Z) = Z^2$ then a solution of (4.11) and (4.12) is

$$g(u, \nu, \omega) = \frac{\nu(u - \omega) - k\omega(u - \nu)}{u - \omega - k(u - \nu)}, \quad k \text{ constant} \tag{4.14}$$

the classical Riccati result. There may be other, more general solutions. It is not known if (4.11) and (4.12) possess a solution for any $f(Z)$.

V. A "MIXED" SUPERPOSITION FOR THE RICCATI EQUATION

While we know the Riccati equation

$$z' + Q(x)z + R(x)z^2 = P(x) \tag{5.1}$$

has the superposition (4.14) it is difficult to construct three solutions of (5.1). As an alternative, if one solution, U, is available, make the transformation

$$z = y^{-1} + U \tag{5.2}$$

to obtain

$$y' + f(x)y = R \tag{5.3}$$

$$f = -2RU - Q$$

Equation (5.3) has an infinity of nonlinear superpositions given by (4.13) where u, ν, ω are solutions of (5.3). Thus a "mixed" superposition for (5.1) is

$$z = U + \frac{1}{u + (\nu - u)H\left(\frac{\omega - u}{\nu - u}\right)},$$

where $H \in C^1$ is arbitrary and u, ν, ω are solutions of the linear equation (5.3).

VI. NONLINEAR DIFFUSION EQUATION

As a way of illustrating how those second order equations possessing nonlinear superpositions can be characterized we study

$$u_{xx} = f(u, u_x, u_t) \tag{6.1}$$

for possible compositions

$$u = g(\nu,\omega) \tag{6.2}$$

of two solutions ν and ω of (6.1). Setting (6.2) into (6.1) g must satisfy

$$\begin{aligned} &g_{\nu\nu}\,\nu_x^2 + 2g_{\nu\omega}\,\nu_x\omega_x + g_{\omega\omega}\,\omega_x^2 \\ &\quad - f[g, g_\nu\nu_x + g_\omega\omega_x, g_\nu\nu_t + g_\omega\omega_t] \\ &\quad + g_\nu f[\nu,\nu_x,\nu_t] + g_\omega f[\omega,\omega_x,\omega_t] = 0 \end{aligned} \tag{6.3}$$

Since one of our goals is the characterization of admissible f it is clear that a possibility for (6.3) occurs if the three right hand terms are equal to

$$A(\nu,\omega,g,g_\nu,g_\omega)\nu_x^2 + B(\nu,\omega,g,g_\nu,g_\omega)\nu_x\omega_x + C(\nu,\omega,g,g_\nu,g_\omega)\omega_x^2 \;. \tag{6.4}$$

If this is the case the g can be determined from the simultaneous solution of

$$\begin{aligned} g_{\nu\nu} + A &= 0 \\ g_{\nu\omega} + \tfrac{1}{2}B &= 0 \\ g_{\omega\omega} + C &= 0 \end{aligned} \tag{6.5}$$

if that exists.

For simplicity in the analysis we set

$$\alpha = g_\nu, \quad \beta = g_\omega, \quad \lambda = \nu_x, \quad \delta = \omega_x, \quad \varepsilon = \nu_t \quad \text{and} \quad \eta = \omega_t$$

whereupon the three right hand terms of (6.3) set equal to (6.4) become

$$\alpha f(\nu,\lambda,\varepsilon) + \beta f(\omega,\delta,\eta) - f[g,\alpha\lambda + \beta\delta, \alpha\varepsilon + \beta\eta] = A\lambda^2 + B\lambda\delta + C\delta^2. \tag{6.6}$$

Upon calculating the third partial derivative with respect to λ we find

$$\frac{\partial^3 f}{\partial\lambda^3} - \alpha^2 \frac{\partial^3 f}{\partial(\alpha\lambda + \beta\delta)^3} = 0$$

With $\beta = 0$

$$\alpha \frac{\partial^3 f}{\partial\lambda^3} = \frac{\partial^3 f}{\partial\lambda^3}$$

for all α. Hence f must be quadratic in λ. The first partial derivative with respect to f yields the requirement that f be linear in ε.

Our candidate f is now

$$f(u,u_x,u_t) = a(u)u_t + b(u)u_x^2 + c(u)\ u_x\ . \tag{6.7}$$

Setting (6.7) into (6.6) and examining the coefficients we easily see that a and c must be constant with b arbitrary. Further the same analysis gives

$$A = b(\nu)g_\nu - b(g)g_\nu^2$$

$$B = -\ 2b(g)\ g_\nu g_\omega$$

$$C = b(\omega)g_\omega - b(g)g_\omega^2$$

so that the study of (6.5) may proceed.

As an example consider the transformed Burgers' equation

$$u_{xx} = u_t + \frac{1}{2}\ u_x^2 \tag{6.8}$$

Then to find the nonlinear superposition we must integrate

$$2g_{\nu\nu} - g_\nu^2 + g_\nu = 0$$

$$2g_{\nu\omega} - g_\nu g_\omega = 0.$$

Thus (6.8) has the nonlinear superposition

$$g(\nu,\omega) = -2 \ln [c_1 + c_2 e^{-\frac{1}{2}\nu} + c_3 e^{-\frac{1}{2}\omega}] \tag{6.9}$$

a result first observed by Jones and Ames [9] in a different setting.

Perhaps we should have anticipated that

$$u_{xx} = au_t + b(u)\, u_x^2 + cu_x, \tag{6.10}$$

with a and c constant, had a nonlinear superposition because of the following "cheap trick" (see Ames [31] for a discussion). When the linear equation

$$\nu_{xx} = a\, \nu_t + c\nu_x \tag{6.11}$$

is subjected to the dependent variable transformation

$$\nu = k(u) \tag{6.12}$$

there results

$$u_{xx} = au_t - [k''(u)/k'(u)]u_x^2 + cu_x$$

which is equation (6.10). The known linear superposition of (6.11) induces a nonlinear superposition on (6.10).

All the analysis of this section is easily generalized to include coefficients of (x,t).

VII. BÄCKLUND TRANSFORMATIONS

These classical transformations (Forsyth [23], Anderson-Abragimov [25] relate two solutions of the same equation, or different equations. In the first instance it was already known by Darboux [32] that the transformation lead to a relation among four solutions of the sine-Gordon equation which was called a theorem of permutability in the early literature. In reality it is a nonlinear superposition. Here we give a few of these relations and omit the details of Backlund transformation construction.

A. Sine-Gordon Equation

An invariance Bäcklund transformation (not unique) for the sine-Gordon equation

$$(z_1)_{xy} = \sin z_1, \tag{7.1}$$

which relates two solutions z_1 and z of (7.1) is

$$\begin{aligned} z_y + (z_1)_y &= 2a^{-1}\sin [(z - z_1)/2] \\ z_x - (z_1)_x &= 2a \sin [(z + z_1)/2] \end{aligned} \tag{7.2}$$

Beginning with a known solution z_1, generate z_2 through a_2 in (7.2), generate z_3 through a_3 and, finally, generate z_4 from z_2 through a_3 and also from z_3 through a_2. This permutability was expressed by Bianchi [33] in the following diagram (Figure 1).

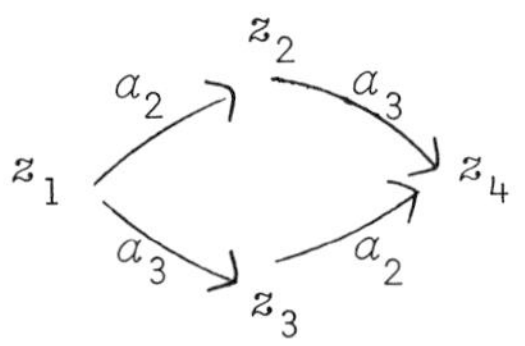

Figure 1

From (7.2) the analytic expressions for these solutions can be written and then it can be shown that they are related by the expression

$$\tan\left(\frac{z_4 - z_1}{4}\right) = \frac{a_2 + a_3}{a_2 - a_3}\tan\left(\frac{z_2 - z_3}{4}\right) \qquad (7.3)$$

which is the classical theorem of permutability. The knowledge of three solutions enables one to recursively generate an infinite sequence of particular solutions to the Sine-Gordon equation.

Perhaps the nonlinear superposition (7.3) should be expected since each of the equations in (7.2) is transformable into a Riccati equation. By setting $r = \tan[(z + z_1)/4]$ into the second equation of (7.2) then results

$$r_x + ar - 1/2p(1 + r^2) = 0,$$

a Riccati equation!

B. Korteweg de-Vries Equation (Lamb [24])

As a second example an invariance Bäcklund transformation of a modified KdV equation

$$u_y + 6uu_x + u_{xxx} = 0 \qquad (7.4)$$

is examined here. Upon setting

$$z(x,y) = \int_{-\infty}^{x} u(x',y)dx$$

equation (7.4) becomes

$$z_y + 3z_x{}^2 + z_{xxx} = 0 \qquad (7.5)$$

after integration and the discard of an arbitrary function of integration. An invariance Bäcklund transformation of (7.4) can be recovered from that for (7.5) by using the Lie-Bäcklund transformation $z_x = u$.

An invariance Bäcklund transformation for (7.5) is

$$z_x + (z_1)_x = m - \frac{1}{2}(z - z_1)^2 \tag{7.6}$$

$$z_y + (z_1)_y = (z - z_1)[z_{xx} - (z_1)_{xx}] - 2[z_x^2 + z_x(z_1)_x + (z_1)_x^2] \tag{7.6a}$$

where m is an arbitrary constant.

A theorem of permutability (nonlinear superposition) which permits the iterative construction of an infinite sequence of particular solutions also exists here, providing we know three solutions. To obtain the composition relation (7.6) is interpreted as a transformation from a known solution z_1 of (7.5) to another solution z_m which is obtained by the use of the constant m. The solution obtained from z_{m_1}, using (7.6) with m_2 is denoted by z_{m_1,m_2}. Writing four such transformations, using $p = z_x$, gives

$$p_{m_1} + p_1 = m_1 - 1/2(z_{m_1} - z_1)^2$$

$$p_{m_2} + p_1 = m_2 - 1/2(z_{m_2} - z_1)^2$$

$$p_{m_1,m_2} + p_{m_1} = m_2 - 1/2(z_{m_1,m_2} - z_{m_1})^2$$

$$p_{m_1,m_2} + p_{m_2} = m_1 - 1/2(z_{m_1,m_2} - z_{m_2})^2$$

where the two already calculated solutions z_{m_1} and z_{m_2} are used in the last two equations. The last two equations must yield the same final solution (i.e., $z_{m_1,m_2} = z_{m_2,m_1}$, as is already written) as demonstrated in the diagram (Figure 2) of the type used by Bianchi. To obtain a nonlinear superposition subtract

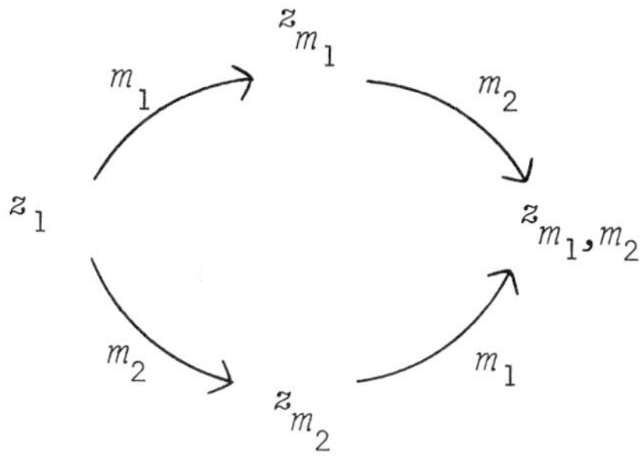

Figure 2

the second equation from the first, the fourth from the third and eliminate $p_{m_1} - p_{m_2}$ from the two resulting equations to obtain

$$z_{m_1,m_2} = z_1 + \frac{2(m_1 - m_2)}{z_{m_1} - z_{m_2}} , \qquad (7.7)$$

which is a remarkable relation in its simplicity! Equation (7.7) can be used to recursively generate an infinite sequence of particular solutions for (7.5).

The second relation, (7.6a), does not yield as elegant a result.

A relation to the Riccati equation is obtainable in a manner similar to the preceding section by setting $r = z - z_1$ in (7.6) whereupon that equation becomes

$$r_x - 1/2\, r^2 + (m - 2p) = 0$$

Thus we have demonstrated that the nonlinear superposition is again associated with a Riccati equation!

VIII. REFERENCES

[1] Davis, H. T., *Introduction to Nonlinear Differential and Integral Equations*, Dover, New York, 1962.

[2] Reid, W. T., *Riccati Differential Equations*, Academic Press,

New York, 1972.

[3] Abel, N. H., *Recherche de fonctions de deux quantitiés variable independantes* x *et* y, *telles que* $f(x,y)$, *qui ont la proprieté que* $f(z,f(x,y))$ *est une fonction symétrique de* z, x *et* y, J. Reine Angew. Math. 1, p. 11, 1826 (Oeuvres complètes Vol. I, pp. 61-65. Christiania 1881).

[4] Aczél, J., *Lectures on Functional Equations and their Applications*, Academic Press, New York, 1966.

[5] Vessiot, M. E., *Sur une classe d'equations différentielles*, Ann. Sci. École Norm. Sup. 10, p. 53, (1893).

[6] Guldberg, A., *Sur les équations différentielles ordinaire qui possèdent un système fundamental d'integrales*, Compt. Rend. Acad. Sci. 116, p. 964, (1893).

[7] Temple, G., *A superposition principle for ordinary nonlinear differential equations*, in *Lectures on Topics in Nonlinear Differential Equations*, pp. 1-15, Rept. 1415, David Taylor Model Basin, Carderock, Md., 1960.

[8] Inselberg, A., "*On classification and superposition principles for nonlinear operators*," AF Grant 7-64, Tech. Rept. 4, Elect. Eng. Res. Lab., Univ. of Illinois, Urbana, 1965.

[9] Jones, S. E. and Ames, W. F., "*Nonlinear superposition*," Jour. Math. Ann. Appl, 17, p. 484, (1967).

[10] Oppenheim, A. V., "*Superposition in a class of nonlinear systems*," Rept. #432, Res. Lab. Elect. MIT, Cambridge, Mass., 1965.

[11] Foerster, H. von, Inselberg, A. and Weston, P., "*Memory and inductive inference*," in "*Bionics Symposium*" (H. Oestreicher and D. Moore, eds.) pp. 31-68, Gordon and Breach, New York, 1968.

[12] Inselberg, A., "*Linear solvability and the Riccati operator*," Jour. Math. Anal. Appl. 22, p. 577, (1968).

[13] Inselberg, A., *Phase plane solutions of Langmuir's equation*," Jour. Math. Anal. Appl. 26, p. 438, (1969).

[14] Inselberg, A., "*Noncommutative superpositions for nonlinear operators*," Jour. Math. Anal. Appl. 29, p. 294, (1970).

[15] Inselberg, A., "*Superpositions for nonlinear operators*," Jour. Math. Anal. Appl. 40, p. 494, (1972).

[16] Weston, P., "*Counting binary rooted trees arboreal numbers of the first and second kind*," *in* "*Rooted Trees*," Rept. 2.0 Biological Computer Lab, Univ. of Illinois, Urbana, Ill., 1966.

[17] Peterson, L., "*Cascades of transformations*," Rept. 5.9, Biological Computer Lab., Univ. of Illinois, Urbana, Ill., 1969.

[18] Levin, S. A., "*Principles of nonlinear superposition*," Jour. Math. Anal., Appl. 30, p. 197, (1970).

[19] Spijker, M. N., "*Superposition in linear and nonlinear ordinary differential equations*," Jour. Math. Anal. Appl. 30, p. 206, (1970).

[20] Kečkić, J. C., *On Nonlinear Superposition*, Math. Balk. 2, p. 88, (1972).

[21] Kečkić, J. D., *On Nonlinear Superposition II*, Math. Balk, 3, p. 206, (1973).

[22] Ames, W. F., *Nonlinear Partial Differential Equations in Engineering*, Vol. II, Academic Press, New York, 1972.

[23] Forsyth, A. R., *Theory of Differential Equations*, Vol. VI, Dover, New York, 1959. (Original Edition 1906).

[24] Lamb, G. L., Jr., *Bäcklund transformations for certain nonlinear evolution equations*, J. Math. Phys. 15, p. 2157, (1974).

[25] Anderson, R. L. and Ibragimov, N., "*Lie-Bäcklund Transformations and Their Applications*," SIAM Monograph, to appear in 1978.

[26] Klamkin, M. S., *Transformation of boundary value into initial value problems*, J. Math. Anal. Appl. 32, p. 308, (1970).

[27] Ames, W. F. and Adams, E., *Monotonically convergent two-*

sided bounds for some invariant parabolic boundary value problems, Zeit. Ang. Math. Mech. 56, T240, (1976).

[28] Bizell, G. D. and Slattery, J. C., *Non-Newtonian boundary layer flow*, Chem. Eng. Sci. 17, S. 777, (1962).

[29] Ames, W. F. and Adams, E., *Exact shooting and eigenparameter problems*, J. Nonlinear Analysis, Theory, Methods and Appl., 1, p. 75, (1976).

[30] Moroney, R. M., *A class of characteristic-value problems*, Trans. Amer. Math. Soc. 102, p. 446, (1962).

[31] Ames, W. F., *Nonlinear Partial Differential Equations in Engineering*, Vol. I, Academic Press, New York, 1965.

[32] Darboux, G., *Lecons sur la Theorie Générale des Surfaces*, Gauthier-Villars, Paris, Troisième Partie, p. 432, (1894).

[33] Bianchi, L., *Lezione di Geometria Differenziale*, Enrico Spoerri, Pisa, p. 743, (1922).

RANDOM FIXED POINT THEOREMS

Heinz W. Engl*
Johannes-Kepler-Universität

I. INTRODUCTION

The study of random operator equations was initiated by the Prague school of probabilists around Špaček and Hanš in the 1950's. As it seems to be a current trend to use stochastic models rather than deterministic ones it is not surprising that the interest in random operator equations has been revived in the last years. The basic questions one might ask about random operator equations contain of course all problems which are interesting for deterministic operator equations, such as existence, uniqueness, stability and approximation of solutions. But the randomization leads to several new questions such as the measurability of solutions and their statistical properties. In this paper we deal with the question of measurability of fixed points of single- and multivalued random operators on randomly varying domains of definition. A complete survey about the "state of the art" in this area up to late 1976 can be found in [1].

Let throughout this paper (unless stated otherwise) X be a real separable Banach space, (Ω, A, μ) a σ-finite measure space. We will use the words "stochastic" and "random" interchangeably also if μ is not a probability measure. B denotes the σ-algebra on X generated by the open sets. By 2^X we denote

* Supported in part by a Auslandsstipendium des österreichischen Bundesministeriums fur Wissenschaft und Forschung (Austria).

ISBN 0-12-434160-8

$\{A/A \subseteq X \wedge A \neq \phi \wedge A$ closed$\}$, by $CB(X)$ $\{A/A \in 2^X \wedge A$ bounded$\}$ and by $CC(X)$ $\{A/A \in 2^X \wedge A$ convex$\}$.

Definition 1. Let $C\colon \Omega \to 2^X$ be a set-valued map. We call C "measurable" iff for all open $D \subseteq X$ $\{\omega \in \Omega/C(\omega) \cap D \neq \phi\} \in A$; we call C "separable" iff C is measurable and there exists a countable set $Z \subseteq X$ such that for all $\omega \in \Omega$ $cl(Z \cap C(\omega)) = C(\omega)$. The "graph of C" is defined as $Gr\ C: = \{(\omega,x) \in \Omega \times X/x \in C(\omega)\}$.

Definition 2. Let $C\colon \Omega \to 2^X$ be measurable. A mapping $T\colon Gr\ C \to 2^X$ is called "set-valued random operator with stochastic domain C" iff for all $x \in X$ and open $D \subseteq X$ $\{\omega \in \Omega/x \in C(\omega) \wedge T(\omega,x) \cap D \neq \phi\} \in A$. A mapping $x\colon \Omega \to X$ will be called "wide-sense fixed point of T" iff for all $\omega \in \Omega$ $x(\omega) \in C(\omega)$ and $x(\omega) \in T(\omega,x(\omega))$; it is called "random fixed point" iff it fulfills the properties of a wide-sense fixed point for μ-almost all $\omega \in \Omega$ and is measurable. A "random operator with stochastic domain C" is a function $T\colon Gr\ C \to X$ such that $\tilde{T}\colon Gr\ C \to 2^X$ defined by $\tilde{T}(\omega,x): = \{T(\omega,x)\}$ is a set-valued random operator with stochastic domain C. Of course in this case our definition coincides with [3, Definition 1].

The following example, which is a modification of an example in [5], shows that the problem of existence of a random fixed point of a random operator is not equivalent to the (non-stochastic!) problem of existence of wide-sense fixed points:

Example 3. $\Omega: = \mathbb{R}$, $A: = \sigma$-algebra of all at most countable subsets of $\mathbb{R}$ and their complements, for all $A \in A$ $\mu(A): = \text{card}(A)$. (Ω, A, μ) is a (not σ-finite) measure space. Let $T\colon \Omega \times \mathbb{R} \to \mathbb{R}$ be defined by

$$T(\omega,x): = \begin{cases} 1 & (x = 1 \wedge \omega = 1) \vee (x \neq 1 \wedge \omega \neq x) \\ 0 & x = 1 \wedge \omega \neq 1 \\ x & x \neq 1 \wedge \omega = x \end{cases}$$

For each $x \in \mathbb{R}$ $T(\cdot,x)$ has the form $c_1 + c_2 \cdot i(A)$, where $\text{card}(A) = 1$ and $c_1, c_2 \in \mathbb{R}$ (i is the indicator function). Therefore $T(\cdot,x)$ is $\mathcal{A}$-measurable, which means that T is a random operator. Obviously $x\colon \Omega \to \mathbb{R}$ defined by $x(\omega) := \omega$ is a wide-sense fixed point of T. It can be easily checked (note that $\mu(A) = 0$ only if $A = \phi$) that x is unique. But x is not measurable, since $\{\omega \in \Omega / x(\omega) < 0\} \notin \mathcal{A}$.

This example tells us that a random operator need not have a random fixed point, even if it has a unique wide-sense fixed point. We will see later that this situation cannot occur if $(\Omega, \mathcal{A}, \mu)$ is σ-finite and each $T(\omega,\cdot)$ is continuous.

II. CONTINUOUS RANDOM OPERATORS

Definition 3. Let $C\colon \Omega \to 2^X$ be measurable and $T\colon Gr\, C \to CB(X)$ a set-valued random operator with stochastic domain C. T is called "continuous" iff for all $\omega \in \Omega$ $T(\omega,\cdot)$ is a continuous mapping from $C(\omega)$ into $(CB(X), D)$, where D is the Hausdorff-metric on $CB(X)$ (i.e. for $A,B \in CB(X)$ $D(A,B) := \max\{\rho(A,B), \rho(B,A)\}$ with $\rho(A,B) := \sup_{a\in A} d(a,B)$ and $d(a,B) := \inf_{b\in B} \|a - b\|$). If $T\colon Gr\, C \to X$ is a random operator with stochastic domain C, then T is called "continuous" iff each $T(\omega,\cdot)$ is continuous as a function from $C(\omega)$ into X (which is of course equivalent with the assumption that the set-valued random operator $\tilde{T}(\omega,x) := \{T(\omega,x)\}$ is continuous in the above sense).

As all our results will be formulated for a separable "domain-function" C we should give some examples of separable functions $C\colon \Omega \to 2^X$. If Ω is countable then every measurable C is trivially separable (remember that X was assumed to be separable). A class of examples of separable "domain-functions" is described by the following

Proposition 4. Let $C\colon \Omega \to CC(X)$ be measurable with solid values (i.e. for all $\omega \in \Omega$ $\text{int}\, C(\omega) \neq \phi$). Then C is separable.

Expecially: If x_0: $\Omega \to X$ and R: $\Omega \to \mathbb{R}_0^+$ are measurable, then K: $\Omega \to CC(X)$ defined by $K(\omega) := \{x \in X / \|x - x_0(\omega)\| \leq R(\omega)\}$ is measurable; if for all $\omega \in \Omega$ $R(\omega) > 0$, then K is separable.

Proof: Let Z be a countable set with $\text{cl } Z = X$ (X is separable!), $\omega \in \Omega$, $x \in C(\omega)$ and $k \in \mathbb{N}$ arbitrary, but fixed. As $C(\omega) = \text{cl}(\text{int } C(\omega))$, there exists an $x_k \in \text{int } C(\omega)$ with $\|x - x_k\| < (2k)^{-1}$. By definition of $\text{int } C(\omega)$ there exist an integer $n_k \geq k$ such that $\{y \in X/\|y - x_k\| < n_k^{-1}\} \subseteq C(\omega)$. By definition of Z there exists a $z_k \in Z$ with $\|x_k - z_k\|$ $< (2n_k)^{-1}$. Because of the choice of n_k we have $z_k \in Z \cap C(\omega)$. $\|x - z_k\| \leq \|x - x_k\| + \|x_k - z_k\| < (2k)^{-1} + (2n_k)^{-1} \leq k^{-1}$. This proves that $\text{cl}(C(\omega) \cap Z) = C(\omega)$.

Now let x_0, R and K be chosen as in the formulation of the proposition, $x \in X$ fixed, but arbitrary. Then $\omega \to d(x, K(\omega))$ $= \max\{0, \|x - x_0(\omega)\| - R(\omega)\}$ is measurable. By [7, Theorem 3.3] K is measurable. If all $R(\omega)$ are positive, then for all $\omega \in \Omega$ $\text{int } K(\omega) \neq \phi$, which yields together with the first part of the proof the separability of K. □

Remark 5. From the proof can be seen that every measurable C: $\Omega \to 2^X$ with $C(\omega) = \text{cl}(\text{int } C(\omega))$ for all $\omega \in \Omega$ is separable.

Now we will prove the main result of this chapter:

Theorem 6. Let C: $\Omega \to 2^X$ be separable and T: $Gr\ C$ $\to CB(X)$ a continuous set-valued random operator with stochastic domain C. Assume that T has a wide-sense fixed point. Then T has a countable set of random fixed points $\{x_1, x_2, x_3, \ldots\}$ such that for μ-almost all $\omega \in \Omega$ $\{x_1(\omega), x_2(\omega), x_3(\omega), \ldots\}$ is dense in the set of all fixed points of $T(\omega, \cdot)$.

If in addition (Ω, A, μ) is complete, then "μ-almost all" can be replaced by "all" here and in the definition of a random fixed point.

Proof: a) We assume first that (Ω, A, μ) is complete. We define $N\colon Gr\ C \to \mathbb{R}$ by $N(\omega,x)\colon = d(x,T(\omega,x))$.

Let $\omega \in \Omega$ and $x,y \in C(\omega)$ be arbitrary, but fixed. Then

$$|N(\omega,x) - N(\omega,y)| = |d(x,T(\omega,x)) - d(y,T(\omega,y))| =$$

$$= |D(\{x\},T(\omega,x)) - D(\{x\},T(\omega,y)) + D(\{x\},T(\omega,y)) - D(\{y\},T(\omega,y))|$$

$$\leqslant |D(T(\omega,x),\{x\}) - D(\{x\},T(\omega,y))| + |D(\{x\},T(\omega,y)) - D(T(\omega,y),\{y\})|$$

$$\leqslant D(T(\omega,x),T(\omega,y)) + d(x,y).$$

This shows together with the continuity of T that each $N(\omega,\cdot)$ is continuous on $C(\omega)$.

Now let $x \in X$ and $r \geqslant 0$ be chosen arbitrarily, but fixed. Then $\{\omega \in \Omega/x \in C(\omega) \wedge N(\omega,x) < r\} = \{\omega \in \Omega/x \in C(\omega) \wedge T(\omega,x) \cap \{y \in X/\|y - x\| < r\} \neq \phi\} \in \mathcal{A}$. Because of this, the continuity of each $N(\omega,\cdot)$ and the fact, that $Gr\ C$ is a measurable subset of $\mathcal{A} \times \mathcal{B}$ ([7, Theorem 3.3]) we have (with Z as a countable set as it appears in Definition 1):

$$\{(\omega,x) \in Gr\ C/N(\omega,x) \leqslant r\} =$$

$$= \{(\omega,x) \in \Omega \times X/x \in C(\omega) \wedge N(\omega,x) \leqslant r\} =$$

$$= \bigcap_{n\in\mathbb{N}} \bigcup_{z\in Z} (\{\omega \in \Omega/z \in C(\omega) \wedge N(\omega,z) < r + n^{-1}\} \times$$

$$\times \{x \in X/\|x - z\| < n^{-1}\}) \cap Gr\ C] \in \mathcal{A} \times \mathcal{B}.$$

So N is a measurable function from the measurable space $(Gr\ C,\ Gr\ C \cap (\mathcal{A} \times \mathcal{B}))$ into $\mathbb{R}$. Especially $\{(\omega,x) \in Gr\ C/N(\omega,x) = 0\} \in \mathcal{A} \times \mathcal{B}$.

Now we define $F\colon \Omega \to \{A/A \subseteq X\}$ by $F(\omega)\colon = \{x \in X/x \in C(\omega) \cap T(\omega,x)\}$. Because T has a wide-sense fixed point, each $F(\omega)$ is non-empty. Now choose $\omega \in \Omega$ arbitrary, but fixed and let $(x_k) \in F(\omega)^{\mathbb{N}}$ such that $(x_k) \to x \in X$. Then $x \in C(\omega)$ and because of

$$d(x,T(\omega,x)) \leqslant \|x - x_k\| + d(x_k,T(\omega,x_k)) + D(T(\omega,x_k),T(\omega,x))$$

and the continuity of T we have $x \in F(\omega)$. So each $F(\omega)$ is closed.

$Gr\ F = \{(\omega,x) \in \Omega \times X / x \in F(\omega)\} = \{(\omega,x) \in Gr\ C / N(\omega,x) = 0\} \in A \times B$. This implies together with Theorem 3.5 (iii) of [7] that F is measurable in the sense of Definition 1 (note that our "measurable" is called "weakly measurable" in [7]!). Because of the Kuratowski-Ryll-Nardzewski-Selection-Theorem ([10, p. 398]) there exists a measurable $x: \Omega \to X$ such that for all $\omega \in \Omega\, x(\omega) \in F(\omega)$, which means $x(\omega) \in C(\omega)$ and $x(\omega) \in T(\omega, x(\omega))$. Especially x is a random fixed point of T.

The assertion about the countable family of random fixed points follows now directly from [7, Theorem 5.6], applied to our function F.

b) Let now (Ω, A, μ) be not necessarily complete; by (Ω, A^*, μ^*) we denote the completion of (Ω, A, μ). Since $A \subseteq A^*$, the assumptions of the theorem are also fulfilled with (Ω, A^*, μ^*) instead of (Ω, A, μ). Because of a) there exists a countable set $\{x_1^*, x_2^*, x_3^*, \ldots\}$ of A^*-measurable functions from Ω to X such that for all $\omega \in \Omega\ \{x_k^*(\omega)/k \in \mathbb{N}\}$ is a dense subset of $F(\omega)$. Let $\mathcal{D}$ be a countable generator of B which exists because of the separability of X. Then for all $D \in \mathcal{D}$ and $k \in \mathbb{N}\ x_k^{*-1}(D) = A_{D,k} \cup L_{D,k}$ with $A_{D,k} \in A$, $L_{D,k} \subseteq N_{D,k}$, $N_{D,k} \in A$ and $\mu(N_{D,k}) = 0$. Let $M := \bigcup_{k \in \mathbb{N}} \bigcup_{D \in \mathcal{D}} N_{D,k}$ and define for each $k \in \mathbb{N}$

$$x_k: \Omega \longrightarrow X$$
$$\omega \longrightarrow \begin{cases} x_k^*(\omega) & \omega \notin M \\ 0 & \omega \in M. \end{cases}$$

Then each x_k is A-measurable and for all $\omega \in \Omega \setminus M\ \{x_k(\omega)/k \in \mathbb{N}\}$ is a dense subset of $F(\omega)$. Because of $\mu(M) = 0$ this proves the theorem. □

Corollary 7. Let $C: \Omega \to 2^X$ be separable and $T: Gr\ C \to X$ a continuous random operator with stochastic domain C such that for all $\omega \in \Omega\ \{x \in C(\omega)/T(\omega,x) = x\} \neq \phi$. Then the conclusions of Theorem 6 hold.

Remark 8. Corollary 7 generalizes Theorem 8 of [3]. There it had to be assumed that C had convex and solid values and that a measurable function $x_0\colon \Omega \to X$ with $x_0(\omega) \in \text{int } C(\omega)$ exists; in [3] sufficient conditions for that to hold were given. These conditions forced to assume that C has uniformly bounded values and that X is reflexive in addition. Using our Corollary 7 instead of [3, Theorem 13] we can of course formulate the applications in [3] in the more general situation. On the other hand Theorem 6 generalizes [4, Theorem 13] and therefore the main result of [8] (cf. [4]). The usefulness of Theorem 6 seems to lie in the fact that unlike many other results about measurability of fixed points of random operators (cf. e.g. [1], [11]) which use special methods for each problem it allows us to neglect the randomness of the operator in the sense that it tells us that a random fixed point (even a dense set of them) always exists as soon as a wide-sense fixed point exists. In other words: For every fixed point theorem for continuous operators on separable Banach spaces the stochastic analogue is immediately available as a consequence of Theorem 6. As an example we give the following Krasnoselski-type theorem which includes stochastic versions of the fixed point theorems of Banach and Schauder:

Corollary 9. Let $C\colon \Omega \to CC(X) \cap CB(X)$ be separable, $L\colon \Omega \to [0,1[$ and T and S random operators with stochastic domain C such that for all $\omega \in \Omega$

(1) for all $x \in C(\omega)$ $T(\omega,x) + S(\omega,x) \in C(\omega)$

(2) for all $x,y \in C(\omega)$ $\|S(\omega,x) - S(\omega,y)\| \leq L(\omega)\cdot\|x - y\|$

(3) $T(\omega,\cdot)$ is completely continuous.

Then there exists a countable set of measurable functions $x_k\colon \Omega \to X$ such that for μ-almost all $\omega \in \Omega$ $x_k(\omega) \in C(\omega)$ and $T(\omega,x_k(\omega)) + S(\omega,x_k(\omega)) = x_k(\omega)$ for all $k \in \mathbb{N}$ and $\{x_1(\omega), x_2(\omega), \ldots\}$ is dense in the set of all fixed points of $T(\omega,\cdot) + S(\omega,\cdot)$. If (Ω,A,μ) is complete, then "μ-almost all" can be replaced by "all".

Proof: Follows from Corollary 7 and [12, p. 339]. □

If we assume $T \equiv 0$ in Corollary 9 then it is obviously sufficient to assume "$C: \Omega \to 2^X$".

As an example we prove a theorem about existence of random solutions of a nonlinear differential equation with random vector field subject to random initial conditions. Other applications can be found in [3]; using our Corollary 7 they can be reformulated as to the existence of a countable dense set of random solutions. Let $\mathbb{R}^n$ be endowed with an arbitrary norm $\| \ \|_n$ and the Borel-σ-algebra; for $a < b$ we denote by $C_n[a,b]$ $\{x: [a,b] \to \mathbb{R}^n / x \text{ continuous}\}$ endowed with the norm $\|x\| := \sup_{s\in[a,b]} \|x(s)\|_n$.

Lemma 10. Let $x: \Omega \times [a,b] \to \mathbb{R}^n$ be such that $x(\omega,\cdot)$ is continuous for all $\omega \in \Omega$. Then $x(\cdot,s)$ is measurable for all $s \in [a,b]$ iff $\tilde{x}: \Omega \to C_n[a,b]$ defined by $\tilde{x}(\omega)(s) := x(\omega,s)$ is measurable. We will therefore identify x and $\tilde{x}$.

Proof: ⇐: For all $s \in [a,b]$, $i \in \{1, \ldots, n\}$ and $x = (x_1, \ldots, x_n) \in C_n[a,b]$ let $f_s^i(x) := x_i(s)$. Then $f_s^i \in C_n[a,b]^*$, therefore $\omega \to f_s^i(\tilde{x}(\omega)) = x_i(\omega,s)$ is measurable; as the Borel-σ-algebra on $\mathbb{R}^n$ is the n-fold product-σ-algebra of the Borel-σ-algebra on $\mathbb{R}$, this means that $\omega \to x(\omega,s)$ is measurable for all $s \in [a,b]$.

⇒: Let $x_0 \in C_n[a,b]$ and $r > 0$ be arbitrary, but fixed and let Z be a countable dense subset of $[a,b]$. For all $s \in Z$ $\omega \to \|x(\omega,s) - x_0(s)\|_n$ is measurable, therefore $\omega \to \sup_{s\in Z} \|x(\omega,s) - x_0(s)\|_n = \|x(\omega,\cdot) - x_0\|$ is measurable, which means $\{\omega \in \Omega / \tilde{x}(\omega) \in \{y \in C_n[a,b] / \|y - x_0\| < r\}\} \in A$. Because of the separability of $C_n[a,b]$ this implies that for all open $D \subseteq C_n[a,b]$ $\{\omega \in \Omega / \tilde{x}(\omega) \in D\} \in A$; therefore $\tilde{x}$ is measurable.□

Theorem 11. Let $U \subseteq \mathbb{R}$ be an open neighborhood of 0, $f: U \times \mathbb{R}^n \times \Omega \to \mathbb{R}^n$ such that $f(\cdot,\cdot,\omega)$ is continuous for all $\omega \in \Omega$ and $f(s,x,\cdot)$ is measurable for all $(s,x) \in U \times \mathbb{R}^n$. Let $x_0\colon \Omega \to \mathbb{R}^n$ and $K\colon \Omega \to]0,+\infty[$ be measurable. We assume that there exists an $\varepsilon > 0$ such that for all $\omega \in \Omega$ and $(s,x) \in U \times \mathbb{R}^n$ with $|s| \leqslant \varepsilon$ and $\|x - x_0(\omega)\|_n \leqslant \varepsilon$: $\|f(s,x,\omega)\|_n \leqslant K(\omega)$ and that there is a $K_0 > 0$ with $K(\omega) \leqslant K_0$ for all $\omega \in \Omega$. Let $d_0 := \min\{\varepsilon, \varepsilon \cdot K_0^{-1}\}$.

Then there exists a countable set of functions $x_k\colon \Omega \times [0,d_0] \to \mathbb{R}^n$ such that for all $k \in \mathbb{N}$

(1) $x_k(\cdot,s)$ is measurable for all $s \in [0,d_0]$.

(2) $x_k(\omega,\cdot)$ is continuously differentiable for all $\omega \in \Omega$.

(3) For μ-almost all $\omega \in \Omega$

$$\frac{dx_k(\omega,s)}{ds} = f(s,x_k(\omega,s),\omega) \text{ and } x_k(\omega,0) = x_0(\omega).$$

(4) For all $\omega \in \Omega$ and $s \in [0,d_0]$

$$\|x(\omega,s) - x_0(\omega)\|_n \leqslant \min\{\varepsilon, \varepsilon \cdot K(\omega)\}.$$

Furthermore:

(5) For μ-almost all $\omega_0 \in \Omega$ every solution on $[0,d_0]$ of the (deterministic!) initial value problem

$$\frac{dx(s)}{ds} = f(s,x(s),\omega_0),\ x(0) = x_0(\omega_0)$$

is the uniform limit of a sequence of elements of the set $\{x_1(\omega_0,\cdot), x_2(\omega_0,\cdot), x_3(\omega_0,\cdot), \ldots\}$.

If (Ω, A, μ) is complete, then "μ-almost all" can be replaced by "all".

Proof: Let for all $\omega \in \Omega$ $d(\omega) := \min\{\varepsilon, \varepsilon \cdot K(\omega)^{-1}\}$, $C(\omega) := \{x \in C_n[0,d_0] /$ for all $s \in [0,d_0] \|x(s) - x_0(\omega)\|_n \leqslant d(\omega) \cdot K(\omega)\}$. We define $T\colon Gr\ C \to C_n[0,d_0]$ by $T(\omega,x)(s) := x_0(\omega) + \int_0^s f(\tau,x(\tau),\omega)d\tau$. Because of Proposition 4 C is separable (in the sense of Definition 1). For all $x \in C_n[0,d_0]$ and $s \in [0,d_0]$ $\omega \to x_0(\omega) + \int_0^s f(\tau,x(\tau),\omega)d\tau$ is measurable as the limit of a sequence of finite sums of measurable functions. Together with Lemma 10 this implies that for all open $D \subseteq C_n[0,d_0]$ and $x \in C_n[0,d_0]$

$$\{\omega \in \Omega / x \in C(\omega) \wedge T(\omega,x) \in D\}$$
$$= [\bigcap_{n \in N} \{\omega \in \Omega / C(\omega) \cap \{y \in C_n[0,d_0] / \|y - x\| < n^{-1}\} \neq \phi\}]$$
$$\cap \{\omega \in \Omega / x_0(\omega) + \int_0^{\cdot} f(\tau,x(\tau),\omega)d\tau \in D\} \in A.$$

So T is a random operator with stochastic domain C. It is well-known (cf. [13, p. 45]) that for each $\omega \in \Omega$ $T(\omega,\cdot)$ is a completely continuous self-map of $C(\omega)$. The conclusions of the theorem follow now from Corollary 9 (with $S \equiv 0$) and Lemma 10. □

III. UPPER SEMICONTINUOUS RANDOM OPERATORS

Definition 12. Let Y be a Banach space, $K \in 2^Y$. $T: K \to 2^Y$ is called "upper semicontinuous (usc)" iff for all $x \in K$ $T(x)$ is compact and for all closed $A \subseteq Y$ $\{x \in K / T(x) \cap A \neq \phi\}$ is closed. T is called "compact" iff it is usc and $\mathrm{cl}(\bigcup_{x \in K} T(x))$ is compact.

As the definitions of upper semicontinuity vary in literature, we state the following well-known

Lemma 13. Let Y be a Banach space, $K \in 2^Y$, $T: K \to 2^Y$ such that $\mathrm{cl}(\bigcup_{x \in K} T(x))$ is compact. Then the following are equivalent:

(1) $\{(x,y) \in K \times Y / y \in T(x)\}$ is closed in the product-topology.

(2) For all $(x_n) \in K^N$, $x \in K$, $y_n \in T(x_n)$, $y \in Y$ with $(x_n) \to x$ and $(y_n) \to y$ we have $y \in T(x)$.

(3) For all $x \in K$ and all open $D \subseteq Y$ with $T(x) \subseteq D$ there exists a neighborhood U of x such that for all $z \in U \cap K$ $T(z) \subseteq D$.

(4) T is compact in the sense of Definition 12.

Proof: e.g. [4].

The following theorem was proved by Kakutani ([9]) for finite-dimentional spaces and by Bohnenblust and Karlin ([2]) in the general situation:

Theorem 14. Let Y be a Banach space, $K \in CC(Y)$ and T: $K \to CC(K)$ compact. Then T has a fixed point (i.e. there exists an $x \in K$ with $x \in T(x)$).

This result has been applied e.g. to game theory ([13, chp. 9.3]) and differential equations with multivalued right-hand side ([6]) which arise in the study of implicit differential equations, differential inequalities, and control theory. In [1, p. 653] A. T. Bharucha-Reid asks for a stochastic version of Theorem 14. We will be able to give one here.

First we observe that the proof of Theorem 6 cannot be simply transferred to upper semicontinuous maps. This proof was based on the fact that a function of two variables, which is continuous in one and measurable in the other variable (satisfies a "Caratheodory-condition") is jointly measurable. The following example shows that this need not be true for set-valued maps if "continuous" is replaced by "usc":

Example 15. Let (Ω, A, μ) be the interval $[-1,1]$ with the Borel-σ-algebra and the Lebesgue-measure, $X: = \mathbb{R}$, $C: = [-1,1]$. X is endowed with the usual topology and the Borel-σ-algebra $\mathcal{B}$. Let $S^1 \subseteq \Omega \times C$ be the unit circle and let E be a subset of S^1 which is not in $A \times \mathcal{B}$. Such a set E exists if we assume the validity of the axiom of choice.

Now we define T: $\Omega \times C \to 2^{\mathbb{R}}$

$$(\omega, x) \to \begin{cases} \{0\} & (\omega, x) \notin E \\ [0,1] & (\omega, x) \in E \end{cases}.$$

For all $x \in C$ and all open $D \subseteq X$

$$\{\omega \in \Omega / T(\omega,x) \cap D \neq \phi\} = \begin{cases} \phi & \text{if } D \cap [0,1] = \phi \\ \Omega & \text{if } 0 \in D \\ B_1(x) & \text{if } 0 \notin D \text{ and } D \cap [0,1] \neq \phi \end{cases} \text{, where}$$

$B_1(x) = \{\omega \in \Omega / (\omega,x) \in E\}$. Because of card $B_1(x) \leq 2$ $\{\omega \in \Omega / T(\omega,x) \cap D \neq \phi\} \in A$.

For all $\omega \in \Omega$ and all closed $A \subseteq X$

$$\{x \in C / T(\omega,x) \cap A \neq \phi\} = \begin{cases} \phi & \text{if } A \cap [0,1] = \phi \\ [-1,1] & \text{if } 0 \in A \\ B_2(\omega) & \text{if } 0 \notin A \text{ and } A \cap [0,1] \neq \phi \end{cases} \text{, where}$$

$B_2(\omega) = \{x \in C / (\omega,x) \in E\}$. Because of card $B_2(\omega) \leq 2$ we have $\{x \in C / T(\omega,x) \cap A \neq \phi\}$ is closed.

So T is measurable as a function of ω and usc as a function of x. But because of $\{(\omega,x) \in \Omega \times X / T(\omega,x) \cap \,]2^{-1}, 1[\neq \phi\} = E \notin A \times B$ T is not jointly measurable.

Nevertheless we can generalize the idea of the proof of Theorem 6 to obtain a stochastic version of Theorem 14 by considering an auxiliary map H instead of T. Theorem 16 is a generalization of [4, Theorem 8]. As the proof is essentially a combination of the proof there and the proof of our Theorem 6 we will only sketch the proof here.

Theorem 16. Let $C: \Omega \to CC(X)$ be separable and $T: Gr\, C \to CC(X)$ be a set-valued random operator with stochastic domain C such that for each $\omega \in \Omega$: $T(\omega,\cdot)$ is compact and for all $x \in bd\, C(\omega)$ $T(\omega,x) \subseteq C(\omega)$. Then T has a random fixed point.

Proof: Let without loss of generality (cf. the proof of Theorem 6) (Ω, A, μ) be complete and let Z be a countable set as it appears in Definition 1. We define $H: Gr\, C \to CC(X)$ by $H(\omega,x) := \bigcap_{n \in \mathbb{N}} \text{cl conv} \bigcup_{\substack{z \in Z \cap C(\omega) \\ \|z-x\| < n^{-1}}} T(\omega,z)$. All properties of H we will use below can be found in [4, Proposition 5].

H is (jointly) measurable as a function on $(Gr\ C, (\mathcal{A} \times \mathcal{B}) \cap Gr\ C)$, and so is the function $\tilde{x}$: $Gr\ C \to X$ defined by $\tilde{x}(\omega,x) := x$. Therefore (cf. [4, Lemma 6]) the function N: $Gr\ C \to X$

$$(\omega,x) \to d(x,H(\omega,x))$$ is (jointly) measurable

$(N(\omega,x) = d(\tilde{x}(\omega,x),H(\omega,x))!)$.

Now we define F: $\Omega \to \{A/A \subseteq X\}$ by

$$F(\omega) := \{x \in X/x \in C(\omega) \wedge x \in H(\omega,x)\}.$$

Because of the Rothe variant of Theorem 14 ([13, Theorem 9.2.4]) $F(\omega) \neq \phi$ for all $\omega \in \Omega$. As $H(\omega,\cdot)$ is usc for all $\omega \in \Omega$, each $F(\omega)$ is closed.

$$Gr\ F = \{(\omega,x) \in \Omega \times X/x \in F(\omega)\} = \{(\omega,x) \in Gr\ C/N(\omega,x) = 0\} \in \mathcal{A} \times \mathcal{B}.$$

We complete the proof analogously to the proof of Theorem 6, using that for all $(\omega,x) \in Gr\ C\ H(\omega,x) \subseteq T(\omega,x)$. □

Remark 17. Again the words "μ-almost all" in the definition of a random fixed point can be replaced by "all" if $(\Omega,\mathcal{A},\mu)$ is complete.

The conclusion about the existence of a dense set of random fixed points, however, cannot be drawn in the same way as in the proof of Theorem 6, because in the proof of Theorem 16 each $F(\omega)$ is the fixed point set of $H(\omega,\cdot)$, which might be a proper subset of the fixed point set of $T(\omega,\cdot)$. If we take $Z := Q$ in Example 15, then for all $(\omega,x) \in \Omega \times X\ H(\omega,x) = \{0\}$. So for each $\omega \in \Omega$ 0 is the only fixed point of $H(\omega,\cdot)$. But for each $(\omega,x) \in E$ with $x \geqslant 0$ 0 and x are fixed points of $T(\omega,\cdot)$.

IV. REFERENCES

[1] Bharucha-Reid, A. T., *Fixed point theorems in probabilistic analysis*, Bull. of the Amer. Math. Soc. 82 (1976), pp. 641-657.

[2] Bohnenblust, H. F. - Karlin, S., *On a theorem of Ville*, in:

H. W. Kuhn - A. W. Tucker (ed.), *Contributions to the theory of games*, Princeton University Press, Princeton 1950.

[3] Engl, H. W., *A general stochastic fixed point theorem for continuous random operators on stochastic domains*, J. Math. Anal. Appl. (to appear).

[4] Engl, H. W., *Random fixed point theorems for multivalued mappings*, Pacific J. Math. (to appear).

[5] Hanš, O., *Random fixed point theorems*, Trans. 1st Prague Conf. on Information Theory, Statist. Decision Functions, and Random Processes (Liblice, 1956), Czechoslovak Acad. Sci., Prague 1957, pp. 105-125.

[6] Hermes, H., *The generalized differential equation* $\dot{x} \in R(t,x)$, Advances in Mathematics 4(1970), pp. 149-169.

[7] Himmelberg, C. J., *Measurable relations*, Fundamenta Math. LXXXVII (1975), pp. 53-72.

[8] Itoh, S., *A random fixed point theorem for a multivalued contraction mapping*, Pacific J. Math. 68 (1977), pp. 85-90.

[9] Kakutani, S., *A generalization of Brouwer's fixed point theorem*, Duke Math. J. 8 (1941), pp. 457-459.

[10] Kuratowski, K. - Ryll-Nardzewski, C., *A general theorem on selectors*, Bull. Acad. Pol. Sc. (Sér. math., astr. et phys.) XIII (1965), pp. 397-403.

[11] Mukherjea, A., *Contractions and completely continuous mappings*, Nonlinear Analysis 1 (1977), pp. 235-247.

[12] Reinermann, J., *Fixpunktsätze vom Krasnoselski-Typ*, Math. Zeitschrift 119 (1971), pp. 339-344.

[13] Smart, D. R., *Fixed point theorems*, Cambridge University Press, London 1974.

DELAY EQUATIONS OF PARABOLIC TYPE IN BANACH SPACE

W. E. Fitzgibbon
University of Houston

I. INTRODUCTION

In this paper we shall consider functional differential equations in Banach space. We shall also provide examples of partial functional differential equations to which our abstract theory may be applied. Specifically we consider equations of the form:

$$\dot{x}(\phi)(t) + A(t)x(\phi)(t) = F(t,x_t(\phi)), \quad t > \tau$$

$$x_\tau(\phi) = \phi \in C.$$

Here, X is a Banach space and $\{A(t) | t \in [0,T]\}$ is a family of densely defined linear operators on X. If I denotes an interval of the form $[-r,0]$ or $(-\infty,0]$ then $C = C(I,X)$ will denote the space of functions which are uniformly continuous from I to X and which is endowed with the supremum norm, $\|\cdot\|$. If u is a function from $I \cup [0,T]$ to X, u_t is that element of C having pointwise definition $u_t(\theta) = u(t+\theta)$ for $\theta \in I$. F is a function mapping $R^+ \times C$ to X. We shall take up two cases which we designate quasilinear and semilinear respectively. The quasilinear case requires that F be Hölder continuous with respect to fractional powers of $A(t)$ and the semilinear case requires that F be continuous.

Our approach will be to apply the theory of abstract parabolic equations developed by Sobolevskii in [15] to obtain

ISBN 0-12-434160-8

existence and then to apply the theory of classical ordinary differential equations to study the behavior of the semilinear case. Abstract functional differential equations have been the subject of much recent activity and the interested reader is referred to Brewer [1], Dyson and Villella Bressan [3], Travis and Webb [16], [17], [18], Webb [20], [21], and the author [4], [5], [6], [7], [8], [9].

II. THE QUASILINEAR CASE

In this section we obtain local existence of solutions to the quasilinear equation. We place the following restrictions on a family of linear operators $\{A(t) | t \in [0,T]\}$ which map X to X and on our family of nonlinear operators $\{F(t,\cdot) | t \in [0,T]\}$ which map $R^+ \times C$ to X.

(S_1) The domain D_A of $\{A(t) | t \in [0,T]\}$ is dense and independent of t. Each $A(t)$ is a closed linear operator.

(S_2) For each $t \in [0,T]$, the resolvent $R(\lambda,A(t))$ exists for all $\lambda \leqslant 0$ and there exists $C > 0$ so that $\|R(\lambda,A(t))\| \leqslant C/|\lambda| + 1$.

(S_3) There exists an α $(0 < \alpha < 1)$ so that for all $t,s,\tau \in [0,T]$, $\|(A(t) - A(s))A^{-1}(\tau)\| \leqslant C\ |t - \tau|^{\alpha}$.

(S_4) For each $\tau \in [0,T]$, $A^{-1}(\tau)$ is completely continuous.

(S_5) There exists a ρ $(0 < \rho < 1)$ so that for all $s,t,\tau \in [0,T]$ and $\phi,\psi \in C$ we have,

$$\|F(t,A^{-\alpha}(s)\phi) - F(\tau,A^{-\alpha}(s)\psi)\| \leqslant C(|t - \tau|^{\rho} + \|\phi - \psi\|_C^{\rho})$$

We remark that Condition (S_2) implies each $A(t)$ is the infinitesimal generator of an analytic semigroup. Moreover for $\alpha > 0$ the fractional power $A^{-\alpha}(t)$ is defined by the equation;

$$A^{-\alpha}(t) = \frac{1}{\Gamma(\alpha)} \int_0^\infty e^{-sA(t)} s^{\alpha-1}\, ds$$

$A^{-\alpha}(t)$ may be shown to be a bounded, one-to-one linear operator and we obtain $A^{\alpha}(t) = (A^{-\alpha}(t))^{-1}$. If conditions (S_1) through (S_3) are satisfied there exists an operator valued function $W(t,\tau)$ defined for $0 \leq \tau \leq t \leq T$ which has values in $B(X)$. $W(t,\tau)$ is jointly continuous in t and τ and maps X to D_A for $t > \tau$. The derivative belongs to $B(X)$, for $t > \tau$ and is strongly continuous in t. Finally $W(t,\tau)$ satisfies the functional equation $W(t,s)W(s,\tau) = W(t,\tau)$ and the differential equation:

$$\partial W(t,\tau)/\partial t + A(t)W(t,\tau) = 0, \quad t > \tau$$

$$W(\tau,\tau) = I.$$

Moreover, if $f\colon\ [0,T] \to X$ is uniformly Hölder continuous then the unique solution to the inhomogeneous linear differential equation;

(2.1)
$$du(t)/dt + A(t)u(t) = f(t), \quad t > \tau$$
$$u(\tau) = u_0$$

has variation of parameters representation:

(2.2)
$$u(t) = W(t,\tau)u_0 + \int_\tau^t W(t,s)f(s)ds.$$

We have the following local existence result.

Theorem 1. Let X be a Banach space and suppose that $\{A(t) \mid t \in [0,T]\}$ and $F\colon\ [0,T] \times C \to X$ satisfy (S_1) through (S_5). Let $\phi \in C$ have the properties that there exists an $R > 0$ and $\beta > \alpha > 0$ such that $\phi(\theta) \in D(A^{\beta}(0))$, $A^{\alpha}(0)\phi \in C$ is uniformly Hölder continuous and $\|A^{\alpha}(0)\phi(\theta)\| \leq R$. There exists a $t_1 = t_1(\phi)$ and a function $x(\phi)\colon\ [0,t_1] \to X$ having properties that $x_0(\phi) = \phi$; $x(\phi)$ is continuously differentiable for $t > 0$ and satisfies,

$$\dot{x}(\phi)(t) + A(t)x(\phi)(t) = F(t, x_t(\phi)). \tag{2.2}$$

Outline of Proof: Let v be a Hölder continuous function mapping an interval $[0, t^*]$ to X having initial value $v(0) = A^{\alpha}(0)\phi(0)$. We define a function $\hat{v}$: $I \cup [0, t^*] \to X$ by the equation

$$\hat{v}(s) = \begin{cases} A^{\alpha}(0)\phi(s) & s \in I \\ v(s) & s \in [0, t^*] \end{cases}$$

It is clear that for $t \in [0, t^*]$, $\hat{v}_t \in C$. Furthermore the function $F_v(t)$ defined by

$$F_v(t) = F(t, A^{-\alpha}(0)\ \hat{v}_t)$$

is Hölder continuous. We can therefore refer to the inhomogeneous linear theory to guarantee unique solutions to,

$$\begin{aligned} z'(t) &= A(t)z(t) + F_v(t) \\ z(0) &= A^{\alpha}(0)\phi(0). \end{aligned} \tag{2.3}$$

We designate a set $Q(t^*, k, \eta)$ as the set of Hölder continuous functions from $[0, t^*]$ to X with Hölder constant K and order η. The properties of the evolution operator $W(t, \tau)$ may be utilized to see that for appropriate choice of K, t_0, and η the operator T defined pointwise by the equation:

$$T_v(t) = w_v(t) = A^{\alpha}(0)W(t,0)\phi(0) + A^{\alpha}(0)\int_0^t W(t,s)(F_v(s))ds$$

maps $Q(t_0, K, \eta)$ to a subset of $Q(t_0, K, \eta)$. One can then observe $z = A^{-\alpha}(0)w$ is the solution (2.3). This may be used to show that the operator T is continuous when $Q(t_0, K, \eta)$ is considered as a subset of the Banach space $C([0, t_0], X)$ and that the image of this subset is precompact. We can now use the Schauder fixed point theorem to insure that T has a fixed point $y \in Q(t_0, K, \eta)$ thus

$$y(t) = A^{\alpha}(0)W(t,0)\phi(0) + A^{\alpha}(0)\int_0^t W(t,s)F_y(s)ds.$$

We then observe $x(\phi)(t) = A^{-\alpha}(0)\hat{y}(t)$ satisfies

$$x(\phi)(t) = W(t,0)\phi(0) + \int_0^t W(t,s)F(s,x_s(\phi))ds \tag{2.4}$$

and the Hölder continuity of $F(t,x_t(\phi))$ in t will guarantee the differentiability. Our argument is a direct adoptation of the abstract quasilinear parabolic theory of Sobolevskii [15] and the reader is referred there for details.

We remark that one can obtain unique existence of solutions if Lipschitz conditions are placed on $F(t,A^{-\alpha}(0)\phi)$. In this case the proof relies on the Banach fixed point theorem rather than the Schauder theorem. Much simplier techniques will yield existence results in the case of positive definite delays. A delay is said to be positive definite if there exists a $d > 0$ so that $F(t,\phi) = F(t,\psi)$ whenever $\phi(\theta) = \psi(\theta)$ for $\theta < -d$. In this case the value of $F(t,\phi)$ depends on values of $\phi(\theta)$ for $\theta < -d$ and if $t < d$ (2.4) can be solved by direct integration. It would be desirable to be able to obtain global existence theorems for solutions. However, in the nondelay case these results appear to depend on apriori bounds of solutions which are unsatisfactory for delay equations.

III. THE SEMILINEAR CASE

In this section, we discuss the case where $F: R \times C \to X$ is required to continuous. Here we are able to obtain local existence of solutions and then apply classical techniques of ordinary differential equations to discuss the behavior of these solutions. The details of the theorems to be presented in this section appear in [8]. We replace Condition (S_4) with the following:

(S_4') For each $t \in [0,T]$ and some $\lambda \in \rho(A(t))$, the resolvent set of $A(t)$, $R(\lambda, A(t))$ is a compact completely continuous operator.

Condition (S_4') together with condition (S_2) implies that for each $\tau \in [0,T]$, $A(\tau)$ is the infinitesimal generator of a semigroup $T(t)$ which is compact for $t > 0$. We are now in a position to present the following local existence theorem:

Theorem 2. Let $\{W(t,s) \mid 0 \leqslant s \leqslant t \leqslant T\}$ be a family of linear evolution operators generated by a family of linear operators, $\{A(t) \mid t \in [0,T]\}$ which satisfy conditions (S_1) through (S_3) and (S_4'). If F is a continuous function from $R \times C$ to X then for each $\phi \in C$ there exists a $t_1 = t_1(\phi)$ and a function $x(\phi)$: $I \cup [0,t_1] \to X$ such that

$$x(\phi)(t) = W(t,0)\phi(0) + \int_0^t W(t,s)F(s,x_s(\phi))ds. \tag{3.1}$$

Theorem 2, is proved by employing a representation for $W(t,\tau)$ which appears in [15] to show that the operator $W(t,\tau)$ is compact whenever $t - \tau > 0$. This permits use of a Schauder fixed point argument to solve (3.1).

Once we have local existence we can utilize classical methods to discuss the extendability of these solutions. Our first result in this direction is:

Theorem 3. Let $\{A(t) \mid t \in [0,T]\}$ and F: $R \times C \to X$ satisfy the conditions of Theorem 2, with α and C independent of t. If F maps bounded subsets to bounded subsets then for every $\phi \in C$, then (3.1) has a maximal interval of existence $[0,t_{max})$.

If $t_{max} < \infty$ then $\overline{\lim_{t \to t_{max}}} \| x(\phi)(t) \| = \infty.$

This result allows us to provide conditions to guarantee the global existence of solutions to (3.1). We basically require that our nonlinear term is linearly bounded.

Theorem 4. Let $\{A(t) | t \in [0,T]$ and $F: R \times C \to X$ satisfy the conditions of Theorem 2. If there exist locally integrable functions K_1 and K_2 so that $\|F(t,\phi)\| \leq K_1(t) + K_2(t)\|\phi\|_C$ for $t \in [0,\infty)$ and $\phi \in C$, then equation (3.1) admits global solutions.

The above result is proved by using a Gronwall argument to establish the existence of a function $M(t) > \| x(\phi)(t)\|$ and applying the previous theorem.

Equation (3.1) need not be differentiable. We therefore turn our attention to the question of regularity and provide conditions which allow the differentiation of (3.1) and insure that the solutions satisfy the functional differential equation. We drop condition (S_4') and assume that the integral equation is well-posed.

Theorem 5. Let $\{W(t,s) |\ 0 \leq s \leq t \leq T\}$ be a linear evolution operator generated by a family of linear operators which satisfy conditions (S_1) through (S_3). Let $x(\phi)$ be a solution to (3.1) on $[0,T]$. If $\phi \in C$ is uniformly Hölder continuous and F is uniformly Hölder continuous in both variables, i.e. there exists an $L > 0$ and $0 < \beta < 1$ so that

$$\|F(t,\phi) - F(\tau, \psi)\| \leq L(|t - \tau|^{\beta} + \|\phi - \psi\|_C^{\beta})$$

then $x(\phi)$ is differentiable for $t > 0$ and satisfies

$$\dot{x}(\phi)(t) + A(t)x(\phi)(t) = F(t,x_t(\phi)). \tag{3.2}$$

We have assumed the existence of a function $x(\phi)(t)$ which satisfies the integral equation. The construction of the evolution operator insures the Hölder continuity of $x(\phi)(t)$. This together with the conditions on ϕ and F will make the function $f(t) = F(t,x_t(\phi))$ Hölder continuous and the theory of inhomogeneous linear equations allows the differentiation of the integral equation to provide solutions to $\dot{x}(\phi)(t) + A(t)x(\phi)(t) = F(t,x_t(\phi))$.

We remark that in the case of initial functions with finite domain we have finite propogation of singularities for non Hölder ϕ. Specifically, if $I = [-r,0]$ then for $t > r$, $\dot{x}(\phi)(t)$ exists and satisfies (3.2). We can also use the theory of linear equations to provide information on the convergence of solutions as $t \to \infty$, cf. [8].

Unique solutions to (3.1) on $[0,T]$ give rise to nonlinear evolution operators. Clearly our existence theory will provide solutions eminating from any $\phi \in C$ at time τ. If $x(\phi)$ is the solution to (3.1) satisfying $x_\tau(\phi) = \phi$ we define $U(t,\tau): C \to C$ by the equation

$$U(t,\tau)\phi = x_t(\phi).$$

It is not difficult to see that $U(t,\tau)$ is a nonlinear evolution operator on C. Moreover its infinitesimal generator

$$\hat{A}(t)\phi = \lim (U(t + h,t)\phi - \phi)/h$$

can be characterized by the equation

$$\hat{A}(t)\phi\ (\theta) = \dot{\phi}(\theta)$$

where

$$D(\hat{A}(t)) = \{\phi |\ \dot{\phi} \in C,\ \dot{\phi}(0) \in D_A \text{ and}$$

$$\dot{\phi}^{-}(0) = -A(t)\phi(0) + F(t,\phi)\}$$

In general $\hat{A}(t)$ is nonlinear by virtue of its nonlinear domain and it can be shown to vary in t. In certain cases it can be used to provide a product integral of these solutions cf. [3], [6].

IV. EXAMPLES

In this section we provide examples with which will hopefully illustrate the applicability of our theory to nonlinear parabolic equations with delay.

We first adapt an example of Sobolevskii [15]. Let Ω be a bounded open region with boundary Γ in n-dimensional complex Euclidean space. We consider the problem:

$$(4.1)\quad \partial v(x,t)/\partial t - \sum_{i,k=1}^{m} a_{ik}(t,x)\partial^2 v(t,x)/\partial x_i \partial x_k + a(x)v(t,x)$$
$$= f(t,x,v(\omega(t)),\partial v(\omega_1(t),x)/\partial x,)$$
$$\ldots\ v(\omega_m(t),x)/\partial x_m)$$
$$t \in [0,T], x = (x_1, \ldots, x_m) \in \Omega$$

where

$$\begin{array}{lll} v(t,x) = 0 & x \in \Gamma, & 0 \leqslant t \leqslant T \\ v(s,x) = \psi(s,x) & x \in \Omega, & s \in [-r,0] \\ t - r \leqslant \omega(t) \leqslant t & t \in [0,T] & \\ t - r \leqslant \omega_i(t) \leqslant t & t \in [0,T]. & \end{array}$$

We require that the function $a(x)$, $a_{ik}(x,t)$, and $f(t,x,v,v_1, \ldots, v_m)$ be continuous in all variables and satisfy a Hölder condition in all places except perhaps x. The functions $a_{ik}(x,t)$ be continuously differentiable in x, the coefficient matrix $[a_{ik}(x,t)]$ be Hermitian and have the property that there is a $c > 0$ so that

$$\sum_{i,k=1}^{m} a_{ik}(x,t)\ \gamma_i\bar{\gamma}_k > c \sum_{i=1}^{m} |\gamma_i|^2.$$

The functions ω, ω_i are required to be Hölder continuous and the function Ψ is required to be Hölder continuous and twice continuously differentiable in x. We further assume that

$a(x) > a_0 > 0$ for $x \in \Omega$. We formally define the operator

$$L(t)v = \sum_{i,k=1}^{m} a_{ik}(x,t)\partial v/\partial x_i \partial x_k - a(x)v$$

and define the operator $A(t)v$ on the Banach space $X = L^p(\Omega)$ by $A(t)v = - L(t)v$ for all $v \in W_p^2(\Omega)$ which vanish on $L^p(\Omega)$. We let $I = [-r,0]$ and $C = C(I,X)$. We define F: $I \times C \to X$ pointwise for $x \in \Omega$ by

$$F(t,\phi)(x) = f(t,x,\phi(\omega(t) - t,x), \partial\phi(\omega_1(t) - t,x)/ \partial x_1$$

$$\ldots, \partial\phi(\omega_m(t) - t,x)/\partial x_m).$$

From [15, p. 60] we see that for sufficiently large p, and $\beta \in (^1/_2, 1)$ the operators $(A(0))^{-\beta}$ and $\partial/\partial x\ (A(0))^{-\beta}$ act completely continuously from $L^p(\Omega)$ into the space of functions satisfying some Hölder condition. The operator $F(t,A^{-\beta}(0)\phi)$ can be shown to be Hölder continuous in t, and ϕ. Thus, if $\phi \in D(A(0))$ we can view (4.1) as the Banach space equation

$$\partial v(t)/\partial t + A(t)v(t) = F(t,v_t)$$

and if $\phi(\theta) \in D(A(0))^{-\alpha}\alpha > \beta$ our existence theorem guarantees a local solution with $v_0 = \phi$.

Our second example deals with a semilinear problem with infinite delay. The linear portion of the problem is developed in [10]. Let Ω be a bounded domain in R^n and Q_t be the cylinder

$$\{(x,t)\ x \in \Omega, \quad 0 \leqslant t \leqslant T\},$$

S_T is the lateral boundary $\{(x,t)|x \in \partial\Omega, \quad 0 \leqslant t \leqslant T\}$ and Ω_s is the cross section of Ω_s at s. We consider a problem of the form

$$(4.2) \quad \partial v(x,t)/\partial t + \sum_{|\alpha|\leqslant m} a_\alpha(x,t)D^\alpha v(x,t) = \int_{-\infty}^{t} g(t-s)f(v(x,s)ds,$$

where the following hold:

(i) $L(t)u = \partial u/\partial t + \sum_{|\alpha| \leqslant m} a_\alpha(x,t)D^\alpha u$ is uniformly parabolic in the sense of Friedman [10].

(ii) $\partial^i v/\partial \nu_j = 0$ on S_T $(0 \leqslant j \leqslant m-1)$; ν is the outward normal

(iii) $v(x,s) = \phi(x,s)$, $x \in \Omega_0$, $s \leqslant 0$

(iv) f is continuous, g is continuous and $\int_{-\infty}^{0} |g(-s)| < \infty$.

We rewrite (4.2) as an abstract semilinear functional equation in $X = L^2(\Omega)$.

(4.3) $$\dot{x}(t) = A(t)x(t) + F(t,x_t)$$

$A(t)$ is that operator defined in X by the action of the partial differential operator on $D_A = H^{2m}(\Omega) \cap H_0^m(\Omega)$. We let I denote the interval $(-\infty, 0)$ and define F from C to X by the equation

$$F(\phi)(x) = \int_{-\infty}^{0} g(-s)f(\phi(x,s))ds$$

In [10], $A(t)$ is shown to satisfy conditions (S_1) through (S_3). The compactness condition on the resolvent is met because it maps X to $D(A)$. Theorem 2 insures the local existence of mild solutions to (4.3). If we further require that ϕ is Hölder continuous we can differentiate the integral equation to obtain (4.3).

V. REFERENCES

[1] Brewer, D. W., "*A nonlinear semigroup for a functional differential equation,*" Dissertation, University of Wisconsin, 1975.

[2] Browder, F., "*Nonlinear equations of evolution*", Ann. Math. 80 (1964), pp. 485-523.

[3] Dyson, J. and Bressan, R. Villella, "*Functional differential equations and evolution operators*", Edinburgh J. Math. (to appear).

[4] Fitzgibbon, W., "*Abstract nonlinear Volterra operations with infinite delay*", Nonlinear Systems and Applications, Academic Press, New York (1977), pp. 513-525.

[5] _______, "*Nonlinear Volterra equations with infinite delay*", Monat. fur Math. (to appear).

[6] _______, "*Representation and approximation of solutions to semilinear Volterra equations with delay*", (to appear).

[7] _______, "*Abstract functional differential equations*", Proceedings of Seventh Annual USL Mathematics Conference, Department of Publications, University of Southwestern Louisiana, Lafayette, Lousiana (1976), pp. 1-12.

[8] _______, *Semilinear functional differential equations in Banach space*, Journal of Differential Equations, (to appear).

[9] _______, "*Stability for abstract nonlinear Volterra equations with finite delay*", Journal Mathematical Analysis and Applications, 60 (1977), pp. 429-434.

[10] Friedman, A., *Partial Differential Equations*, Holt, Rinehart and Winston, New York, 1969.

[11] Hale, J., *Functional Differential Equations*, Appl. Math. Series, Vol. 3, Springer - Verlag, New York, 1969.

[12] Krein, S. G., *Linear Differential Equations in Banach Space*, American Mathematical Society, Providence, Rhode Island, 1971.

[13] Ladas, G. and Lakshmikantham, V., *Differential Equations in Abstract Space*, Academic Press, New York, 1972.

[14] Pazy, A., "*A class of semilinear equations of evolution*", Israel Journal of Mathematics 20 (1975), pp. 23-35.

[15] Sobolevskii, P. E., "*Equations of parohlic type in Banach space*", American Mathematical Society Translations, Series

2, 49 (1966), pp. 1-62.

[16] _______, *"On the differentiability and compactness of semigroups of linear operators"*, J. Math. Mech. 17 (1968), pp. 1131-1139.

[17] Travis, C. and Webb, G., *"Existence, stability and compactness in the* α*-norm for partial functional differential equations"*, (to appear).

[18] _______, *"Existence and stability for partial functional differential equations"*, Trans American Mathematical Society 200 (1976), pp. 395-418.

[19] _______, *Partial differential equations with diviating arguments in the time variable*, J. Math. Anal. Appl., (to appear).

[20] Webb, G., *"Autonumous nonlinear functional differential equations"*, J. Math. Annal. Appl. 46 (1974), pp. 1-12.

[21] _______, *"Asymptotic stability for abstract nonlinear functional differential equations"*, Proc. Amer. Math. Soc., (to appear).

[22] Yosida, K., *Functional Analysis*, Springer - Verlag, New York, 1968.

THE EXACT AMOUNT OF NONUNIQUENESS FOR SINGULAR ORDINARY DIFFERENTIAL EQUATIONS IN BANACH SPACES WITH AN APPLICATION TO THE EULER-POISSON-DARBOUX EQUATION

Jerome A. Goldstein*
Tulane University

I. INTRODUCTION

The classical Euler-Poisson-Darboux equation is

$$\frac{\partial^2 u}{\partial t^2} + \frac{\rho}{t}\frac{\partial u}{\partial t} = \frac{\partial^2 u}{\partial x^2} \qquad (t > 0, \quad -\infty < x < \infty),$$

It has been known for a long time that uniqueness holds for the Cauchy problem (i.e. specifying $u(0,x)$) provided $\rho \geqslant 0$, whereas uniqueness fails when ρ is negative. One of the purposes of this paper is to show quantitatively how much nonuniqueness occurs when $\rho < 0$; it turns out that $-\rho$ is a measure of the amount of nonuniqueness. This result will be deduced as a consequence of a new uniqueness theorem for nonlinear differential equations in Banach spaces. This illustrates the principle that new and interesting linear results can be obtained from nonlinear analysis.

Let X be a Banach space and let f map $D \subset [0,\infty) \times X$ into X. There is a substantial literature devoted to the question of uniqueness for the initial value problem

(1) $$du/dt = f(t,u), \quad u(0) = u_0.$$

If $f(t,u)$ is well-behaved at and near $t = 0$, then (1) has a unique solution u and one can find the derivatives of u at $t = 0$ from (1);

$$u'(0) = f(0,u_0), \; u''(0) = \frac{\partial f}{\partial t}(0,u_0) + \frac{\partial f}{\partial u}(0,u_0)u'(0), \text{ etc.}$$

* Partially supported by an NSF grant.

ISBN 0-12-434160-8

We are interested in the case when f is singular at $t = 0$, i.e. $f(t,u)$ "blows up" in some sense as $t \to 0^+$ and $f(0,u)$ is not defined. Many interesting equations which arise in the applications have this form; cf. [3]. In this case it is well-known that nonuniqueness can occur for (1). In particular, in this case (1) does not determine the derivatives $u^{(k)}(0)$ for $k \geqslant 1$. Two questions now arise quite naturally. Is the solution of $u' = f(t,u)$ uniquely determined by $u^{(k)}(0) = u_k$, $k = 0, 1, \ldots, n$ for some value of n? If so, what is the minimum value of n?

The classical Nagumo criterion involves three assumptions: (i) X is finite dimensional, (ii) f is continuous, (iii) $f(t,u)$ satisfies a modulus of continuity condition involving $1/t$. Recently several authors have shown that (i) and (ii) can be dispensed with and (iii) can be replaced by a certain one-sided bound, namely, a dissipative (or accretive) condition on $f(t,u)$ involving $1/t$. (See Goldstein [8], Deimling [4], and the bibliographies of these papers.)

It is well-known that the Nagumo type results are sharp in that $1/t$ cannot be replaced by α/t for $\alpha > 1$. For example, Donaldson and Goldstein [5] showed this in the context of the Euler-Poisson-Darboux (EPD) equation.

On the other hand, in the context of (i) and (ii), (iii) can be weakened by replacing $1/t$ with α/t if certain additional assumptions are made on f; see Bounds and Metcalf [2], Bernfield, Driver and Lakshmikantham [1], and Gard [6].

The purpose of this paper is to give (in Section 2) sharp criteria for uniqueness when f satisfies a dissipative condition involving α/t. In this case it is necessary to specify $u^{(k)}(0)$ for $k \leqslant n$ where n is the largest integer satisfying $n < \alpha + 1$. This abstract theorem is applied (in Section 3) to an abstract EPD equation; the result is a precise description of the minimum number of initial conditions needed to determine a solution uniquely. These results are new even in the classical (one space dimensional) case. Moreover, we explicitly construct solutions which show that these results are best possible.

II. THE ABSTRACT UNIQUENESS THEOREM

Let X be a real or complex Banach space with norm $\|\cdot\|$ and dual space X^*. The image of $x \in X$ under $\psi \in X^*$ will be denoted by $\langle x,\psi\rangle$. For each $x \in X$ let $J(x)$ be the (nonempty) set of all $\psi \in X^*$ for which $\langle x,\psi\rangle = \|x\|^2 = \|\psi\|^2$. J is called duality mapping of X. $\mathcal{D}(f)$ will denote the domain of an operator f. Let $f\colon \mathcal{D}(f) \subset X \to X$. f is called dissipative if, for $x,y \in \mathcal{D}(f)$, there is a $\psi \in J(x - y)$ such that $\mathrm{Re}\langle f(x) - f(y),\psi\rangle \leqslant 0$. We could equally well treat multi-valued functions f; this would be useful in an existence context, but it only complicates the notation unnecessarily as far as uniqueness is concerned. For a brief discussion of J when $X = L^p$, see [8, p. 94]. When X is a Hilbert space, J is the identity map I and f dissipative means $\mathrm{Re}\langle f(x) - f(y), x - y\rangle \leqslant 0$ for $x,y \in \mathcal{D}(f)$, where $\langle\cdot,\cdot\rangle$ is the inner product in X. The readers who are primarily interested in the application to the EPD equation should take X to be Hilbert and $J = I$ in this section; this will simplify the proofs. However, dissipative conditions can sometimes be shown to hold only with respect to special non-Hilbert norms. Thus, for instance, when dealing with equations related to conservation laws [resp. the Hamilton-Jacobi equation], we expect the L^1 norm [resp. the L^∞ norm] to be the appropriate norm; cf. [9]. These considerations justify our treating the general Banach space case.

Our final preliminary observation is the following simple but useful observation due to Kato [10, p. 510].

Lemma 1. Suppose u has a weak derivation $u'(s) \in X$ at $t = s$, and suppose $\|u(\cdot)\|$ is differentiable at $t = s$. Then

$$\|u(s)\|\,(d/dt)\|u(t)\|\;\Big|_{t=s} = \mathrm{Re}\,\langle u'(s),\psi\rangle$$

for each $\psi \in J(u(s))$.

For completeness we quote Kato's short proof.

$$\mathrm{Re}\,\langle u(t),\ \psi\rangle \leqslant \|u(t)\|\ \|\psi\| = \|u(t)\|\ \|u(s)\|$$

(for $\psi \in J(u(s))$ and $\mathrm{Re}\,\langle u(s), \psi \rangle = \|u(s)\|^2$ together imply

$$\mathrm{Re}\,\langle u(t) - u(s), \psi \rangle \leqslant \|u(s)\| \, (\|u(t)\| - \|u(s)\|).$$

Dividing both sides by $t - s$ and letting t approach s from above and from below yields

$$\mathrm{Re}\,\langle u'(s), \psi \rangle \lesseqgtr \|u(s)\| \, (d/dt) \, \|u(t)\| \big|_{t=s}. \qquad \square$$

We can now state the uniqueness theorem.

Theorem 1. Let $f\colon \mathcal{D}(f) \subset [0,c) \times X \to X$, $0 < c \leqslant \infty$. Suppose that $f(t,\cdot) - (n/t)I$ is dissipative for some positive integer n and all t, $0 < t \leqslant c$. Then given $u_0, u_1, \ldots, u_n \in X$ there is at most one function $u\colon [0,c] \to X$ such that

(i) u is locally strongly absolutely continuous and has a strong derivative a.e. in $(0,c)$,

(ii) $u'(t) = f(t,u(t))$ a.e. on $(0,c)$,

(iii) $u^{(k)}(0)$ exists for $k = 0, \ldots, n$,

(iv) $u^{(k)}(0) = u_k$ for $k = 0, \ldots, n$.

The case $n = 1$ of this theorem was proved in 1971 by Goldstein [8; cf. Footnote 1 on p. 91]; see also Deimling [4]. Note that $f(t,\cdot) - (n/t)I$ is dissipative is equivalent to: for $x,y \in \mathcal{D}(f(t,\cdot))$ there is a $\psi \in J(x - y)$ so that

$$\mathrm{Re}\,\langle f(t,x) - f(t,y), \psi \rangle \leqslant (n/t) \, \|x - y\|^2.$$

Proof of Theorem 1. Let u,v satisfy (i), (ii), (iii), and (iv) with the same data, i.e., $u^{(k)}(0) = v^{(k)}(0) = u_k$, $k = 0, 1, \ldots, n$. We must show that $u(t) = v(t)$ for $0 \leqslant t < c$.

For almost all $t \in (0,c)$, there is a ψ in $J(u(t) - v(t))$ such that

$$\|u(t) - v(t)\| \frac{d}{dt} \|u(t) - v(t)\| = \mathrm{Re}\,\langle f(t,u(t)) - f(t,v(t)), \psi \rangle \tag{2}$$

$$\leqslant \frac{n}{t} \, \|u(t) - v(t)\|^2; \tag{3}$$

the equality in (2) holds by Lemma 1 (for all $\psi \in J(u(t) - v(t))$), and (3) holds (for a particular ψ in $J(u(t) - v(t))$) by the dissipative assumption. Let

$$\phi(t) = t^{-n}\|u(t) - v(t)\|, \quad 0 < t < c.$$

ϕ is locally absolutely continuous on $(0,c)$ and for a.e. $t \in (0,c)$,

$$\|u(t) - v(t)\|\frac{d\phi(t)}{dt} = \|u(t) - v(t)\|^2(-nt^{-n-1})$$

$$+ t^{-n}\|u(t) - v(t)\|\frac{d}{dt}\|u(t) - v(t)\| \leqslant 0$$

by (3). Thus $\frac{d\phi}{dt}(t) \leqslant 0$ for each t for which $u(t) \neq v(t)$. But $\frac{d\phi}{dt}(t) = 0$ for a.e. t for which $(\phi(t) =)\|u(t) - v(t)\| = 0$. It follows that the locally absolutely continuous function ϕ is nondecreasing on $(0,c)$. (Cf. [8, p.93].) We CLAIM that $\lim_{t\to 0^+} \phi(t) = 0$.

This claim, together with the monotonicity and nonnegativity of ϕ, implies that $\phi \equiv 0$ or $[0,c)$, hence $u \equiv v$ on $[0,c)$ and the proof is finished.

Thus it only remains to prove the claim. By Taylor's theorem of the calculus,

$$u(t) = \sum_{k=0}^{n} \frac{t^k u^{(k)}(0)}{k!} + o(t^n) = \sum_{k=0}^{n} \frac{t^k}{k!} u_k + o(t^n)$$

as $t \to 0^+$, by (iv); similarly

$$v(t) = \sum_{k=0}^{n} \frac{t^k}{k!} u_k + o(t^n)$$

as $t \to 0^+$. Consequently,

$$\|u(t) - v(t)\| = o(t^n)$$

as $t \to 0^+$. This clearly implies the claim. □

III. APPLICATION TO SINGULAR SECOND ORDER LINEAR EQUATIONS

Let S be a self-adjoint operator on a complex Hilbert space Y. (More generally, we could replace iS by B, the infinitesimal generator of a (C_0) group of bounded linear operators acting on a real or complex Banach space Y. We could even work on a much more general setting [7], but the case of self-adjoint

S is simpler and is adequate for the applications in this section.)

Consider the equation

(4) $$v''(t) + \alpha(t)\, v'(t) + S^2 v(t) = 0 \quad (t > 0)$$

together with the initial conditions

(5;m) $$v^{(k)}(0) = v_k, \quad k = 0, 1, \ldots, m.$$

α is assumed to be a complex-valued function on $(0,\infty)$. Here and for the remainder of the paper, m and n are nonnegative integers.

Theorem 2. The initial value problem (4), (5;n+1) has at most one solution on $[0,c)$ provided that $\mathrm{Re}(\alpha(t)) \geqslant -n/t$ for $0 < t < c$.

Proof: By a solution of (4), (5;m) we mean an m-times continuously differentiable $u\colon [0,c) \to Y (u \in C^m([0,c);\, Y)$ for short) satisfying (4) for $0 < t < c$ and (5;m). Let $\mathcal{D}(S)$, ker (S) denote the domain and kernel of S. Let X be the completion of $(\mathcal{D}(S)/\ker(S)) \times Y$ in the norm $\| \binom{x}{y} \|_X = (\|Sx\|_Y^2 + \|y\|_Y^2)^{1/2}$.

Let

$$u(t) = \begin{pmatrix} v(t) \\ v'(t) \end{pmatrix}, \quad A = \begin{pmatrix} 0 & I \\ -S^2 & 0 \end{pmatrix},$$

$$P(t) = \begin{pmatrix} 0 & 0 \\ 0 & -\alpha(t)I \end{pmatrix}, \quad u_k = \begin{pmatrix} v_k \\ v_{k+1} \end{pmatrix},$$

$k = 0, 1, \ldots, m - 1$. Then one can readily check that iA, with domain $\mathcal{D}(S\ ) \times \mathcal{D}(S)$, is a self-adjoint operator on the Hilbert space X, and the problem (4), (5;m) becomes equivalent to the problem

(6) $$u'(t) = f(t,u(t)) \quad \text{a.e. on } (0,c)$$

(7;n) $$u^{(k)}(0) = u_k, \quad k = 0, 1, \ldots, n$$

under the above identifications together with

$$f(t,u) = (A + P(t))u, \quad m - 1 = n/$$

The condition that $\mathrm{Re}(\alpha(t)) \geqslant - n/t$ now becomes equivalent to the dissipativity of $f(t,\cdot) - (n/t)I$. Theorem 2 thus follows from Theorem 1. □

We now apply Theorem 2 to the abstract Euler-Poisson-Darboux (EPD) equation

$$(8;\rho) \qquad v''(t) + \frac{\rho}{t} v'(t) + S^2 v(t) = 0.$$

Here S is a self-adjoint operator on a complex Hilbert space Y and ρ is a complex constant.

For a comprehensive bibliography of the literature related to the EPD equation, see the recent book of Carroll and Showalter [3].

Let n be a nonnegative integer. By an n-<u>null solution</u> of (8;ρ) we mean a solution v of (8;ρ) satisfying

$$v \in C^{n+1}([0,\infty);\ Y) \cap C^2((0,\infty);\ Y) \text{ and } v^{(k)}(0) = 0 \text{ for } k = 0,\ 1,\ \ldots,\ n+1.$$

Saying that an n-null solution of (8;ρ) is zero is equivalent to saying that the data $v^{(k)}(0)$ $(k = 0, 1, \ldots, n+1)$ uniquely determines a solution of (8;ρ).

Theorem 3. An n-null solution of (8;ρ) is zero if $\mathrm{Re}(\rho) \geqslant - n$. □

Proof: This is an immediate consequence of the preceding remark and Theorem 2; take $\alpha(t) = \rho/t$. □

Next we show that the hypotheses of Theorem 3 (and hence those of Theorem 1 and 2 as well) are best possible.

Theorem 4. Let n be a nonnegative integer. For each non-integral negative number $\rho < - n$, there is a nonzero solution of (8;ρ) which is n-null.

In other words, Theorem 3 becomes false if $\mathrm{Re}(\rho) \geqslant - n$ is replaced by $\mathrm{Re}(\rho) \geqslant - (n - \varepsilon)$ for any $\varepsilon > 0$.

Proof of Theorem 4. For $\sigma > 0$, a solution of the EDP equation (8;σ) is given by

$$(9) \qquad v_\sigma(t) = \frac{\Gamma(\sigma+1)}{\Gamma(\sigma/2)\Gamma(1/2)} \int_{-1}^{1} (1-x^2)^{\frac{\sigma}{2}-1} e^{itxS}\phi \, dx$$

(cf. [5, p. 151]). Here $\{e^{irS}: r \in \mathcal{R}\}$ denoted the (C_0) unitary group generated by iS and $\phi \in \mathcal{D}(S^2)$. v_σ satisfies $v_\sigma(0) = \phi$ and $v_\sigma'(0) = 0$. Now let $\rho \in \mathcal{R}$ and let w_ρ be any solution of (8;ρ). Then for ρ negative and nonintegral,

$$u_\rho(t) = w_\rho(t) + t^{1-\rho} w_{2-\rho}(t)$$

is a solution of (8;ρ) on $(0,\infty)$ (cf. [5, p. 151]).

Now let n be a nonnegative integer and let $0 \neq \phi \in \mathcal{D}(S^m)$ when m is the larger of 2 and $n+1$. For $\rho < -n$, ρ nonintegral, let

$$u_\rho(t) = t^{1-\rho} v_{2-\rho}(t)$$

where $v_{2-\rho}$ is defined by (9). A straightforward computation shows that

$$u_\rho \in C^{n+1}([0,\infty); Y) \cap C^2((0,\infty); Y),$$

u_ρ satisfies (8;ρ), and $u_\rho^{(k)}(0) = 0$ for $k = 0, 1, \ldots, n+1$.

Thus u_ρ is an n-null solution of (8;ρ), but u_ρ is nonzero since $\phi \neq 0$. □

The results of Theorems 3 and 4 are new even in the classical case $(Y = L^2(\mathcal{R}^q),\ S^2 = -\Delta = -\sum_{j=1}^{q} \partial^2/\partial x_j^2)$, even in the one dimensional case $(q = 1)$. The case of ρ being a negative integer is not covered by Theorem 4, but we believe that Theorem 4 is valid for these values of ρ as well; see [5, p. 152, Remark 4].

We thank Professors V. Lakshmikantham and S. Leela for pointing out that some ideas related to our Theorem 1 in the classical (finite dimensional) case go back to S. Krein and M. Krasnoselskii in the mid-fifties. Cf. V. Lakshmikantham and S. Leela, *Differential and Integral Inequalities*, Vol. I, Academic, New York

(1969), Chapter 2.

IV. REFERENCES

[1] Bernfeld, S. R., Driver, R. D., and Lakshmikantham, V., *Uniqueness for ordinary differential equations*, Math. Systems Theory 9 (1976), pp. 359-367.

[2] Bounds, J. M., and Metcalf, F. T., *An extension of the Nagumo uniqueness theorem*, Proc. Amer. Math. Soc. 27 (1971), pp. 313-316.

[3] Carroll, R. W., and Showalter, R. E., *Singular and Degenerate Cauchy Problems*, Academic, New York, 1976.

[4] Deimling, K., *On existence and uniqueness for differential equations*, Ann. Mat. Pure Appl. 106 (1975), pp. 1-10.

[5] Donaldson, J. A., and Goldstein, J. A., *Some remarks on uniqueness for a class of singular abstract Cauchy problems*. Proc. Amer. Math. Soc. 54 (1976), pp. 149-153.

[6] Gard, T. C., *A generalization of the Nagumo uniqueness theorem*, to appear.

[7] Goldstein, J. A., *A perturbation theorem for evolution equations and some applications*, Ill. J. Math. 18 (1974), pp. 196-207.

[8] Goldstein, J. A., *Uniqueness for nonlinear Cauchy problems in Banach spaces*, Proc. Amer. Math. Soc. 53 (1975), pp. 91-95.

[9] Goldstein, J. A., *Nonlinear semigroups and nonlinear partial differential equations*, Proc. Colóquio Brasileiro de Matemática (Poços de Caldas, 1975), to appear.

[10] Kato, T., *Nonlinear semigroups and evolution equations*, J. Math. Soc. Japan 19 (1967), pp. 508-520.

ON THE EQUATION $Tx = y$ IN BANACH SPACES WITH WEAKLY CONTINUOUS DUALITY MAPS

Athanassios G. Kartsatos
University of South Florida

I. INTRODUCTION

Let X be a reflexive Banach space with dual X^*, and let T be an operator with domain $D(T) \subset X$ and values in X. In this paper, we are seeking solutions $x \in D(T)$ of the equation $Tx = y$, where y is a known element of X. The operator T will be assumed maximal monotone on its domain with respect to a weakly continuous duality map which will also be assumed to exist for the space X.

We show here that if, in addition to the above hypotheses, $D(T)$ contains zero and

(*)
$$\|T(0)\| < r \leqslant \liminf_{\substack{x \in D(T) \\ \|x\| \to \infty}} \|Tx\|,$$

then $Tx_0 = 0$ for some $x_0 \in D(T)$ Here r is a fixed positive constant. It can easily be seen that in the case of a Hilbert space the above condition is implied by coerciveness:

$$\lim_{\substack{x \in D(T) \\ \|x\| \to \infty}} \frac{\langle Tx, x \rangle}{\|x\|} = +\infty.$$

Conditions of the type (*) have been considered by Lange in [5], and the present results are extensions of those in [5]. Lange considered operators T defined on a Hilbert space H with values in H or on a Banach space B with values in the

ISBN 0-12-434160-8

dual space B^*. For several results in the spirit of this paper, the reader is referred to the works of Minty [6-7], Zarantonello [9] and Browder [2-3]. The book of Brezis [1] is an excellent reference of relevant onto-ness results in Hilbert spaces. Besides the above mentioned result, we provide conditions under which a neighborhood of zero belongs to the range of T (defined on the whole of X), or a neighborhood of the point x_0 is mapped onto a neighborhood of the point y_0 where x_0, y_0 are known points with the property $Tx_0 = y_0$.

II. PRELIMINARIES

In what follows, X will be a real reflexive Banach space. For $f \in X^*$, $x \in X$ we write $f(x) = \langle x, f \rangle$. Following Browder [2, p. 261], let μ: $R_+ \to R_+$ be continuous, strictly increasing and such that $\mu(0) = 0$. A "duality map J: $X \to X^*$ with gauge function μ" satisfies the following two conditions:

(i) $\langle x, Jx \rangle = \|Jx\| \, \|x\|$ for every $x \in X$,

(ii) $\|Jx\| = \mu(\|x\|)$ for every $x \in X$.

An operator T: $D(T) \to X$ is "J-monotone" if for every $x_1, x_2 \in D(T)$ we have

$$\langle Tx_1 - Tx_2, J(x_1 - x_2) \rangle \geq 0.$$

The operator T: $D(T) \to X$ $(D(T) \subset X)$ is said to be "demi-continuous" if $x_n, x \in X$ with $x_n \to x$ imply $Tx_n \rightharpoonup Tx$. Here $\to$ $(\rightharpoonup)$ denotes strong (weak) convergence.

T: $D(T) \to X$ is said to be "maximal J-monotone" if it is J-monotone and $(T + \lambda I)D(T) = X$ for every $\lambda > 0$. Here K is the identity operator on X. In what follows, J will be a weakly continuous duality map on X corresponding to a gauge function μ.

III. MAIN RESULTS

Lemma 1. Let $\{x_n\}$ be a bounded sequence in $D(T)$, y a fixed point in X and

(1) $$Tx_n + (1/n)x_n = y, \quad n = 1, 2, \ldots ,$$

where T is a maximal J-monotone operator with J weakly continuous. Then $\{x_n\}$ contains a subsequence $\{x_j\}$ such that $x_j \to x_0 \in D(T)$ and $Tx_0 = y$.

Proof: Letting $t_n = 1/n$ and using the monotonicity of T we obtain

(2) $$\langle t_m x_m - t_n x_n, J(x_m - x_n)\rangle = -\langle Tx_m - Tx_n, J(x_m - x_n)\rangle \leqslant 0.$$

Now, since X is reflexive, we may assume that $x_n \rightharpoonup$ (some) $x_0 \in X$. Since J is weakly continuous, taking the limit of the left hand side of (2) as $n \to \infty$ we obtain

(3) $$t_m \langle x_m, J(x_m - x_0)\rangle \leqslant 0, \quad m = 1, 2, \ldots .$$

From this we obtain

(4) $$\langle x_m - x_0, J(x_m - x_0)\rangle + \langle x_0, J(x_m - x_0)\rangle \leqslant 0,$$

or

(5) $$\mu(\|x_m - x_0\|)\|x_m - x_0\| \leqslant -\langle x_0, J(x_m - x_0)\rangle .$$

Since the right hand side of (5) tends to zero as $m \to \infty$ and μ is strictly increasing with $\mu(0) = 0$, it follows that $x_n \to x_0$. Since T is maximal, there exists $x \in D(T)$ such that

(6) $$(T + I)x = Tx + x = y + x_0.$$

Now we have

(7) $$\langle Tx - y, J(x_0 - x)\rangle = \langle x_0 - x, J(x_0 - x)\rangle = \mu(\|x_0 - x\|)\|x_0 - x\| \geqslant 0.$$

On the other hand,

(8) $$\langle Tx_n - Tx, J(x_n - x)\rangle \geqslant 0,$$

which, by the weak continuity of J, implies

(9) $$\langle y - Tx, J(x_0 - x)\rangle \geqslant 0.$$

From (7) and (9) we obtain now

(10) $$0 = \langle Tx - y, J(x_0 - x)\rangle = \mu(\|x_0 - x\|)\|x_0 - x\|.$$

This yields $x_0 = x$, or $x_0 \in D(T)$ and $Tx_0 = y$.

Lemma 2. Let $T: D(T) \to X$ be maximal J-monotone and $\{x_n\}$ a bounded sequence of $D(T)$ such that $Tx_n \to y \in X$. Then there exists a subsequence $\{x_j\}$ of $\{x_n\}$ such that $x_j \rightharpoonup x_0 \in D(T)$ and $Tx_0 = y$.

Proof: The conclusion of this lemma follows from the proof of Lemma 1, because the relations (6) - (9) hold also for a weakly convergent sequence $\{x_j\}$ replacing $\{x_n\}$.

Now we are ready for the main result of this paper.

Theorem 1. Let $T: D(T) \to X$ be maximal J-monotone, $0 \in D(T)$, and, for some $r > 0$,

(11) $$\|T(0)\| < r \leqslant \liminf_{\substack{x \in D(T)\\ \|x\| \to \infty}} \|Tx\|.$$

Then $Tx_0 = 0$ for some $x_0 \in D(T)$.

Proof: Since T is maximal, there exists a sequence $\{x_n\}$, $n = 1, 2, \ldots$ in $D(T)$ such that

(12) $$Tx_n + (1/n)x_n = 0, \quad n = 1, 2, \ldots .$$

To show that $Tx_0 = 0$ for some $x_0 \in X$ it suffices (by Lemma 1) to show that the sequence $\{x_n\}$ is bounded. Assume the contrary and let $\{x_j\}$ be a subsequence of $\{x_n\}$ such that

(13) $$\lim_{j\to\infty} \|x_j\| = +\infty .$$

Let t_j denote the corresponding subsequence of $\{\frac{1}{n}\}$ in (12).

Then we have

(14) $$0 \leqslant \langle Tx_j - T(0), J(x_j)\rangle = \langle -t_jx_j - T(0), J(x_j)\rangle,$$

or

(15) $$0 \leq -t_j \langle x_j, J(x_j)\rangle - \langle T(0), J(x_j)\rangle .$$

This implies

(16) $$t_j\|x_j\| \mu(\|x_j\|) \leq \|T(0)\| \mu(\|x_j\|),$$

or

(17) $$t_j\|x_j\| \leq \|T(0)\|.$$

However, (12) implies $\|Tx_j\| \leq t_j\|x_j\|$. Thus,

(18) $$r \leq \liminf_{\|x\|\to\infty} \|Tx\| \leq \liminf_{j\to\infty} \|Tx_j\| \leq \|T(0)\|.$$

This contradicts (11). Thus $T(x_0) = 0$ for some $x_0 \in X$.

Now let L be a class of operators $T\colon X \to X$ with the following properties:

(a) If $T \in L$ then $T_1 \in L$, $T_2 \in L$ where

$$T_1x = T(x + x_0) \qquad (x_0 \in X \text{ fixed})$$

$$T_2x = Tx + y_0 \qquad (y_0 \in X \text{ fixed}).$$

(b) If $T \in L$ and for some $r > 0$

$$\|T(0)\| < r \leq \liminf_{\|x\|\to\infty} \|Tx\|,$$

then there exists $x_0 \in X$ with $Tx_0 = 0$.

(c) If $T \in L$ and $x_n \in X$ with $Tx_n \to y_0$, and $\{x_n\}$ is bounded, then $Tx_0 = y_0$ for some $x_0 \in X$.

Lemma 3. Let $T \in L$ and $r > 0$ be such that

$$\|T(0)\| < r \leq \liminf_{\|x\|\to\infty} \|T(x)\|.$$

Then $K_r \subset T(X)$, where

$$K_r = \{x \in X;\ \|x\| < r\}.$$

We omit the proof of this lemma, because it is almost identical to Lange's proof [5, p. 16].

Theorem 2. Let $T\colon X \to X$ be maximal J-monotone and J

weakly continuous. Moreover let $r > 0$ be such that

(19) $$\|T(0)\| < r \leqslant \liminf_{\|x\| \to \infty} \|Tx\|$$

Then $K_r \subset T(X)$.

If $r > 0$, $s > 0$ are such that

(20) $$r \leqslant \inf_{\|x - x_0\| \geqslant s} \|Tx - Tx_0\|,$$

and $Tx_0 = y_0$, then $K_r(y_0) \subset T(K_s(x_0))$, where

$$K_r(u) = \{x \in X;\ \|x - u\| < r\}.$$

Proof: Let $r > 0$ be such that (19) holds

Then, by Theorem 1, there exists $x_0 \in X$ such that $Tx_0 = 0$. Now let L be the class of all maximal J-monotone operators. This class satisfies (a), (b) and (c) above. If fact, (b), (c) follow from Theorem 1 and Lemma 2. We only show that if $T \in L$ then $T_1 \in L$, where $T_1 u = T(u + x_0)$ for some fixed $x_0 \in X$. T_1 is obviously monotone. To show that it is maximal, let $\lambda > 0$ and $y \in X$. Also, because of the maximality of T. let $v \in X$ be such that $Tv + \lambda v = y + \lambda x_0$. Then $T_1(u_0) + \lambda u_0 = y$ where $u_0 = v - x_0$. Thus, T_1 is maximal. It follows from Lemma 3 that $K_r \subset T(X)$. The rest of the proof follows exactly as in Lange [5] and is therefore omitted.

IV. DISCUSSION

As it was mentioned in the introduction, Lange considered in [5] multi-valued, demicontinuous and monotonic operators $T\colon X \to X^*$, where X is a Hilbert, or a reflexive Banach space. Lange's proof of a result analogous to Theorem 1 does not carry over to the present case, because he makes use of a result of Browder [2], referring to strongly monotone operators $(\langle Tx - Ty, x - y\rangle \geqslant \alpha\|x - y\|^2, \alpha > 0)$ on a Banach or Hilbert space. Browder has shown in [2] that all the ℓ^p-spaces $(1 < p < +\infty)$ possess weakly continuous duality maps, but L^4

does not have one. Opial showed in [8] that no space L^p, $p > 1$, $p \neq 2$ possesses a weakly continuous duality map. It would therefore be very interesting to have results similar to those of Theorem 1 or 2 for spaces, which do not necessarily have weakly continuous duality maps. Theorem 1 suggests a fixed point theorem which we do now state:

"Let $S: D(S) \to X$ satisfy the following assumptions:

(i) $0 \in D(S)$ and $(S + \lambda I)D(S) = X$ for every $\lambda > 0$;

(ii) S is strongly J-monotone: $x, y \in D(S)$ imply

$$\langle S(x) - S(y), J(x - y)\rangle \geqslant \mu(\|x - y\|)\|x - y\|;$$

(iii) for some $r > 0$,

$$\|S(0)\| < r \leqslant \liminf_{\substack{\|x\| \to \infty \\ x \in D(S)}} \|Sx - x\|. \tag{21}$$

Then S has a fixed point in $D(S)$

To prove this theorem it suffices to show that the operator $T = Sx - x$ has a zero in $D(S)$. First, it is easy to see that T is J-monotone. Moreover $(T + 2I)D(S) = (S + I)D(S) = X$. This however implies that $(T + \lambda I)D(S) = X$ for every $\lambda > 0$ (cf. Kato [4, p. 511]). Thus, T satisfies all the assumptions of Theorem 1.

It should also be noted that if $D(T)$ and $D(S)$ in Theorem 1 and the above fixed point theorem are bounded, then both of these results hold without, of course, the conditions (11) and (21). This is due to the fact that the sequence $\{x_n\}$ in the proof of Theorem 1 belongs to $D(T)$ and is thus bounded. Actually, in this case Lemma 1 implies that T is an onto operator. This is Corollary 2.2 in [1] if X is a Hilbert space.

V. REFERENCES

[1] Brézis, H., *Operateurs maximaux monotones et semi-groupes de contractions dans les espaces de Hilbert*, North Holland

(1973).

[2] Browder, F. E., *Fixed point theorems for nonlinear semicontractive mappings in Banach spaces*, Arch. Rational Mech. Anal., 21 (1966), pp. 259-269.

[3] Browder, F. E., *Nonlinear maximal monotone operators in Banach spaces*, Math. Anallen, 175 (1968), pp. 89-113.

[4] Kato, T., *Nonlinear semigroups and evolution equations*, J. Math. Soc. Japan, 19 (1967), pp. 508-520.

[5] Lange, H., *Abbildungssatze fur monotone operatoren in Hilbert-und Banach-Raumen*, Dissert. Albert-Ludwigs-Universitat, Freiburg, Germany (1973).

[6] Minty, G. J., *Monotone (nonlinear) operators in Hilbert space*, Duke Math. J., 29 (1962), pp. 341-346.

[7] Minty, G. J., *On a monotonicity method for the solution of nonlinear equations in Banach spaces*, Proc. Nat. Acad. Sci., U.S.A., 50 (1963), pp. 1038-1041.

[8] Opial, Z., *Weak convergence of the sequence of successive approximations for non-expansive mappings*, Bull. Amer. Math. Soc., 73 (1967), pp. 591-597.

[9] Zarantonello, F. H., *Solving functional equations by contractive averaging*, Techn. Rep. Nr. 160, U.S. Army Research Center, Madison Wisconsin, 1960.

NONLINEAR EVOLUTION OPERATORS IN BANACH SPACES

Yoshikazu Kobayashi
Faculty of Engineering
Niigata University

We are concerned in this lecture with the nonlinear evolution operator associated with a system of time-dependent evolution equations

$$(\mathrm{DE})_s \qquad u'(t) \in A(t)u(t), \quad s < t < T; \quad u(s) = x,$$

which are formulated for $0 \leqslant s < T$ and a one-parameter family $\{A(t); \ 0 \leqslant t \leqslant T\}$ of nonlinear operators in a real Banach space X. Following a paper by K. Kobayasi, S. Oharu and me, we shall introduce a notion of generalized solution of the above equation and discuss the construction of an evolution operator which provides the generalized solutions.

Suppose for the moment that the initial-value problem for the evolution equation has a unique solution on $[s,T]$ for every x in a subset C of X and s in $[0,T]$ such that $u(t) \in C$ for $t \in [s,T]$. Defining the operator $U(t,s)\colon C \to C$ by $U(t,s) = u(t)$ where $u(t)$ is the solution of $(\mathrm{DE})_s$ associated with the initial-value x, we obtain the relation

(E1) $\quad U(t,t) = I/C$ (the identify operator on C) and

$$U(t,s)U(s,r) = U(t,r) \text{ on } C \text{ for } 0 \leqslant r \leqslant s \leqslant t \leqslant T.$$

One also will have that $U(t,s)$ is continuous on $[s,T]$ with respect to t for fixed s and $x \in C$ under most definitions of a "solution" of $(\mathrm{DE})_s$. Usually a stronger continuity of $U(t,s)$ with respect to the parameters s and t is obtained, namely:

ISBN 0-12-434160-8

(E2) $U(t,s)x$ is continuous over the triangle $0 \leqslant s \leqslant t \leqslant T$ with respect to s and t.

A family of operators $U(t,s)$ having properties (E1) and (E2) is called an evolution operator on C.

Ph. Bénilan has recently introduced in [1] a notion of integral solution as one of a generalized solution for the time-independent evolution equation. We first extend the notion of integral solution to the time-dependent case.

Next we consider the generation of evolution operators which provide the generalized solutions. A fundamental result on the construction of an evolution operators in general Banach spaces has been established by Crandall-Pazy [2]. We shall discuss the generation of evolution operators which provide the integral solutions mentioned above and extend their result.

There are two significances in our argument. First we think of a proper, lower semi-continuous functional $p(\cdot)$: $X \to [0,+\infty]$ to specify the t-dependence of $A(t)$, stability, etc. The t-dependence discussed in Crandall-Pazy [2] will be treated as a special case of ours by taking appropriate functional $p(\cdot)$.

Secondly, we discuss the construction of evolution operators through the discrete approximation of $(\mathrm{DE})_s$:

$$(\mathrm{DS})_s \quad \begin{cases} \dfrac{u_k^n - u_{k-1}^n}{t_k^n - t_{k-1}^n} = v_k^n + \varepsilon_k^n, \quad [u_k^n, v_k^n] \in A(t_k^n) \\ \lim_{n\to\infty} u_0^n = x, \quad k = 1, 2, \dots, N_n; \quad n = 1, 2, 3, \dots . \end{cases}$$

where

$$\Delta_n: \quad s = t_0^n < t_1^n < \dots < t_{N_n-1}^n < T \leqslant t_{N_n}^n, \quad \|\Delta_n\| = \max_k (t_k^n - t_{k-1}^n)$$

$$\lim_{n\to\infty} \|\Delta_n\| = 0, \quad \lim_{n\to\infty} \sum_{k=1}^{N_n} (t_k^n - t_{k-1}^n) \|\varepsilon_k^n\| = 0,$$

and x is the initial value. Defining step functions $u_n(t)$ by setting $u_n(s) = u_0^n$ and $u_n(t) = u_k^n$ for $t \in (t_{k-1}^n, t_k^n] \cap [s,T]$, the evolution operator $U(t,s)$ is obtained as $U(t,s)x = \lim_{n\to\infty} u_n(t)$. Recently Evans discussed in [3] the convergence of the approximate solutions $u_n(t)$ and treated the construction of evolution operators from the same point of view in Crandall-Pazy [2]. Here we treat a modified version of the estimation by the author [4] to treat the convergence of $u_n(t)$ and our argument turns out to be rather simpler than that of Evans [3].

REFERENCES

[1] Bénilan, Ph., *Équations d'évolution dans un espace de Banach quelconque et applications*, Thèse Orsay, 1972.

[2] Crandall, M. and Pazy, A., *Nonlinear evolution equations in Banach spaces*, Israel J. Math., 11 (1972), pp. 57-94.

[3] Evans, L. C., *Nonlinear evolution equations in an arbitrary Banach space*, Israel J. Math., 26 (1977), pp. 1-42.

[4] Kobayashi, Y., *Difference approximation of Cauchy problems for quasi-dissipative operators and generation of nonlinear semigroups*, J. Math. Soc. Japan, 27 (1975), pp. 640-665.

ABSTRACT BOUNDARY VALUE PROBLEMS

V. Lakshmikantham
The University of Texas at Arlington

I. INTRODUCTION

Consider the abstract boundary value problem

(1.1) $$x'' = H(t,x,x'), \quad 0 < t < 1,$$

(1.2) $$B^i x = \alpha_i x(i) + (-1)^{i+1}\beta_i x'(i) = b_i, \quad i = 0, 1.$$

Here $H \in C\Big[[0,1] \times B \times B, B\Big]$, B being a real Banach space, $\alpha_0,\alpha_1 \geqslant 0$ and $\beta_0,\beta_1 > 0$. In case $B = R^n$ existence was proved by first obtaining a priori bounds for $\|x(t)\|$, $\|x'(t)\|$ of a solution of (1.1) and (1.2) and then employing a theorem of Scorza-Dragoni [7, 11]. The methods involve assuming inequalities in terms of the second derivative of Lyapunov like functions relative to H, using comparison theorems for scalar second order equations and utilizing Leray Schauder's alternative or equivalently the modified function approach [2, 3, 7, 8].

To extend this fruitful method to the case when B is an arbitrary Banach space, it is first necessary to extend the basic result of Scorza-Dragoni. This can be achieved by imposing compactness-type condition on H and using Darbo's fixed point theorem. To extend the modified function approach, one needs a new comparison result, Lypunov like functions and an argument similar to the case in R^n with appropriate modifications. A general comparison result is proved which includes known results and is also more flexible in applications. Finally, it is shown how to extend monotone iterative methods to generate two-sided

ISBN 0-12-434160-8

point wise bounds on solutions of (1.1) and (1.2)

II. GENERAL COMPARISON RESULT

Let B be a real Banach space with $\|\cdot\|$ and let B^* denote the set of all continuous linear functionals on B. Let K be a cone in B which induces a partial ordering $\leqslant$ as follows: $x \leqslant y$ if and only if $y - x \in K$. A linear functional $\phi \in B^*$ is called a positive linear functional if $\phi(x) \geqslant 0$ whenever $x \in K$. Let K^* denote the set of all positive linear functionals. Note that K is contained in the closed half-space $C_\phi = [x \in B: \phi(x) \geqslant 0, \phi$ a positive linear functional]. Thus the positive linear functionals are support functionals for K and since K is a cone in B, K is the intersection of all the closed half-spaces which support it. If $S \subseteq K^*$ and $K = \cap [C_\phi: \phi \in S]$, then S is said to generate K. We suppose that S generates K and let $S_u = [\phi \in S: \|\phi\| = 1]$. We denote by K^0, the interior of K and $\overline{S_u}$ the closure of S_u in the weak star topology. If $K^0 \neq 0$, then K is called a solid cone.

Let S_u generate the cone K. We shall say that $H(t,x,x')$ is quasimonotone nonincreasing in x with respect to K if $x \leqslant y$, $\phi(x) = \phi(y)$, $\phi(x') = \phi(y')$ implies

$$\phi(H(t,x,x')) \geqslant \phi(H(t,y,y')) \quad \text{for} \quad \phi \in S_u.$$

Let us consider a family of functions $\{z_\lambda(t) = z(t,\lambda), \lambda > 0, t \in [0,1]\}$. We shall say that the family $z_\lambda(t)$ is admissible if $z(t,\lambda) \in C^2[I,B]$, $I = [0,1]$ for each λ, $z(t,\lambda)$ is continuous in λ for each $t \in I$ and $z(t,0) \equiv 0$. The family is said to satisfy a uniformity condition at $\lambda = 0 (\lambda = \infty)$ if for each $\phi \in S_u$, $\phi(z_\lambda(t)) \to 0$ as $\lambda \to 0$ uniformly in $t \in I$ ($\phi(z_\lambda(t)) \to \infty$ as $\lambda \to \infty$ uniformly in $t \in I$).

We then have the following comparison result [3a].

Theorem 2.1. Assume that

(i) $H(t,x,x')$ is quasimonotone nonincreasing in x relative to the cone K;

(ii) $V, W \in C^2[I,B]$ and for $t \in I$, $i = 0, 1$,

$V'' \geq H(t,V(t),V'(t))$, $B^i V \leq b_i$,

$W'' \leq H(t,W(t),W'(t))$, $B^i W \geq b_i$;

(iii) $\{z_\lambda(t)\}$ is an admissible family satisfying a uniformity condition at $\lambda = 0$ and $\lambda = \infty$ such that for $t \in I$ and for $\phi \in Su$,

$$\phi(z''_\lambda(t)) < \phi(H(t,W(t) + z_\lambda(t), W'(t) + z'_\lambda(t))) - \phi(H(t,W(t), W'(t)))$$

and $B^0 z_\lambda(0) > 0$, $B^1 z_\lambda(1) > 0$.

Then $V(t) \leq W(t)$ on I.

This theorem includes a result in [10] where the case $B = R$ is considered. Moreover this result also includes the comparison theorem in [6] which was concerned with the case $B = R^n$, $1 \leq n \leq \infty$. Observe that theorem 2.1 has enough flexibility since for a given $H(t,x,x')$ one may have several cones in which H is quasimonotone. Furthermore, we do not require the cone K to be solid and this is an asset in applications.

III. EXISTENCE RESULTS

For a bounded set $A \subset B$, let $\alpha(A)$ denote the Kuratowski's measure of noncompactness. To prove Scorza-Dragoni's Theorem in the abstract set up, we need the following assumptions:

(H_1) $H \in C[I \times B \times B, B]$ and for bounded set $A_1, A_2 \subset B$,

$$\alpha(H(I \times A_1 \times A_2)) \leq \beta \max [\alpha(A_1), \alpha(A_2)];$$

(H_2) $\|H(t,x,y)\| \leq L$ for $(t,x,y) \in I \times B \times B$.

Theorem 3.1. Let the assumptions (H_1) and (H_2) hold. Then there exists a solution $x \in C^2[I,B]$ of the problem (1.1) and (1.2) provided $\beta < \frac{1}{2}p$ where $p = \max [1, \sup_{I\times I} |G(t,s)|]$, $G(t,s)$

being the Green's function associated with the scalar boundary value problem $y'' = h(t)$, $B^i y = 0$, $i = 0, 1$.

If the assumption (H_2) is dispensed with then one can only prove a result which gives existence in the small. See for details [7]. To remove the condition (H_2) which is very restrictive and to prove existence in the large, we need additional conditions which are listed below.

(H_3) $f \in C[I \times R \times R, R]$, $W \in C^2[I,R]$ with $W(t) \geqslant 0$ and

$$W''(t) \leqslant f(t, W(t), W'(t)), \quad t \in (0,1).$$

such that $B^0 W(0) \geqslant b_0$, $B^1 W(1) \geqslant b_1$;

(H_4) $z \in C^2[I,R]$ with $z(t) > 0$ and for each $\lambda > 0$,

$$\lambda z'' < f(t, W + \lambda z, W' + \lambda z') - f(t, W, W')$$

such that $B^0 z(0) > 0$ and $B^1 z(1) > 0$,

(H_5) The left maximal solution $r(t, 1, \eta_1)$ and the right minimal solution $\phi(t, 0, \eta_0)$ of $v' = \hat{f}(t,v)$ exist on I where

$$\hat{f}(t,v) = \min_{0 \leq u \leq B_0} f(t,u,v), \; B_0 = \max_I W(t), \quad \eta_0 = \frac{-b_0}{\beta_0} \text{ and } \eta_1 = \frac{b_1}{\beta_1},$$

(H_6) $V \in C^2[I \times B, R^+]$ such that $V(t,x) \to \infty$ as

$$x \to \infty \text{ uniformly in } t \in I,$$

(H_7) for $0 \leqslant \delta \leqslant 1$,

$$V''_{\delta H}(t,x) \geqslant f(t, V(t,x), V'(t,x)) + \sigma\delta \| H(t,x,x') \|,$$

where $\sigma > 0$, $V''_{\delta H}(t,x) = U(t,x,x') + \delta V_x(t,x) H(t,x,x')$, and $U(t,x,x') = V_{tt}(t,x) + 2V_{tx}(t,x)x' + V_{xx}(t,x)(x',x')$, where $V_{xx}(t,x)$ is the bilinear operator mapping $B \times B$ into $L(B \times B, R)$.

(H_8) The boundary conditions (1.2) imply that

$$B^0 V(0, x(0)) \leqslant b_0 \quad \text{and} \quad B^1 V(1, x(1)) \leqslant b_1.$$

Theorem 3.2. Suppose that the assumptions (H_1), (H_3) to (H_8) hold. Then there exists a solution $x \in C^2(I,B)$ of the problem

(1.1) and (1.2) provided $\beta < \frac{1}{2}p$.

The proof of this Theorem utilizes the scalar version of the comparison theorem 2.1 and the modified function approach. For details see [7].

IV. MONOTONE ITERATIVE METHOD

Let us briefly describe the monotone method for the problem (1.1) and (1.2). The first step is to prove existence of solutions of (1.1) and (1.2) by the modified function approach combined with a Nagumo condition assuming the existence of upper and lower solutions $V(t)$, $W(t)$ such that $V(t) \leq W(t)$. This modified function approach is different than the one described in Section 3 which is not suitable to obtain monotone sequences which converge uniformly to maximal and minimal solutions of (1.1) and (1.2). See [3, 9] for details of such a method in case $B = R^n$. As before assume that $H(t,x,x')$ is quasimonotone relative to the cone K. Suppose also that $H(t,x,y)$ is continuously differentiable in x and y for $t \in I$ and $V(t) \leq x \leq W(t)$ and $\|y\| \leq N$, where N is obtained in view of Nagumo type condition, and the Frechet derivative $H_x(t,x,y)$ satisfies, for $z \in B$,

$$H_x(t,x,y)(z) \leq Q(z)$$

where $Q(z)$ is quasimonotone nonincreasing in z with respect to K. We then define for any $\eta(t)$ such that $V(t) \leq \eta(t) \leq W(t)$,

$$F(t,x,x') = H(t,\eta(t),x') + Q(x - \eta(t)).$$

one can show that assuming that $Q(z)$ is α-contraction, for each $\eta(t)$, the problem (1.2) and

$$x'' = F(t,x,x')$$

has a unique solution $x \in C^2(I,B)$ such that $V(t) \leq x(t) \leq W(t)$ on I and the mapping A defined by $A\eta = x$ is a monotone operator on the segment $\langle V,W \rangle = [u \in B\colon\ V(t) \leq u \leq W(t),\ t \in I]$. We then define the sequences $V_n = AV_{n-1}$, $W_n = AW_{n-1}$ with

$V_0 = V$ and $W_0 = W$ and show that the sequences $\{V_n\}$, $\{W_n\}$ converge uniformly and monotonically to minimal and maximal solutions of (1.1) and (1.2) respectively. If there is a unique solution to (1.1) and (1.2), then the sequences $\{V_n\}$, $\{W_n\}$ converge to that unique solution. We thus have a constructive method to obtain solutions of the problem (1.1) and (1.2).

It is clear that the results described above include the sequence of papers that have appeared in this direction [1, 4, 5]. See also other references given in these papers.

V. REFERENCES

[1] Bernfeld, S. and Chandra, J., *Minimal and maximal solutions of nonlinear boundary value problems*, Pacific J. of Math, (to appear).

[2] Bernfeld, S., Ladde, G., and Lakshmikantham, V., *Existence of solutions of two point boundary value problems for nonlinear systems*, Jour. Diff. Eq. 18(1975).

[3] Bernfeld, S. and Lakshmikantham, V., *An Introduction to Nonlinear Boundary Value Problems*, Academic Press, New York, 1974.

[3a] Bernfeld, S. R., Lakshmikantham, V., and Leela, S., *A generalized comparison principle and monotone method for second order boundary value problems in Banach spaces*, (to appear).

[4] Chandra, J. and Davis, P., *A monotone method for quasilinear boundary value problems*, Arch. Rat. Mech. Anal., 54(1974), pp. 257-266.

[5] Chandra, J., Lakshmikantham, V., and Leela, S., *A monotone method for infinite system of nonlinear boundary value problems*, Arch. Rat. Mech. Anal., (to appear).

[6] Chandra, J., Lakshmikantham, V., and Mitchell, A., *Existence of solutions of boundary value problems for nonlinear second order systems in a Banach space*, J. Nonlinear Analysis, (to appear).

[7] Hartman, P., *Ordinary Differential Equations*, Wiley Interscience, New York, 1964.

[8] Hartman, P., *On two point boundary value problems for nonlinear second order systems*, SIAM J. Math. Anal. 5 (1974).

[9] Thompson, R. C., *Differential inequalities for infinite second order systems and an application to the method of lines*, Jour. Diff. Eq. 17 (1975), pp. 421-434.

[10] Schroder, Johann, *Upper and lower bounds for solutions of generalized two point boundary value problems*, (to appear).

[11] Scorza-Dragoni, G., *Sul problema dei valori ai limiti per i systemi di equazioni differenziali del secondo ordine*, Boll. Un Mat. Ital. 14 (1935), pp. 225-230.

EXISTENCE THEORY OF DELAY DIFFERENTIAL EQUATIONS IN BANACH SPACES

V. Lakshmikantham
University of Texas at Arlington

S. Leela
SUNY College at Genesco

V. Moauro
University di Napoli

The study of the abstract Cauchy problem

(1) $$x' = f(t,x), \quad x(t_0) = x_0,$$

where $f \in C([t_0, t_0 + a] \times D, E)$, $D \subset E$ and E is a real Banach space, has been carried out mainly along two directions: (a) finding compactness-type conditions to obtain the existence of solutions of (1); (b) finding dissipative-type conditions to assure the existence as well as uniqueness of solutions of (1). See [1-7, 11-14]. In general, the steps involved in showing the existence of a solution of (1) are

(i) constructing suitable approximate solutions for (1);

(ii) proving the convergence of the sequence of approximate solutions and

(iii) showing that the limit of the sequence of approximate solution is actually a solution of (1).

It is in step (ii) that either the compactness-type condition is used to obtain the relative compactness of the set of approximate solutions in order to apply Ascoli-Arzela theorem or the dissipative-type condition is used to show that the sequence of approximate solutions is uniformly Cauchy. In the case when D is a closed (or locally closed) subset of E, an additional condition called "boundary condition" given by

ISBN 0-12-434160-8

(2) $\liminf_{h\to 0^+} \frac{1}{h}[d(x + hf(t,x), D)] = 0$ for every $x \in D$,

is needed. See [7, 12].

Till recently, there was not much progress in extending this existence theory to delay differential equations in Banach spaces. We shall now focus on the delay differential equation

(3) $$x'(t) = f(t,x_t), \quad x_{t_0} = \phi_0,$$

where $f \in C([t_0,t_0 + a] \times C, E)$, $C = C([-\tau,0], E)$, $\tau > 0$, and for every function $x(t)$ defined in $[t_0 - \tau, t_0 + a]$, $x_t \in C$ is defined by $x_t(\theta) = x(t + \theta)$, $\theta \in [-t,0]$. In [8], assuming that f satisfies a dissipative-type condition

$$\liminf_{h\to 0^-} \frac{1}{h}[\|\phi(0) - \psi(0) + h(f(t,\phi) - f(t,\psi))\| - \|\phi(0) - \psi(0)\|]$$
$$\leq g(t,\|\phi(0) - \psi(0)\|)$$

whenever $\phi,\psi \in \Omega = \{\phi,\psi \in C: \|\phi(\theta) - \psi(\theta)\| \leq \|\phi(0) - \psi(0)\|$ for every $\theta \in [-\tau,0]\}$, where $g \in C(R_+ \times R_+, R_+)$ is such that $g(t,0) \equiv 0$ and $u(t) \equiv 0$ is the unique solution of

$$u' = g(t,u), \; u(t_0) = 0,$$

an existence and uniqueness result is proved for the Cauchy problem (3).

In order to consider the Cauchy problem (3) in a closed subset F of the Banach space E, the boundary condition (2) needs to be appropriately modified. The two possibilities are to require (i) that

(4) $$\liminf_{h\to 0^+} \frac{1}{h}[d(\phi(0) + hf(t,\phi), F)] = 0$$

holds for those $\phi \in C$ which satisfy the condition $\phi(\theta) \in F$ for every $\theta \in [-\tau,0]$ or (ii) that (4) holds for $\phi \in C$ for which only $\phi(0) \in F$. However, as shown by the following counter-example [10], the boundary condition (i) is not enough to yield the existence of solutions in a closed set.

Example. Let $E = R$ and $F = (-\infty,0] \cup \{\frac{1}{n}: \; n \geq 1\}$. Consider the function $T\colon \; R \to R$ defined by

$$T(x) = \begin{cases} -x & \text{if } x < 0 \\ 0 & \text{if } x \geq 0, \end{cases}$$

and the function $f\colon C([-1,0], R) \to R$ defined by

$$f(\phi) = T(\phi(-1)).$$

Consider the Cauchy problem

$$(5) \qquad x'(t) = f(x_t), \; x_{t_0} = \phi_0.$$

It is easy to verify that f satisfies the boundary condition (i). But if we consider $\phi_0 \in C([-1,0], R$ defined by $\phi_0(s) = s$, $s \in [-1,0]$ the Cauchy problem (5) does not have a solution in the closed set F, since we have, for a solution $x(t)$ of (5) $x'(t_0) = f(\phi_0) = T(-1) = 1 > 0$ and $x(t_0) = 0$. On the otherhand, it is also easy to check that the function f in (5) does not satisfy the boundary condition (ii), by considering a continuous function ϕ such that $\phi(0) = 1$ and $\phi(-1) = -1$.

For the convenience of summarizing some of the recent results obtained for the Cauchy problem (3) in [9, 10], let us introduce the following notations and assumptions.

Let $\|\phi\|_0 = \sup_{\theta \in [-\tau,0]} \|\phi(\theta)\|$, for any $\phi \in C$, where $\|\cdot\|$ denotes the norm in E. Let $C_F = \{\phi \in C\colon \phi(0) \in F\}$ and $\hat{C}_F = C_F \cap \{\phi \in C\colon \phi(\theta) \in \overline{\text{Co}}(F)$ for every $\theta \in [-\tau,0]\}$, where F is a closed subset of E and $\overline{\text{co}}F$ is the closed convex hull of F. For $a > 0$, $t_0 \in R_+$ and $\phi_0 \in C_F$, let the function $y \in C([t_0 - \tau, t_0 + a], E)$ be defined by

$$y(t) = \begin{cases} \phi_0(t - t_0), & \text{if } t_0 - \tau \leq t \leq t_0, \\ \phi_0(0), & \text{if } t_0 \leq t \leq t_0 + a. \end{cases}$$

For $b > 0$ and $t \in [t_0, t_0 + a]$, define the set $C^t_F(b)$ by

$$C^t_F(b) = C_F \cap \{\phi \in C\colon \|\phi - y_t\|_0 \leq b\},$$

If $f \in C([t_0,t_0 + a] \times C_F, E)$, it is possible to show [6] that there exists a $b > 0$ such that the function f is bounded on the set

$$C_0(b) = \bigcup_{t \in [t_0, t_0 + a]} \{t\} \times C_F^t(b).$$

In the sequel, we need the following hypotheses:

(A_1) Let $t_0 \in R_+$, $\phi_0 \in C_F$, $f \in C([t_0, t_0 + a] \times C_F, E)$, a, b, and M are such that $\|f(t,\phi)\| \leq M - 1$ $(M \geq 1)$ on $C_0(b)$;

(A_2) $\liminf_{h \to 0^+} \frac{1}{h}[d(\phi(0) + h\, f(t,\phi), F)] = 0$ for every $\phi \in C_F$ and $t \in [t_0, t_0 + a]$.

The following two results proved in [10] guarantee the existence of a sequence of polygonal approximate solutions and the fact that the limit function, if it exists, is a solution of (3).

Lemma 1. Let (A_1) and (A_2) hold. If $\{\varepsilon_n\} \subset (0,1)$ is a non-increasing sequence with $\lim_{n\to\infty} \varepsilon_n = 0$, then there exists a sequence of ε_n - approximate solutions for (3), that is, for every n, there exists a function $x_n(t): [t_0 - \tau, t_0 + \gamma] \to E$, where $\gamma = \min(a, \frac{b}{M})$, satisfying the following properties:

(i) there exists a sequence $\{t_i^n\}_{i=1}^{\infty}$ in $[t_0, t_0 + \gamma]$ such that $t_0^n = t_0$, $t_{i+1}^n = t_i^n + \delta_i^n$, $\delta_i^n > 0$ if $t_i^n < t_0 + \gamma$, $\lim_{i\to\infty} t_i^n = t_0 + \gamma$;

(ii) $x_n(t) = \phi_0(t - t_0)$ for $t \in [t_0 - \tau, t_0]$, $\|x_n(t) - x_n(s)\| \leq M|t - s|$, for $t, s \in [t_0, t_0 + \gamma]$;

(iii) for each $i \geq 0$, $(t_i^n, x_{n,t_i^n}) \in C_0(b)$ and $x_n(t)$ is linear on each of the intervals $[t_i^n, t_{i+1}^n]$;

(iv) if $t \in (t_i^n, t_{i+1}^n)$ and $t_i^n < t_0 + \gamma$, then $\|x_n'(t) - f(t_i^n, x_{n,t_i^n})\| \leq \varepsilon_n$;

(v) δ_i^n can be chosen less than $\min\{\varepsilon_n, \delta_{\phi_0}(\frac{\eta_n}{2}), \frac{\eta_n}{2M}\}$ where $\delta_{\phi_0}(\eta_n)$ is the number associated with η_n by the uniform continuity of ϕ_0 on $[-\tau, 0]$ and η_n is such that $|t - t_i^n| \leq \eta_n$, $\|\phi - x_{n,t_i^n}\|_0 \leq \eta_n$

imply that $\|f(t,\phi) - f(t_i^n, x_{n,t_i^n})\| \leq \varepsilon_n$.

It should be noted that Lemma 1 remains valid when in the assumptions (A_1) and (A_2), the set C_F is replaced by $\hat{C}_F$.

Lemma 2. Let the assumptions of Lemma 1 hold. If the sequence $\{x_n(t)\}$ of ε_n - approximate solutions of (3) converges to $x(t)$ uniformly on $[t_0 - \tau, t_0 + \gamma]$, then $x(t)$ is a solution of (3) such that $x(t) \in F$ for $t \in [t_0, t_0 + \gamma]$.

Now in order to prove the uniform convergence of a subsequence of ε_n - approximate solutions constructed in Lemma 1, in [10] a compactness-type condition is employed. The main existence result under a compactness-type condition in terms of Kuratowski's measure of noncompactness α proved in [10] is given below. For properties of α and existence results of this type for ordinary differential equations in a Banach space, see [11, 14].

Theorem 1. Suppose that (A_1) and (A_2) hold. Assume that f is uniformly continuous on $[t_0, t_0 + a] \times C_F$. Let f satisfy the following compactness-type condition: for $t \in [t_0, t_0 + a]$ and $\Phi^t \subset C_F^t(b)$,

$$\liminf_{h \to 0+} \frac{1}{h}[\alpha(\Phi^t(0)) - \alpha(\{\phi(0) - hf(t,\phi): \ \phi \quad \Phi^t\})]$$
$$\leq g(t, \alpha(\Phi^t(0)))$$

whenever Φ^t is such that $\alpha(\Phi^t(\theta)) \leq \alpha(\Phi^t(0))$ for every $\theta \in [-\tau, 0]$, where $g \in C([t_0, t_0 + a] \times [0, 2b], R)$ is such that $g(t,0) \equiv 0$ and $u(t) \equiv 0$ is the unique solution of

$$u' = g(t,u), \quad u(t_0) = 0. \tag{6}$$

Then the Cauchy problem (3) has a solution existing on $[t_0 - \tau, t_0 + \gamma]$, where $\gamma = \min\{a, b/M\}$.

The proof of Theorem 1 consists of first showing that $\alpha(\{x_n(t): \ n \geq 1\}) = 0$ by means of theory of differential inequalities and the properties of α and then obtaining a uniformly convergent subsequence as a consequence of Ascoli-Arzela theorem. The existence of a solution $x(t)$ of (3) such that

$x(t) \in F$ follows then by Lemma 2. An existence result under a general compactness-type condition in terms of a Lyapunov-like function is also given in [10].

Existence and uniqueness of solutions of (3) in the closed set F, under dissipative-type conditions are established in [9] for the two cases: (a) when F is convex and (b) when F is not convex. We shall only state the main existence and uniqueness theorems. The details of the proofs will appear in [9].

Theorem 2. Suppose F is convex and that (A_1), (A_2) hold. Let f satisfy the following dissipative-type condition:

(A_3) For $t \in [t_0, t_0 + a]$ and $\phi, \psi \in C_F^t(b)$ such that

$$\|\phi(\theta) - \psi(\theta)\| \leq \|\phi(0) - \psi(0)\| \quad \text{for every} \quad \theta \in [-\tau, 0],$$

$$\|f(t,\phi) - f(t,\psi)\| \leq g(t, \|\phi(0) - \psi(0)\|), \tag{7}$$

where $g \in C[R_+ \times R_+, R_+]$ is such that $g(t,0) \equiv 0$ and $u(t) \equiv 0$ is the unique solution of (6). Then the Cauchy problem (3) has a unique solution $x(t)$ existing on $[t_0 - \tau, t_0 + \gamma]$ such that $x(t) \in F$ for $t \in [t_0, t_0 + \gamma]$.

If no convexity of F is assumed, then the dissipative-type condition (A_3) has to be strengthened as follows:

(A_4) For $t \in [t_0, t_0 + a]$, and $\phi, \psi \in C_F^t(b)$,

$$\|f(t,\phi) - f(t,\psi)\| \leq g(t, \|\phi(0) - \psi(0)\|, \|\phi(\cdot) - \psi(\cdot)\|) \tag{8}$$

where $g \in C(R_+ \times R_+ \times C^+, R_+)$, $C^+ = C([-\tau, 0], R_+)$ is such that $g(t,0,0) \equiv 0$, $g(t,u,u_t)$ is nondecreasing in u and u_t for each (t,u_t) and (t,u) respectively and $u(t) \equiv 0$ is the unique solution of the scalar functional differential equation

$$u'(t) = g(t,u,u_t), \quad u_{t_0} \equiv 0. \tag{9}$$

Theorem 3. Suppose F is not convex. Let (A_1), (A_2) and (A_4) hold. Then the conclusion of Theorem 2 is valid.

Remark. The positive invariance of the set F with respect to the equation $x'(t) = f(t,x_t)$ is an easy consequence of Theorem 2 whenever the dissipativity condition (7) is satisfied for $\phi,\psi \in B[y_t,b] = \{\sigma \in C\colon \|\sigma - y_t\|_0 \leq b\}$ such that $\|\phi(\theta) - \psi(\theta)\| \leq \|\phi(0) - \psi(0)\|$. Similarly if (8) is satisfied for $\phi,\psi \in B[y_t,b]$, Theorem 3 also yields the positive invariance of the set F with respect to the equation $x'(t) = f(t,x_t)$. Theorem 3 with the set C_F replaced by the set $\hat{C}_F$ gives an existence and uniqueness result for (3) whenever $\phi_0 \in \hat{C}_F$ Therefore Theorem 2 in [15], in the case of a constant delay, can be generalized in two different ways, namely, either by eliminating the convexity hypothesis on F or by relaxing the dissipative condition. Existence and uniqueness results under dissipative conditions in terms of a Lyapunov-like function when the set F is convex are also given in [9].

The problem of proving an existence and uniqueness result when the set F is not convex and f satisfies a general dissipative-type condtion (weaker than (8)) presents several difficulties as in the case of differential equations without delay. See [7, 12]. Nonetheless, in [9], it is shown that if (i) f satisfies a dissipative-type condition in terms of a Lyapunov function belonging to a certain class and (ii) the boundary condition (A_2) holds, we can construct certain auxiliary sequences of continuous functions which are in some sense close to the polygonal approximate solutions obtained in Lemma 1. Then, by applying a new comparison result which involves a sequence of functions satisfying a certain functional differential inequality, the existence and uniqueness of solutions of (3) in the closed set F is established. See [9], for details of this general existence and uniqueness result.

REFERENCES

[1] Cellina, A., *On the local existence of solutions of ordinary differential equations*, Bull. Acad. Pol. Sci. III Math. 20 (]972), pp. 293-296.

[2] Ladas, G. E., and Lakshmikanthan, V., *Differential*

Equations in Abstract Spaces, Academic Press, New York, 1972.

[3] Lakshmikantham, V., *Stability and asymptotic behavior of solutions of differential equations in a Banach space*, Lecture notes CIME, Italy, 1974.

[4] Lakshmikantham, V., *Existence and comparison results for differential equations in a Banach space*, Proc. Int. Conf. on Diff. Eqns., pp. 459-473, Academic Press, New York, 1975.

[5] Lakshmikantham, V. and Eisenfeld, J., *On the existence of solutions of differential equations in a Banach space*, Rev. Math. Pures et Appl. (to appear).

[6] Lakshmikantham, V. and Leela, S., *Differential and Integral Inequalities*, Vol. I and II, Academic Press, New York, 1969.

[7] Lakshmikantham, V., Mitchell, A. R., and Mitchell, R. W., *Differential equations on closed subsets of a Banach space*, Trans. Amer. Math. Soc., 220 (1976), pp. 103-113.

[8] Lakshmikantham, V., Mitchell, A. R., and Mitchell, R. W., *On the existence of solutions of differential equations of retarded type in a Banach space*, Ann. Pol. Math. (to appear).

[9] Lakshmikantham, V., Leela, S., and Moauro, V., *Existence and uniqueness of solutions of delay differential equations on a closed subset of a Banach space*, J. Nonlinear Anal. (to appear).

[10] Leela, S. and Moauro, V., *Existence of solutions in a closed set for delay differential equations in Banach spaces*, J. Nonlinear Anal. (to appear).

[11] Li Tien-Yien., *Existence of solutions for ordinary differential equations in Banach spaces*, J. Diff. Eqns., 18 (1975), pp. 29-40.

[12] Martin, R. H., Jr., *Differential equations on closed subsets of a Banach space*, Trans. Amer. Math. Soc., 179 (1973), pp. 399-414.

[13] Martin, R. H., Jr., *Nonlinear Operators and Differential*

Equations in Banach spaces, John Wiley, New York, 1976.

[14] Martin, R. H., Jr., *Approximation and existence of solutions to ordinary differential equations in Banach spaces*, Funckcialaj Ekvacioj, 16 (1973), pp. 195-211.

[15] Seifert, G., *Positively invariant closed sets for systems of delay differential equations*, J. Diff. Eqns., 22 (1976), pp. 292-304.

INVARIANT SETS AND A MATHEMATICAL MODEL INVOLVING SEMILINEAR DIFFERENTIAL EQUATIONS*

Robert H. Martin, Jr.
North Carolina State University

The purpose of this note is twofold: first we indicate an abstract result on the existence of and invariant sets for solutions to a semilinear differential equation with inhomogeneous terms in a Banach space; and then we indicate an application of these techniques to a mathematical model of a gas exchange system. In Section I we briefly indicate that the results of Lightbourne and Martin [3] can be extended by allowing inhomogeneous terms. Very general results extending those of [3] can be found in the paper [1] by H. Amann. Here our extension is indicated in relation to the results and techniques in [3]; the reader should also compare with those of [1]. In Section II these abstract results are used to study the existence and behavior of solutions to a gas exchange model, which is also studied by D. Viaud [8] using different techniques.

I. AN ABSTRACT SYSTEM

In this section we use many of the notations and suppositions of [3]. Let X be a real or complex Banach space with norm denoted by $|\cdot|$ and assume that each of the following conditions is satisfied.

* Work supported by U.S. Army Research Office, Research Triangle Park, N. C.

ISBN 0-12-434160-8

(C1) $T = \{T(t): t \geq 0\}$ is an analytic semigroup of bounded linear operators on X (see, e.g., [2, p. 101] or [6, p. 60]), A is the infinitesimal generator of T, and the numbers M and ω are such that $\|T(t)\| \leq Me^{\omega t}$ and $\|AT(t)\| \leq Mt^{-1}e^{\omega t}$ for all $t > 0$.

(C2) $g: [0,\infty) \to X$ is continuous and

$$S(t,s)x \equiv T(t - s)(x - g(s)) + g(t)$$

for all $x \in X$ and $t \geq s \geq 0$.

(C3) $\alpha \in (0,1)$, $\lambda > \omega$, $L^{-\alpha} \equiv (\lambda I - A)^{-\alpha}$, $L^{\alpha} \equiv (L^{-\alpha})^{-1}$ and $C_\alpha > 0$ is such that $\|L^{\alpha}T(t)\| \leq C_\alpha t^{-\alpha}e^{\omega t}$ for all $t > 0$ (see, e.g., [2, p. 158] or [6, p. 71]).

(C4) D is a closed subset of X and

$$D_\alpha(t) \equiv \{x + g(t): x \in D(L^{\alpha})\} \cap D$$

is dense in D for each $t > 0$.

(C5) $D(B)$ is a subset of $[0,\infty) \times X$ such that $D_\alpha(t) \subset \{x: (t,x) \in D(B)\}$ for each $t \geq 0$, and $B: D(B) \to X$ is a function with the property that $(t,x) \to B(t,L^{-\alpha}x + g(t))$ is continuous from $\{(t,x): (t,L^{-\alpha}x + g(t)) \in D(B)\}$ into X.

(C6) $T(t)$ is a compact operator for each $t > 0$.

The compactness assumption (C6) is used by Pazy [7]; this assumption could also be replaced by a Lipschitz type assumption (relative to L^{α}) for the operator B. In this section we consider the existence of solutions to the integral equation.

$$(1.1)\quad u(t) = S(t,a)z + \int_a^t T(t - r)B(r,u(r))dr \quad \text{for } t \geq a \geq 0,$$

where $(a,z) \in [0,\infty) \times D$. A continuous function $u:[a,a + \sigma) \to D$ is said to be a solution to (1.1) on $[a,a + \sigma)$ if $u(t) - g(t) \in D(L^{\alpha})$ (and hence $u(t) \in D_\alpha(t)$) for $t \in (a,a + \sigma)$, the map $t \to L^{\alpha}(u(t) - g(t))$ is continuous on $(a, a + \sigma)$, and u satisfies (1.1) for all $t \in [a, a + \sigma)$. When $z - g(a) \in D(L^{\alpha})$ it is also assumed that $L^{\alpha}(u(t) - g(t)) \to L^{\alpha}(z - g(a))$ as $t \to a^{+}$. One should note also that $r \to B(r, u(r))$ is continuous on $(a, a + \sigma)$ by (C5) and the fact that

$$B(r, u(r)) = B(r, L^{-\alpha}(L^{\alpha}(u(r) - g(r))) + g(r)).$$

For each $y \in X$ define $d(y; D) = \inf\{|y - x|: \ x \in D\}$. Our main results are given by the following two theorems.

Theorem 1. In addition to (C1) - (C6) suppose that

(C7) $S(t,s)x \in D$ for each $x \in D$ and $t \geq s \geq 0$.

(C8) $\liminf_{h \to 0^{+}} d(x + hB(t,x); D)/h = 0$ for all $t \geq 0$ and $x \in D_{\alpha}(t)$.

Then (1.1) has a local solution for each $(a,z) \in [0,\infty) \times D$ with $z - g(a) \in D(L^{\alpha})$.

Theorem 2. In addition to the suppositions of Theorem 1, suppose $Q\colon\ [0,\infty) \to (0,\infty)$ is continuous and that

(C9) $|B(t,x)| \leq Q(t)[1 + |L^{\alpha}(x - g(t))|$ for all $t \geq 0$ and $x \in D$ with $x - g(t) \in D(L^{\alpha})$.

Then for each $(a,z) \in [0,\infty) \times D$, equation (1.1) has a solution $u_{(a,z)}$ on $[a,\infty)$ with the property that

$$|u_{(a,z)}(t) - g(t)| + (t - a)^{\alpha}|L^{\alpha}(u_{(a,z)} - g(t))| \leq M(a, |z|, t)$$

for all $t \geq 0$, where $M\colon\ [0,\infty)^{3} \to (0,\infty)$ is continuous.

Remark 1. Suppose that g is also continuously differentiable and for each $t \geq 0$ define the operator $A^{*}(t)$ on X by $A^{*}(t) = A(x - g(t)) + g'(t)$ for all $x \in D(A^{*}(t))$ $\equiv \{x \in X\colon\ x - g(t) \in D(A)\}$. If $t > a \geq 0$ and $x \in X$, then $S(t,a)x \in D(A^{*}(t))$ and $\frac{d}{dt} S(t,a)x = A^{*}(t)S(t,a)x$. In particular, if a solution u to (1.1) on $[a, a + \sigma)$ is continuously differentiable on $(a, a + \sigma)$, then $u(t) \in D(A^*(t))$ and

$$(1.1)' \quad u'(t) = A^*(t)u(t) + B(t, u(t)) \text{ for all } t \in (a, a + \sigma).$$

Therefore, solutions to the integral equation (1.1) are called mild solutions to the differential equation (1.1)'.

Remark 2. Let $\nu \in (0,1]$ and suppose that for each $b, R > 0$ there is a number $N = N(b,R) > 0$ such that

$$|B(t, L^{-\alpha}x + g(t)) - B(s, L^{-\alpha}y + g(s))| \leqslant N(|t - s|^{\nu} + |x - y|^{\nu})$$

for all $t, s \in [0,b]$ and $x, y \in D$ with $|L^{-\alpha}x|$, $|L^{-\alpha}y| \leqslant R$. If u is a solution to (1.1) on $[a, a + \sigma)$ then $u - g$ is continuously differentiable on $(a, a + \sigma)$ (and hence if g is continuously differentiable then u solves the differential equation (1.1)' in Remark 1). If there exists a constant $N = N(b,R) > 0$ such that

$$|B(t,x) - B(t,y)| \leqslant N|L^{\alpha}(x - y)|$$

for all $t \in [0,b]$ and $x, y \in D$ such that $x - g(t)$, $y - g(t) \in D(L^{\alpha})$ and $|L^{\alpha}(x - g(t))|$, $|L^{\alpha}(y - g(t))| \leqslant R$, then each solution u to (1.1) is unique whenever $z - g(a) \in D(L^{\alpha})$.

Theorems 1 and 2 above are essentially Theorems 1 and 2 of [3] with the inhomogeneous term S in place of the homogeneous term T. The proof techniques needed here are straightforward modification of those used in [3], and so we only indicate briefly the techniques for setting up approximations of solutions to (1.1).

Now suppose that $(a,z) \in [0,\infty) \times D$ with $z - g(a) \in D(L^{\alpha})$. Subtracting $g(t)$ from each side of (1.1) and applying L^{α} to each side of the resulting equation, we are led to the equation

$$(1.2) \quad v(t) = T(t)w + \int_a^t L^{\alpha}T(t - r)B^*(r, v(r))dr \quad \text{for } t \geqslant a,$$

where $w = L^{\alpha}(z - g(a))$ and

$$B^*(r,x) \equiv B(r, L^{-\alpha}x + g(r)) \quad \text{for all} \quad (r,x) \in D(B^*)$$

where $D(B^*) \equiv \{(r,x): \ r \geq 0 \ \text{and} \ L^{-\alpha}x + g(r) \in D\}$.

Note that if $v: \ [a, a + \sigma) \to X$ is continuous with $(t, v(t)) \in D(B^*)$ for all $t \in [a, a + \sigma)$ and v satisfies (1.2) for all $t \in [a, a + \sigma)$, then $u(t) \equiv L^{-\alpha}v(t) + g(t)$ is a solution to (1.1) on $[a, a + \sigma)$.

Let $\varepsilon, \rho > 0$ and let $\{t_i\}_0^n$ be a partition of $[a, a + \rho]$ with $t_i - t_{i-1} \leq \varepsilon$. If $x: \ [a, a + \rho] \to X$ then the pair $(x, \{t_i\}_0^n)$ is a piecewise continuous ε-approximation (written P. C. ε-A) to (1.2) on $[a, a + \rho]$ if each of the following is satisfied:

(ε1) $x(a) = L^{\alpha}(z - g(a))$ and $L^{-\alpha}x(t_i) + g(t_i) \in D$ (and hence $L^{-\alpha}x(t_i) + g(t_i) \in D_{\alpha}(t_i))$ for $i = 1, \ldots, n$.

(ε2) $x(t) = T(t - t_i)x(t_i) + (t - t_i)L^{\alpha}T(t - t_i)B^*(t_i, x(t_i))$ for $t \in [t_i, t_{i+1})$ and $i = 0, \ldots, n - 1$.

(ε3) $x(t_{i+1}) - x(t_{i+1}^-) = L^{\alpha}T(t_{i+1} - t_i)\nu_{i+1}$ where $\nu_{i+1} \in X$ with $|\nu_{i+1}| \leq \varepsilon(t_{i+1} - t_i)$ for $i = 0, \ldots, n - 1$.

Suppose that $(x, \{t_i\}_0^n)$ is a P. C. ε-A to (1.2) on $[a, a + \rho]$ that also satisfies

(ε4) $|T(t - t_i)x(t_i) - x(t_i)| \leq \varepsilon$ for $t \in [t_i, t_{i+1}]$ and $i = 0, \ldots, n - 1$.

(ε5) $|T(t - t_i)B^*(t_i, \gamma(t_i)) - B^*(s, x(t_i))| \leq \varepsilon$ for $s, t \in [t_i, t_{i+1}]$ and $i = 0, \ldots, n - 1$.

We show that $(x, \{t_i\}_0^n)$ can be extended to a P. C. ε-A. to (1.2) on $[a, a + \sigma]$ for some $\sigma > \rho$. Choose $\delta \in (0, \varepsilon)$ such that $|T(h)x(t_n) - x(t_n)| \leq \varepsilon$ for $h \in [0, \delta]$ and

$$|T(t - t_n)B^*(t_n, x(t_n)) - B(s, x(t_n))| \leq \varepsilon \quad \text{for} \quad s, t \in [t_n, t_n + \delta].$$

Also, from (C8) and the denseness of $D_{\alpha}(t_n + \delta)$ in D, δ can be chosen so that

$$L^{-\alpha}x(t_n) + g(t_n) + \delta B^*(t_n, x(t_n)) = \mu_{n+1} - \nu_{n+1}$$

where $\mu_{n+1} \in D_\alpha(t_n + \delta)$ and $|\nu_{n+1}| \leqslant \varepsilon\delta$. Setting $t_{n+1} = t_n + \delta$ and extending x to $[a, a + \rho + \delta]$ by defining

$$x(t) = \begin{cases} T(t - t_n)x(t_n) + (t - t_n)L^{\alpha}T(t - t_n)B^*(t_n, x(t_n)) \\ \text{for } t_n \leqslant t < t_{n+1} \\ L^{\alpha}T(t_{n+1} - t_n)(\mu_{n+1} - g(t_n)) \quad \text{for } t = t_{n+1} \end{cases}$$

we see that $(x, \{t_i\}_0^{n+1})$ is a P. C. ε-A. on $[a, a + \rho + \delta]$ satisfying (ε1) - (ε5) (note that $L^{-\alpha}x(t_{n+1}) + g(t_{n+1})$ $= S(t_{n+1}, t_n)\mu_{n+1} \in D$ by (C7) and the fact that $\mu_{n+1} \in D_\alpha(t_{n+1}) \subset D$). Using arguments similar to those in [3, Proposition 1], one can easily show that (1.2) has a P. C. ε-A. on $[a, a + \rho_\varepsilon]$ that satisfies (ε1) - (ε5) for each $\varepsilon > 0$, where $\inf\{\rho_\varepsilon : \varepsilon > 0\} > 0$.

II. A MATHEMATICAL MODEL

Suppose that ρ, ℓ, P, σ_1, σ_2, σ_3, and k_1, k_2, k_3 are positive numbers; that $\beta = (\beta_i)_1^3$ and $\gamma = (\gamma_i)_1^3$ are continuous functions from $\mathbb{R}^3 \times \mathbb{R}^3$ into $\mathbb{R}^3$, and that $\mu = (\mu_i)_1^3$ and $\gamma = (\nu_i)_1^3$ are continuous and bounded functions from $[0, \infty)$ into $\mathbb{R}^3$. Given $a \geqslant 0$ and initial functions $z = (z_i)_1^3$ and $w = (w_i)_1^3$ from $[0, \ell]$ into $\mathbb{R}^3$, the problem is to determine the existence of solutions $u = (u_i)_1^3$, $v = (v_i)_1^3 : [a, \infty) \times [0, \ell] \to \mathbb{R}^3$ and $c : [a, \infty) \times [0, \ell] \to \mathbb{R}$ to the semilinear parabolic system

$$\frac{\partial}{\partial t} u_i = \rho \frac{\partial^2}{\partial x^2} u_i - \frac{\partial}{\partial x}(cu_i) + \beta_i(u, v)$$

(2.1) for $(t, x) \in (a, \infty) \times (0, \ell)$ and $i = 1, 2, 3$

$$\frac{\partial}{\partial t} v_i = \sigma_i \frac{\partial^2}{\partial x^2} v_i - k_i \frac{\partial}{\partial x} v_i + \gamma_i(u, v)$$

satisfying the initial-boundary values

$$(2.2)\quad \begin{aligned} & u(0,x) = z(x),\ v(0,x) = w(x) \\ & u(t,0) = \mu(t),\ v(t,0) = \nu(t) \\ & \frac{\partial}{\partial x} u(t,\ell) = 0,\ \frac{\partial}{\partial x} v(t,\ell) = 0 \\ & c(t,\ell) = 0 \end{aligned} \qquad \text{for all } (t,x) \in (a,\infty) \times [0,\ell]$$

and also satisfying the side condition

$$(2.3)\qquad \sum_{i=1}^{3} u_i(t,x) \equiv P \qquad \text{for all } (t,x) \in (a,\infty) \times [0,\ell].$$

If $b_i > 0$ and $\beta_i(\xi,\eta) \equiv -\gamma_i(\xi,\eta) \equiv b_i(\eta_i - \xi_i)$ for all $(\xi,\eta) \in \mathbb{R}^3 \times \mathbb{R}^3$, then this equation is the model of a gas exchange system studied by Viaud [8].

Let $p > 1$ and let $L^p(= L^p([0,\ell]; \mathbb{R}^3 \times \mathbb{R}^3))$ denote the space of all measurable functions $(\phi,\psi)\colon\ [0,\ell] \to \mathbb{R}^3 \times \mathbb{R}^3$ (where $\phi = (\phi_i)_1^3$, $\psi = (\psi_i)_1^3\colon\ [0,\ell] \to \mathbb{R}^3$) such that

$$|(\phi,\psi)|_p \equiv [\sum_{i=1}^{3} \int_0^\ell (|\phi_i(x)|^p + |\psi_i(x)|^p)dx]^{1/p} < \infty.$$

Throughout this section we suppose that R_1, R_2, and R_3 are positive numbers and that Λ_1 and Λ_2 are the subsets of $\mathbb{R}^3$ defined by

$$\Lambda_1 = \{\xi \in \mathbb{R}^3\colon\ \sum_{i=1}^{3} \xi_i = P \text{ and } \xi_i \geqslant 0 \text{ for } i = 1,\ 2,\ 3\} \text{ and}$$

$$\Lambda_2 = \{\eta \in \mathbb{R}^3\colon\ 0 \leqslant \eta_i \leqslant R_i \text{ for } i = 1,\ 2,\ 3\}.$$

Also, it is assumed that the following hypotheses are satisfied:

(H1) If $(\xi,\eta) \in \Lambda_1 \times \Lambda_2$ and $j \in \{1,\ 2,\ 3\}$, then $\xi_j = 0$ implies $\beta_j(\xi,\eta) \geqslant 0$; $\eta_j = 0$ implies $\gamma_j(\xi,\eta) \geqslant 0$; and $\eta_j = R_j$ implies $\gamma_j(\xi,\eta) \leqslant 0$.

(H2) $(z,w) \in L^p$ with $(z(x),w(x)) \in \Lambda_1 \times \Lambda_2$ for almost all $x \in [0,\ell]$.

(H3) $(\mu(t),\nu(t)) \in \Lambda_1 \times \Lambda_2$ for all $t \geq 0$, $(\mu(\cdot),\nu(\cdot))$ is continuous and right differentiable on $[0,\infty)$ and $(\mu_+'(\cdot),\nu_+'(\cdot))$ is piecewise continuous on $[0,\infty)$.

Define the linear operator L on $D(L) \subset L^p$ by

$$L\,(\phi,\psi) = ((\rho\phi_i'')_1^3,\ (\sigma_i\psi_i'')_1^3) \text{ for all } (\phi,\psi) \in D(L) \text{ where}$$
(2.4)
$$D(L) \equiv \{(\phi,\psi) \in L^p\colon (\phi,\psi),\ (\phi',\psi') \textit{ are abs. cont. and } (\phi'',\psi'') \in L^p\}$$

and define the linear operator A on $D(A) \quad L^p$ by

(2.5) $A(\phi,\psi) = L(\phi,\psi)$ for $(\phi,\psi) \in D(A) \equiv \{(\phi,\psi) \quad D(L)\colon$

$$(\phi(0),\psi(0)) = (\phi'(\ell),\psi'(\ell)) = 0\}.$$

It is well known that A is the generator of a compact, analytic semigroup $T = \{T(t)\colon\ t \geq 0\}$ on L^p (see, e.g., [5, p. 309]). Moreover, if $\alpha \in (1/2,1)$ then $(I - A)^{-\alpha}$ exists, the range of $(I - A)^{-\alpha}$ contains $C_1[\ \equiv \{(\phi,\psi) \in L^p\colon\ (\phi,\psi)$ is continuously differentiable on $[0,\ell]\}]$ and there is a constant $M_\alpha > 0$ such that

(2.6) $$|[(1 - A)^{-\alpha}(\phi,\psi)]'|_\infty \leq M_\alpha|(\phi,\psi)|_p \quad \text{for all } (\phi,\psi) \in L^p,$$

where $|(\phi,\psi)|_\infty \equiv \text{ess sup}\,\{|(\phi(x),\psi(x))|\colon\ x \in [0,\ell]\}$ for each essentially bounded member (ϕ,ψ) of L^p.

Define the map $m\colon [0,\infty) \to L^p$ by $[m(t)](x) \equiv (\mu(t),\nu(t))$ for all $x \in [0,\ell]$ and note that m is continuous and that $[m_+'(t)](x) \equiv (\mu_+'(t),\nu_+'(t))$ for all $x \in [0,\ell]$ and $t \geq 0$. Now define

(2.7) $$g(t) = m(t) - \int_0^t T(t - r)\, m_+'(r)dr \quad \text{for all } t \geq 0$$

and

(2.8) $$S(t,a)(\phi,\psi) = T(t - a)[(\phi,\psi) - g(a)] + g(t)$$

$$\text{for all } t \geq a \geq 0 \text{ and } (\phi,\psi) \in L^p.$$

Also, define the closed, bounded, convex subset D of L^p by

(2.9) $$D \equiv \{(\phi,\psi) \in L^p\colon \ (\phi(x),\psi(x)) \in \Lambda_1 \times \Lambda_2 \text{ for almost all } x \in [0,\ell]\},$$

and for each $t \geqslant 0$ define

(2.10) $$D_\alpha(t) \equiv \{(\phi,\psi) + g(t)\colon (\phi,\psi) \in D((I - A)^\alpha)\} \cap D$$

we have the following results for the map g and the semigroup S:

Lemma 1. Suppose that $g\colon [0,\infty) \to L^p$ and $S = \{S(t,a)\colon t \geqslant a \geqslant 0\}$ are as above. Then

(i) $g(t) \in C^1$ for each $t \geqslant 0$ and the maps $t \to g(t)$ and $t \to \frac{\partial}{\partial x} g(t)$ are continuous from $[0,\infty)$ into L^p.

(ii) $S(t,a)\colon D \to D$ for all $t \geqslant a \geqslant 0$.

(iii) $D_\alpha(t)$ is dense in D for each $t \geqslant 0$.

Indication of Proof. Since $\frac{\partial}{\partial x} m(t) \equiv 0$ and since

$$t \to \int_0^t (I - A)^\alpha T(t - r)m_+'(r)dr$$

is locally Hölder continuous on $[0,\infty)$ (see, e.g., Pazy [6, p. 129]) it follows from (2.6) that (i) holds. Let $a \geqslant 0$ and define

$$U_a(t)(\phi,\psi) \equiv T(t)[(\phi,\psi) - m(a)] + m(a)$$

for all $t \geqslant 0$ and $(\phi,\psi) \in L^p$.

Note that $t \to U_a(t)(\phi,\psi)$ is the solution (u,v) to the autonomous inhomogeneous linear system

$$\frac{\partial}{\partial t} u_i = \rho \frac{\partial^2}{\partial x^2} u_i;\ \frac{\partial}{\partial t} v_i = \alpha_i \frac{\partial^2}{\partial x^2} v_i \quad \text{for } (t,x) \in (0,\infty) \times [0,\ell]$$

$$u(t,0) = \mu(a);\ v(t,0) = \nu(a);\ \frac{\partial}{\partial x} u(t,\ell) = \Theta; \quad \text{and}$$

$$\frac{\partial}{\partial x} v(t,\ell) = \Theta;\ u(0,x) = \phi(x);\ v(0,x) = \psi(x) \quad i = 1,\ 2,\ 3.$$

Using the maximum principle (along with the smoothness of solutions to this equation for $t > 0$) it is easy to check that $U_a(t)\colon D \to D$ for all $t \geqslant 0$. If $(\phi,\psi) \in D$ then $U_a(h)(\phi,\psi) \to (\phi,\psi)$ as $h \to 0+$. Moreover, if $h > 0$,

$$U_a(h)(\phi,\psi) = T(h)[(\phi,\psi) - g(a)] + g(a) \in D(A) + g(a) \subset D((I-A)^{\alpha}) + g(a)$$

and one sees that (iii) is true. Noting that $m_+'(\cdot)$ is continuous from the right at $t = a$ it follows that

$$\begin{aligned} S(a+h,a)(\phi,\psi) &- U_a(h)(\phi,\psi) \\ &= T(h)[m(a) - g(a)] + m(a+h) - m(a) = \int_0^{a+h} T(a+h-r)m'(r)dr \\ &= m(a+H) - m(a) - \int_a^{a+h} T(a+h-r)m_+'(r)dr \\ &= m(a+h) - m(a) - hm_+'(a) + o(h) \end{aligned}$$

where $h^{-1}|o(h)|_p \to 0$ on $h \to 0+$. Therefore, if $(\phi,\psi) \in D$ and $a \geqslant 0$,

$$\lim_{h\to 0+} d(s(a+h,a)(\phi,\psi);D)/h \quad \lim_{h\to 0+} |S(a+h,a)(\phi,\psi) - U_a(h)(\phi,\psi)|_p/h = 0,$$

and it follows from [4] that (ii) is also true.

The preceeding lemma and discussions show that the inhomogeneous linear part of the system (2.1) - (2.3) satisfy the suppositions of Theorems 1 and 2 in Section I. We now consider the nonlinear part of (2.1) - (2.3).

Summing each side of the first three equations in (2.1) and using the constraint (2.3) one sees that

$$P\frac{\partial}{\partial x}c(t,x) = \sum_{i=1}^{3} \beta_i(u(t,x),v(t,x)) \text{ for all } (t,x) \in [0,\infty) \times [0,\ell].$$

From the boundary condition $c(t,\ell) = 0$ it follows that

$$c(t,x) = -P^{-1}\int_x^{\ell} [\sum_{i=1} \beta_i(u(t,y),v(t,y))]dy$$

$$\text{for all } (t,x) \in [0,\infty) \times [0,\ell]$$

(see [8, p. 729]). Therefore, define the operator C on D by

(2.11) $$[C(\phi,\psi)](x) = -P^{-1}\int_x^{\ell} [\sum_{i=1}^{3} \beta_i(\phi(y),\psi(y))]dy$$

for all $x \in [0,\ell]$ and $(\phi,\psi) \in D$.

and note that the map C as well as the map $(\phi,\psi) \to C(\phi,\psi)'[= \frac{d}{dx} C(\phi,\psi)]$ are continuous and bounded from D into $L^p([0,\ell]; \mathbb{R})$ Set

$$D^1 = \{(\phi,\psi) \in D: \ (\phi,\psi) \text{ is continuously differentiable}\}$$

and define the operators F and G from D^1 into $L^p([0,\ell]; \mathbb{R}^3)$ by

(2.12)
$$[F(\phi,\psi)](x) = (-C(\phi,\psi)(x)\phi_i'(x) - C(\phi,\psi)'(x)\phi_i(x) + \beta_i(\phi(x),\psi(x)))_1^3$$
$$[G(\phi,\psi)](x) = (- k_i\psi_i'(x) + \gamma_i(\phi(x), \psi(x)))_1^3$$

for all $x \in [0,\ell]$ and $(\phi,\psi) \in D^1$. Now define the operator B from D into L^p by

(2.13) $B(\phi,\psi) = (F(\phi,\psi),G(\phi,\psi))$ for all $(\phi,\psi) \in D^1$.

We have the following continuity and boundedness properties for B.

Lemma 2. If $L^\alpha \equiv (I - A)^\alpha$ and B, A and g are as defined above, then B satisfies (C5) and (C9) in Section I.

Indication of Proof. Let $t \geq 0$ and let $(\phi,\psi) \in D^1$ with $(\phi,\psi) - g(t) \in D((I - A)^{-1})$. Since $\Lambda_1 \times \Lambda_2$ is bounded in $\mathbb{R}^3 \times \mathbb{R}^3$ it is easy to see that there are positive numbers N_1 and N_2 such that

$$|B(\phi,\psi)|_p \leq N_1 + N_2 \ |(\phi,\psi)'|_p$$

and hence

$$|B(\phi,\psi)|_p \leq N_1 + N_2|\frac{\partial}{\partial x} g(t)|_p + N_2|[(I-A)^{-\alpha}(I-A)^{\alpha}((\phi,\psi) - g(t))]'|_p$$

and it follows from (i) of Lemma 1 and (2.6) that (C9) holds.

Observe first that there is a continuous function $N\colon [0,\infty) \to (0,\infty)$ such that $|\frac{\partial}{\partial x} g(t)|_\infty \leqslant N(t)$ for all $t \geqslant 0$. For note that

$$(I - A)^\alpha [g(t) - m(t)]|_p \leqslant \int_0^t |(I - A)^\alpha T(t - r) m_+'(r)|_p dr$$

$$\leqslant \sup\{|m_+'(r)|_p\colon \ 0 \leqslant r \leqslant t\} \int_0^t C_1 (t - r)^{-\alpha} dr$$

where C_1 is some constant, and then use (2.6) along with the fact that $\frac{\partial}{\partial x} m(t) \equiv \Theta$. Thus, in order to show that (C5) is valid it suffices (using (2.6), part (i) of Lemma 1 and the fact that $|\frac{\partial}{\partial x} g(t)|_\infty \leqslant N(t)$) to show that if $\{t_n\}_1^\infty$ is a sequence in $[0,\infty)$ and $\{(\phi^n,\psi^n)\}_1^\infty$ is a sequence in D^1 such that

$$\lim_{n\to\infty} |t_n - t| = 0, \quad \lim_{n\to\infty} (\phi^n,\psi^n) - (\phi,\psi)|_p = 0,$$

$$\lim_{n\to\infty} |(\phi^n,\psi^n)' - (\phi,\psi)'|_p = 0 \text{ and } \sup_{n\geqslant 1}\{|(\phi^n,\psi^n)'|_\infty\} < \infty,$$

then $\lim_{n\to\infty} |B(\phi^n,\psi^n) - B(\phi,\psi)|_p = 0$. This follows routinely and we omit the details.

Under the hypotheses (H1) - (H3) and the above notation, we consider the following integral equation is the space L^p:

$$\text{(2.13)} \quad (u(t),v(t)) = S(t,a)(z,w) + \int_a^t T(t - r) B((u(r),v(r))dr$$

for all $t \geqslant a \geqslant 0$ and $(z,w) \in D$. If (u,v) is a solution to (2.13) on $[a,\infty)$ and we define $u,v\colon [0,\infty) \times [0,\ell] \to \mathbb{R}^3$ and $c\colon [0,\infty) \times [0,\ell] \to \mathbb{R}$ by

$$u(t,x) \equiv [u(t)](x),\ v(t,x) \equiv [v(t)](x), \text{ and } c(t,x) \equiv [C(u(t),v(t))](x)$$

Then u, v and c are called mild L^p-solutions to the system (2.1) - (2.3). Our main result is

Theorem 3. Suppose that hypotheses (H1) - (H3) are satisfied, that $a \geqslant 0$, and that $(z,w) \in D$. Then the integral equation (2.13) has a solution on $[a,\infty)$, and hence the system (2.1) -

(2.3) has a mild L^p-solution on $[a,\infty) \times [0,\ell]$.

<u>*Indication of Proof.*</u> From the preceeding results of this section and Theorem 2 in Section I, it suffices to show that $\lim_{h\to 0+} d((\phi,\psi) + hB(\phi,\psi);\ D)/h = 0$ for all $(\phi,\psi) \in D^1$ (i.e., (C8) in Theorem 1 is satisfied). So let $(\phi,\psi) \in D^1$. Using the techniques in [5, p. 386] we need only show that

(2.14) $$\lim_{h\to 0+} d((\phi(x),\psi(x)) + h[B(\phi,\psi)](x);\ \Lambda_1 \times \Lambda_2)/h = 0$$
$$\text{for all } x \in (0,\ell).$$

Assertion (2.14) follows routinely. For let $x \in (0,\ell)$ and for each $h > 0$ and $i \in [1,2,3]$ define

$$\phi_i^h(x) = \phi_i(x) + h[-C(\phi,\psi)'(x)\phi_i(x) - C(\phi,\psi)(x)\phi_i'(x) + \beta_i(\phi(x),\psi(x))]$$

$$\psi_i^h(x) = \psi_i(x) + h[-k_i\psi_i'(x) + \gamma_i(\phi(x),\psi(x))]$$

Since $\sum_{i=1}^{3} \phi_i(y) = P$ for all $y \in [0,\ell]$ it is easy to check directly from the definitions that

$$\sum_{i=1}^{3} [-C(\phi,\psi)'(x)\phi_i(x) - C(\phi,\psi)(x)\phi_i'(x) + \beta_i(\phi(x),\psi(x))] = 0,$$

and hence $\sum_{i=1}^{3} \phi_j^h(x) = P$ for each $h > 0$. If $\phi_j(x) = 0$ then $\phi_j'(x) = 0$ and so $\phi_j^h(x) = h\beta_j(\phi(x),\psi(x)) \geqslant 0$ by (H1). Thus it is easy to see that there is an $h_1 > 0$ such that $\phi^h(x) \in \Lambda_1$ for all $h \in (0,h_1]$. Similarly, if $\psi_j(x) = 0$ (resp., $\psi_j(x) = R_j$) then $\psi_j'(x) = 0$ and it follows from (H1) that $\psi_j^h(x) = h\gamma_j(\phi(x),\psi(x)) \geqslant 0$ (resp., $\psi_j^h(x) = R_j + h\gamma_j(\phi(x),\psi(x)) \leqslant R_j$).

Therefore, there is an $h_2 > 0$ such that $\psi^h(x) \in \Lambda_2$ for all $h \in (0,h_2]$. Setting $\delta = \min\{h_1,h_2\}$ we wee that

$$(\phi(x),\psi(x)) + h[B(\phi,\psi)](x) = (\phi^h(x),\psi^h(x)) \in \Lambda_1 \times \Lambda_2$$

for all $h \in (0,\delta]$, and (2.14) is immediate. Thus (2.14) holds and the proof indication of Theorem 3 is complete.

Remark 3. If β and γ satisfy a Lipschitz condition on $\Lambda_1 \times \Lambda_2$, then the mild L^p-solutions to (2.1) - (2.3) are unique whenever $(z,w) - g(a) \in D((I - A)^\alpha)$ (see Remark 2 in Section I). Note that our techniques do not depend on the assumption that $i \in \{1,2,3\}$ -- we could just as easily considered the system (2.1) - (2.3) for $i \in \{1, \ldots, n\}$ for any $n \geq 2$ (simply replace $\mathbb{R}^3 \times \mathbb{R}^3$ by $\mathbb{R}^n \times \mathbb{R}^n$). However, these techniques depend in a crucial manner on the assumption that the coefficient ρ of $\frac{\partial^2}{\partial x^2} u_i$ in (2.1) is independent of $i \in \{1,2,3\}$.

III. REFERENCES

[1] Amann, H., *Invariant sets and existence theorems for semi-linear parabolic and elliptic systems*, (to appear).

[2] Friedman, A., "*Partial Differential Equations*", Holt, Rinehart and Winston, New York, (1969).

[3] Lightbourne, J. H. and Martin, R. H., *Relatively continuous nonlinear perturbations of analytic semigroups*, J.N.A. - T.M.A. 1(1977), pp. 277-292.

[4] Martin, R. H., *Invariant sets for evolution systems*, Proc. International Conf. Diff. Equations (H. Antosiewicz, Ed.) Academic Press, New York/London, 1975.

[5] Martin, R. H., "*Nonlinear Operators and Differential Equations in Banach Spaces*", Wiley-Interscience, New York, 1976.

[6] Pazy, A., "*Semigroups of Linear Operators and Applications to Partial Differential Equations*", Lecture Notes No. 10, University of Maryland, (1974).

[7] Pazy, A., *A class of semilinear equations of evolution*, Israel J. Math. 20(1975), pp. 23-36.

[8] Viaud, D., *Mathematical model of a gas exchange system*, J. Math. Anal. Appl. 55(1976), pp. 726-742.

TOTAL STABILITY AND CLASSICAL HAMILTONIAN THEORY*

V. Moauro[1], L. Salvadori[2], M. Scalia[3]

I. INTRODUCTION

Let S be a conservative mechanical system and $(\mathcal{H})$ be the system of Hamilton's equations of motion relative to S. In an interesting paper [1] Cetaev examines the following problem: are observable motions well described by the corresponding stable solutions of $(\mathcal{H})$? This question arises because, in the edification of the theoretical scheme, small and non-estimable forces acting upon S are necessarily neglected. The elegant analysis of Cetaev, which is mainly concerned with equilibrium positions, is strictly connected with the known arguments of Kelvin on the "secular" or permanent stability under dissipative forces. The observable equilibrium positions are thus identified with those represented by the static solutions of $(\mathcal{H})$ for which this secular character arises.

We consider in our paper the same problem in the framework of the concept of total stability as defined by Dubosin [2] and subsequently developed in the fundamental works by Malkin [3], Gorsin [4] and Seibert [5]. Some more recent results in this direction are also taken into account [6]. We wish to emphasize that in

---

* Work performed under the auspices of Italian Council of Research (C.N.R.).

1 Istituto Matematico "R. Caccioppoli" dell'Università di Napoli, and The University of Texas at Arlington, Arlington, TX 76019.

2 Istituto Matematico "G. Castelnuovo" dell'Università di Roma, and The University of Texas at Arlington, Department of Mathematics, Arlington, Texas 76019.

3 Istituto Matematico "G. Castelnuovo" dell'Università di Roma.

ISBN 0-12-434160-8

many interesting cases there is no connection between total stability and secular stability: for instance, when the analysis of linear approximation is not sufficient to recognize the stability of equilibrium. We consider also nonobservable steady motions and show that the corresponding solutions of $(\mathcal{H})$ are not totally stable with respect to dissipative forces.

We note that the periodic solutions of $(\mathcal{H})$ are not totally stable (and neither orbitally totally stable) and it is very difficult to get a justification of their observability in terms of some kind of total stability. Thus the general problem remains open.

II. PRELIMINARIES

Consider the differential equation

$$\dot{x} = f(t,x) \tag{2.1}$$

where $x \equiv (x_1,x_2,\ldots,x_n) \in \mathbb{R}^n$. Suppose that $f \in C[\mathbb{R}_+ \times E,\mathbb{R}^n]$, E being an open neighborhood of the origin 0 of $\mathbb{R}^n$, and $f(t,0) \equiv 0$. We shall denote by $x(t,t_0,x_0)$ any non-continuable solution of (2.1) passing through (t_0,x_0). Let $\mathfrak{S}$ be the set of all mappings from $\mathbb{R}_+ \times E$ into $\mathbb{R}^n$ satisfying Caratheodory conditions. We recall the following stability concepts.

2.1 *Definitions.* Let $y \equiv (x_1,x_2,\ldots,x_m)$, $m \leqslant n$. The solution $x \equiv 0$ is said to be:

2.1.1 *y-uniformly stable*, if given any $\varepsilon > 0$ there exists a $\delta = \delta(\varepsilon) > 0$ such that $t_0 \in R_+$ and $\|x_0\| < \delta$ imply $\|y(t,t_0,x_0)\| < \varepsilon$ for all $t \geqslant t_0$;

2.1.2 *y-uniformly asymptotically stable*, if it is y-uniformly stable and there exists $\sigma > 0$ such that $t_0 \in \mathbb{R}_+$ and $\|x_0\| < \sigma$ imply

(i) $x(t,t_0,x_0)$ exists for all $t \geqslant t_0$;

(ii) $y(t,t_0,x_0) \to 0$ as $t \to +\infty$, uniformly in (t_0,x_0).

2.1.3 *$(y,\mathcal{U})$-uniformly totally stable*, with $\mathcal{U} \subset \mathfrak{S}$, if given any $\varepsilon > 0$ there exist $\delta_1 = \delta_1(\varepsilon) > 0$, $\delta_2 = \delta_2(\varepsilon) > 0$, such that if $t_0 \in \mathbb{R}_+$, $\|x_0\| < \delta_1$ and $R \in \mathcal{U}$ with $\|R(t,x)\| < \delta_2$ when $t \geq t_0$ and $\|x_0\| < \delta_1$, then $\|y_R(t,t_0,x_0)\| < \varepsilon \ \forall t \geq t_0$, where $y_R(t,t_0,x_0)$ is the y-part of any solution of the perturbed system

$$\dot{x} = f(t,x) + R(t,x), \tag{2.2}$$

passing through (t_0,x_0).

In the following, the uniform character of the above stability properties will not be explicitly mentioned. Moreover, the reference to the y-part of x, or to $\mathcal{U}$, will be omitted when $m = n$ or $\mathcal{U} = \mathcal{S}$, respectively.

We also need the well-known theorem of Malkin and Gorsin [3,4].

2.2 Theorem. *Suppose that for any compact set $K \subset E$ there exists $\lambda(K) > 0$ such that $\|f(t,x') - f(t,x'')\| \leq \lambda(K)\|x'' - x'\|$ for $t \in \mathbb{R}_+$ and $x',x'' \in K$. Then, (uniform) asymptotic stability of $x \equiv 0$ implies its total stability.*

Suppose now $\mathcal{U} \subset \mathcal{S}$ such that $R \in \mathcal{U}$ implies $R(t,0) \equiv 0$. Thus $x \equiv 0$ is a solution of any perturbed equation (2.2). Consider the following properties:

(A) The solution $x \equiv 0$ of (2.1) is $\mathcal{U}$-totally stable;

(B) For any $R \in \mathcal{U}$, $x \equiv 0$ is a stable solution of (2.2).

We want to emphasize that everyone of these two properties does not imply the other. Indeed, consider the two following families of differential systems in $\mathbb{R}^2$

$$\begin{cases} \dot{y} = -z - y(y^2 + z^2) + \mu y \\ \dot{z} = y - z(y^2 + z^2) + \mu z \end{cases} \tag{2.3}$$

and

$$\begin{cases} \dot{y} = -z - \mu^2 y + \mu y(y^2 + z^2) \\ \dot{z} = y - \mu^2 z + \mu z(y^2 + z^2) \end{cases} \tag{2.4}$$

both depending on a parameter $\mu \in \mathbb{R}_+$. For every $\mu \geq 0$, $(y,z) \equiv 0$ is a solution of (2.3); but for $\mu = 0$ if is asymptotically stable and therefore totally stable, while for $\mu > 0$ it is unstable.

Setting now $V(y,z) = y^2 + z^2$, we have along the solution of (2.4)

$$\dot{V}_\mu(y,z) = 2\mu(y^2 + z^2)[(y^2 + z^2) - \mu].$$

Therefore for $\mu > 0$ the null solution of (2.4) is asymptotically stable (with a region of attraction tending to 0 as $\mu \to 0$), while for $\mu = 0$ it is stable, but not U-totally stable if

$$U = \{R_\mu(x) = x(-\mu^2 + \mu(y^2 + z^2)),\quad \mu \geqslant 0,\quad x = (y,z)\}.$$

2.3 Remark. Let $U \subset \mathcal{S}$ such that $R \in U$ implies $R(t,0) \equiv 0$. If $x \equiv 0$ is U-totally stable for (2.1) and unstable for every equation (2.2), $R \in U \setminus \{0\}$, then we have both

(i) $(\exists\ \varepsilon(R) > 0)(\forall\delta > 0)(\exists t_0 \in R_+)(\exists x_0 : \|x_0\| < \delta)(\exists t_1 \geqslant t_0)$, $\|x_R(t_1,t_0,x_0)\| \geqslant \varepsilon(R)$,

(ii) $(\forall\eta > 0)(\exists\gamma > 0)(\|R(t,x)\| < \gamma$ for $t \in R_+, \|x\| < \eta)$, $\varepsilon(R) < \eta$.

Roughly speaking, the "ε" of instability tends to zero as the perturbation R tends to zero.

III. CRITICAL ANALYSIS OF A CETAEV'S REMARK ON THE VALIDITY OF HAMILTONIAN SCHEME

Let us assume that (2.1) is the equation governing the evolution of a physical system S. Let σ be an observable evolution of S represented by means of the solution σ^* of (2.1) corresponding to the same initial conditions. These conditions are only approximatively known and, moreover, some small and non-estimable forces are necessarily neglected. Then, the validity of our theoretical description of σ will be strictly connected with the total stability of σ^*. We shall suppose that σ^* is the null solution of (2.1), and that it is stable but not totally stable. Then, if (2.1) is assumed to be a valid scheme of evolution law, we have to argue either

(1) the right hand side of (2.1) is known to within small perturbations belonging to a class U, and σ^* is U-totally stable; or

(2) among the neglected forces acting on S, some of them

protect S from deviating actions of arbitrarily small exciting forces.

The point (2) was particularly considered by Cetaev [1], in a discussion concerning the Hamilton's equations for conservative mechanical systems. That was done not in connection with the requirement of total stability property; but in order to insure that for a certain class $\mathcal{U}$ of permanently acting perturbations, $R(t,x)$, $R(t,0) \equiv 0$, $x \equiv 0$ be a stable solution of (2.2). As we have pointed out by comparison of properties (A) and (B), this requirement of Cetaev is not connected with total stability. Moreover, the instability of $x \equiv 0$ for every equation (2.2) with $R \in \mathcal{U} \setminus \{0\}$ does not seem important, from a physical point of view,if $x \equiv 0$ is totally stable. Indeed according to Remark 2.3, $\varepsilon(R)$ is then vanishing as $R \to 0$ in a suitable sense. Conversely, stability of $x \equiv 0$ for (2.2) is not sufficient to preserve the physical system from a drastic change of behavior, if $x \equiv 0$ is not totally stable.

Let us consider a mechanical system S with a finite number of degrees of freedom subjected to conservative forces. Suppose that the constraints are holonomic and independent of time. Choose a t-independent system of coordinates $q \equiv (q_1,q_2,\ldots,q_n)$ to specify the configuration of S, and denote by p the vector momentum associated with q. The Hamilton's equations of motion

$$\dot{p} = -\frac{\partial H}{\partial q}, \quad \dot{q} = \frac{\partial H}{\partial p}, \tag{3.1}$$

are autonomous and $H = T - U$, where T is the kinetic energy and U is the potential of forces. T is expressed by a quadratic form in p whose coefficients are dependent on q only. We suppose that these coefficients and U are of class C^2.

We can now better illustrate the arguments involved in Cetaev's analysis. He observes that "arbitrarily small exciting forces may make stable solutions of (3.1) become unstable". In fact, let $(p,q) \equiv 0$ be an isolated static solution of (3.1) and suppose that U has a maximum for $q = 0$. By virtue of the Lagrange-Dirichlet theorem this solution is stable and the corresponding evolution of S is actually an observable rest motion of S. The eigenvalues of the linear part of (3.1) have all real parts equal to zero. Therefore, arbitrarily small forces can be found so that $(p,q) \equiv 0$ becomes an unstable solution of the corresponding perturbed system. Set $U = U_2 + o(\|q\|^2)$, with U_2 a quadratic form, and suppose that the maximum of U can be recognized on U_2. Then, in the Cetaev analysis, the validity of the Hamiltonian scheme is stated by using the argument that linear dissipative forces, acting upon S, with complete dissipation change (3.1) to new system whose eigenvalues have negative real parts. In this way $(p,q) \equiv 0$ is an asymptotically stable solution of the new equations and this behavior is preserved under small exciting forces.

We notice that the above conclusion cannot be applied to the following two cases:

(C_1) The maximum of U is recognizable on U_2 and the dissipative forces are not linear;

(C_2) The maximum of U is not recognizable on U_2.

Let us examine now the question in its natural framework, that is in connection with concept of total stability. Actually the solution $(p,q) \equiv 0$ of (2.1) is not totally stable according to the following theorem.

3.1 Theorem [6]. *No periodic solution of (3.1) is totally stable.*

Can dissipative forces with complete dissipation cause that $(p,q) \equiv 0$ is a totally stable solution of the modified equation? The answer is positive as stated by the theorem given below. In this theorem, cases (C_1) and (C_2) are also included.

3.2 Theorem. *Let $R \equiv (R_1, R_2, \ldots, R_n)$ be the system of Lagrangian components of dissipative forces with complete dissipation*

(R is locally Lipschitzian and $R \cdot \frac{\partial H}{\partial p} \leqslant 0$, $R \cdot \frac{\partial H}{\partial p} = 0$ if and only if $p = 0$). If $(p,q) \equiv 0$ is an isolated static solution of (3.1) and U has a maximum for $q = 0$, then $(p,q) \equiv 0$ is an asymptotically stable (and therefore totally stable) solution of the differential system:

$$\dot{p} = -\frac{\partial H}{\partial q} + R \qquad \dot{q} = \frac{\partial H}{\partial p} \quad . \tag{3.2}$$

Proof. There exists a neighborhood Ω of $(p,q) = 0$ such that: (i) H is positive definite in Ω; (ii) $\dot{H}(p,q) \leqslant 0$ for all $(p,q) \in \Omega$ and $\dot{H}(p,q) = 0$ iff $p = 0$; (iii) in Ω there exists no positive orbit of (3.2). Then, the statement is a trivial consequence of a theorem of Barbasin-Krasovskii [7] or LaSalle [8].

Therefore the solution $(p,q) \equiv 0$ of (3.1), which is not totally stable, gains this stability property when it is considered as a solution of (3.2). In this sense we can intrepret Cetaev's statement that it is possible to recognize, to within the small forces which are neglected in the Hamiltonian scheme, the existence of "a barrier force which protects the mechanical system against large deviations under the action of arbitrarily small acting forces".

Open question. It seems very difficult to find out a barrier force which makes a periodic solution of (3.1) totally stable (or at least orbitally totally stable). In particular, the question arises for the planets' motions. How can we explain their lack of total stability? We think that the point (1) is particularly important for this problem. Perhaps the forces acting on the planets are exactly known to within an appropriate class $\mathcal{U}$ of small perturbations that are conservative. This seems reasonable and if so, the stable periodic character of these motions could be explained by looking for the $\mathcal{U}$-total stability of the corresponding solution of (3.1).

IV. NON-OBSERVABLE MOTIONS

In this Section we wish to emphasize the fact that certain stable solutions of Hamilton's equations, relative to a mechanical system S, cannot represent observable motions of S. This happens, for instance, when the dissipative forces, far from being "barrier forces", are themselves permanent perturbations which emphasize the lack of total stability.

(i) Suppose that S is acted upon by "generalized" conservative forces [9] $Q^* = Q' + Q$ where Q' depends, in the usual sense, on a potential function $U(q) \in C^2$ and Q (gyrostatic forces) has components given by

$$Q_h(q,\dot{q}) = \sum_1^n {}_k \, g_{hk}(q)\dot{q}_k, \; g_{hk} \in C^1, \; g_{hk} = -g_{kh}, (h,k = 1,2,\ldots,n).$$

The equations of motion could be put in a Hamiltonian form by means of a definition of the vector momentum p involving Q. However, we prefer to write these equations in the following form:

$$(4.1) \qquad \dot{p} = -\frac{\partial H}{\partial q} + Q, \quad \dot{q} = \frac{\partial H}{\partial p},$$

where $H = T - U$ has the same analytical expression as in Sec. 3 and Q is expressed in terms of p,q. It is known that an isolated static solution of (4.1), say $(p,q) \equiv 0$, can be stable even if U has not a maximum for $q = 0$ (gyrostatic stabilization). But, such a solution does not represent an observable rest motion of S. Indeed denote by $\mathcal{U}$ the class of dissipative forces $R \equiv (R_1,R_2,\ldots,R_n)$ with complete dissipation. The following theorem holds.

4.1 <u>*Theorem*</u>. *Suppose that* $(p,q) \equiv 0$ *is an isolated static solution of (4.1) and* U *has not a maximum for* $q = 0$. *Then* $(p,q) \equiv 0$ *is not* $\mathcal{U}$*-totally stable.*

<u>*Proof*</u>. Consider the differential system

$$(4.2) \qquad \dot{p} = -\frac{\partial H}{\partial q} + Q + R, \quad \dot{q} = \frac{\partial H}{\partial p},$$

where $R \in \mathcal{U}$. Denote by $\dot{H}_R$ the derivative of H along the solutions of (4.2). There exists a neighborhood Ω of $(p,q) = 0$, independent of R, such that: (i) $\dot{H}_R(p,q) \leqslant 0$ for every $(p,q) \in \Omega$ and $\dot{H}_R(p,q) = 0$ iff $p = 0$; (ii) in the set $\{(p,q) \in \Omega: p = 0,\ q \neq 0\}$ there is no positive orbit of (4.2).

Let $t_0 \in \mathbb{R}_+$. Since U does not have a maximum in $q = 0$, there exists (p_0,q_0) arbitrarily close to $(0,0)$ with $H(p_0,q_0) < 0$. Then (i) and (ii) imply [8] that $(p_R(t,t_0,p_0,q_0), q_R(t,t_0,p_0,q_0)) \notin \Omega$ for some $t \geqslant t_0$. Thus, the statement is a consequence of Remark 2.3.

(ii) Consider a heavy top S and suppose that a point O of its symmetry axis is fixed without friction. Let $O\xi\eta\zeta$ be an external Cartesian frame with axis η vertically descendent, and $OXYZ$ a Cartesian frame fixed in S with Z coinciding with the symmetry axis oriented from O to the center of mass. Let $q_1 + \frac{\pi}{2}$, q_2, and ϕ the Euler angles of $OXYZ$ with respect to $\Omega\xi\eta\zeta$ (nutation, precession, proper rotation angle, respectively) with $|q_1| < \pi/2$, $|q_2| < \pi$ $|\phi| < \pi$. The Hamilton's function is given by

$$H = \mathcal{T} + \frac{1}{2}\,\frac{k^2}{b} - U$$

where $\mathcal{T} = \frac{1}{2}\left(\frac{p_1^2}{a} + \frac{p_2^2}{a\cos^2 q_1}\right)$ and $U = c(1 - \cos q_1 \cos q_2)$.

Here $a,b,c > 0$ are constants and (p_1,p_2,k) is the vector momentum associated with (q_1,q_2,ϕ). The coordinate ϕ is cyclic so that the determination of $q \equiv (q_1,q_2)$, $p \equiv (p_1,p_2)$ and k can be reduced to the integration of equations

$$\dot{p} = -\frac{\partial H}{\partial q}, \quad \dot{q} = \frac{\partial H}{\partial p}, \quad \dot{k} = 0. \tag{4.3}$$

Let

$$(p,q) \equiv 0 \qquad k \equiv k^* \tag{4.4}$$

be a static solution of (4.4). For any k^*, values (4.4)

correspond to a one parameter family of solutions of the complete (Hamilton's) equations of motion, and these solutions represent the so called permanent rotations of S around Z in the vertically ascendent position. $\frac{k^*}{b}$ is the angular velocity of S in the permanent rotations. Let $\mathcal{U}$ be the set of dissipative forces (R_1,R_2,Q_ϕ) having dissipation reduced to non-cyclic coordinates. That is, $R \equiv (R_1,R_2)$ depends only on (p,q), $Q_\phi \equiv 0$ and

$$R \cdot \frac{\partial H}{\partial p} \leqslant 0, \quad R \cdot \frac{\partial H}{\partial p} = 0 \quad \text{iff} \quad p = 0.$$

We shall prove that the above permanent rotations are not observable motions. Indeed the following theorem holds.

4.2 <u>*Theorem.*</u> *For any k^* the solution (4.4) of (4.3) is not $(q,\mathcal{U})$-totally stable.*

<u>*Proof.*</u> For a fixed value of k, consider the differential system

$$\dot{p} = -\frac{\partial H}{\partial q} + R, \quad \dot{q} = \frac{\partial H}{\partial p}, \tag{4.5}$$

where $(R,0) \in \mathcal{U}$. Set $F = \mathcal{T} - U$ and let $\dot{F}_R = R \cdot \frac{\partial H}{\partial p}$ be its derivative along the solutions of (4.5). For $\varepsilon \in]0,c[$, set $\Omega = \{(p,q) : U(q) \leqslant \varepsilon, \mathcal{T}(p,q) \leqslant \varepsilon\}$. We have: (i) $\dot{F}_R(p,q) \leqslant 0$ for every $(p,q) \in \Omega$ with $\dot{F}_R(p,q) = 0$ iff $p = 0$; (ii) in the set $\{(p,q) \in \Omega : p = 0, q \neq 0\}$ there is no positive orbit of (4.5).

Let $t_0 \in \mathbb{R}_+$. Since U has a minimum in $q = 0$, there exists (p_0,q_0) arbitrarily close to $(0,0)$ with $F(p_0,q_0) < 0$. Let $(p_R(t),q_R(t))$ the solution of (4.5) passing through (t_0,p_0,q_0). Then (i) and (ii) imply [8] that there exists $t_1 \geqslant t_0$ such that $(p_R(t_1),q_r(t_1)) \notin \Omega$, which, in turn, implies, again taking in account (i), $U(q_R(t_1)) > \varepsilon$. By Remark 2.3, the proof of the theorem is complete.

V. REFERENCES

[1] Cetaev, N.G., *Note on classical Hamiltonian theory*, P.M.M.

24, 1(1960).

[2] Dubosin, G. N., *On the problem of stability of a motion under constantly acting perturbations*, Trudy gos. astron. Inst. Sternberg *14*, 1(1940).

[3] Malkin, I. G., *Stability in the case of constantly acting disturbances*, P.M.M. *8* (1944). *Theory of stability of motion*, Gos. Izdat. tekh.-teoret. Lit., Moscow 1952; English transl., AEC tr. 3352, 1958; German transl., Oldenbourg, 1959.

[4] Gorsin, S., *On the stability of motion under constantly acting perturbations*, Izv. Akad. Nauk Kazakh. S.S.R. *56*, Ser. Mat. Mekh., 2 (1948).

[5] Seibert, P., *Estabilidad bajo perturbaciones sostenidas y su generalizacion en flujos continuos*, Acta Mexicana Cienc. y Tecnol., *11*, 3 (1968).

[6] Salvadori, L., and Schiaffino, A., *On the problem of total stability*, Nonlinear Analysis, Theory, Methods, and Applications, *1*, 3(1977).

[7] Barbasin, E. A., and Krasovskii, N. N., *On the stability of motion in the large*, Doklady Akad. Nauk S.S.S.R. *86*, 3 (1952).

[8] LaSalle, J. P., *The extent of asymptotic stability*, Proc. Nat. Acad. Sci. *48* (1960). *Some extensions of Liapunov's second method*, IRE Trans. CT-7 (1960).

[9] Whittaker, E. T., *A treatise on the analytical dynamics of particles and rigid bodies*, Cambridge University Press, 1937; Dover Publications, New York, 1944.

ON SOME MATHEMATICAL MODELS OF SOCIAL PHENOMENA

Elliott W. Montroll
Institute for Fundamental Studies
University of Rochester

I. INTRODUCTION

It is a pleasure to participate in the dedication of your new center of applied mathematics. There has probably never been a more appropriate time for the organization of such centers. With the increased concern for the future consequences of society's present activities and with the availability of a sophisticated low priced computing capacity, an increased demand for applied mathematicians is guaranteed.

Practicers of almost all academic disciplines, as well as interdisciplinary buffs, are using computers in some way, which forces them to characterize the subjects of their investigation more precisely in mathematical terms. Such a characterization will motivate a certain number of investigators to improve their mathematical skills and to be concerned with styles of mathematical model making.

Physical scientists have developed a style of subtle interplay between organization of numerical data and introduction of mathematical models to classify and/or mimic the data. All new branches of mathematical physics start with some simple model. At the later stages of development of the science, more global models (or, as they are called, principles) evolve which imply a wide variety of the more primitive models. Embedded in this is an accountant's conscience displaying a special disposition toward

ISBN 0-12-434160-8

conservation "laws". There is constant search for simplicity and elegance of expression. A tendency exists to first examine isolated portions of a complex situation with an inherent faith that the full complexity might eventually be understood in terms of an appropriate conjunction of the simple portions.

The aim of this lecture is to review a few rate equation models and a few statistical models of social phenomena in the spirit described above. Rate equations will frequently be nonlinear. In view of the short time available for this presentation, no topic will be presented in great detail. Further information is available in a book by Wade Badger and the author[1] and in many references given at the end of this report.

The first topic we wish to consider in some detail is the modeling of population dynamics. Since considerable data exists on this subject and since large books are devoted to it, our discussion should be considered as an introduction to the rate equation style of modeling rather than as an exhaustive review.

We will postulate a certain "perfect population growth process" and take the view that when real populations do not grow according to that process, some social, ecological, or economic force has appeared. Our strategy will then be to investigate the consequences of certain "common sense" mathematical forms for these forces. The consequences in some cases will be favorably compared with "real" population data and in others will be left hanging for future consideration and correction.

The equations in each section will be numbered 1, 2, ... When equations from previous sections are identified, the equation number will be prefixed by the section number, thus: III.2.

II. THE FIRST AND SECOND "LAWS" OF SOCIAL DYNAMICS

This section is a parody of Newton's first and second laws of mechanics adapted to the discussion of certain social phenomena. Newton's first law is the postulate that in the absence of an external force every body in a state of motion will remain in that state of motion; i.e., it will continue to move in a straight line with a constant velocity. Of course this situation never prevails in earthbound experiments, but it is still a good starting point for the construction of mathematical models of dynamical systems.

The first law of population dynamics is here chosen to have two similarly stated parts:

a) In the absence of any social, economic or ecological force the rate of change of the logarithm of a population, $N(t)$, of an organism is constant,

$$d\log N(t)/dt = \text{constant} \tag{1a}$$

Without the prescribed forces this equation is also postulated to be valid for the variation of the population of objects of production (automobiles, radios, etc.).

b) In the absence of any social, economic or ecological forces, the rate of change of the logarithm of the price of maintenance $P(t)$ (per unit time) of an "organism" is also constant

$$d\log P(t)/dt = \text{constant} \tag{1b}$$

In the case of objects of production $P(t)$ is to be interpreted as a unit cost.

Our inclusion of populations of inanimate objects is made so that our population growth models might occasionally be applied to production of and competition between manufactured items.

Equation (1a) is, of course, nothing but the Malthusian law of exponentiation of populations, and (1b) is a statement of the accountants' "discounting" principle and the housewives' observation that things are always getting more expensive. The sign of the constant in (1a) might be negative as well as positive because interest in some items just dies away.

It might be claimed that our first law of social dynamics is more often applicable to the real world than Newton's first law of mechanics is to real dynamical systems, since a considerable effort of social reformers as well as conservative politicians is devoted to finding means of generating forces to induce the violation of laws (1a) and (1b).

Newton was very astute in his employment of the second law as the definition of a force. Who can go wrong by making definitions if he doesn't make too many of them? The second law is just the statement that a force is that which causes the first law to be violated. We will not attempt to outdo the master on this point.

The second law of population dynamics is the postulate: Equations (1a) or (1b) or both are violated when a social, economic or ecological force is applied. How is the force to be chosen or measured? By observing the manner in which the first law is violated! Part of the remainder of this lecture will be devoted to the investigation of the consequences of the application of various postulated forces.

A third law - evolution is a sequence of replacements - should also be considered.

Newton never tried to derive his laws of force from first principles. The restoring force for a displaced mass in a harmonic oscillator was merely the force of greatest mathematical simplicity and one found to be useful in describing many physical phenomenon. The postulation of the inverse square law of the gravitation force was natural for a genius whose geometrical intuition told him it was just what was required to produce Kepler's empirical observations on elliptical planetary orbits.

III. THE LINEAR FORCE AND VERHULST'S LOGISTIC MODEL OF POPULATION GROWTH AND SATURATION

The simplest mathematical form of the force which might replace the constant on the right hand side of (II.1b) is the linear form (with parameters k and θ)

$$F\{N(t)\} = k - \alpha N(t) \equiv k\{1 - [N(t)/\theta]\} \qquad (1)$$

We choose the sign of the $N(t)$ dependent term to be negative so that it represents a deterent to exponential growth. Then our equation of population growth is

$$d \log\{N(t)/\theta\}/dt = k \{1 - [N(t)/\theta]\} \quad (2)$$

Clearly when $N(t) = \theta$ there will be no growth; i.e., the population saturates (if initially $N(t) < \theta$).

The form of equation (2) was first proposed by the Belgian mathematician Verhulst[2]. While the author has heard the rumor that it was known even earlier to Euler, he has not been able to track down any specific reference. The equation was rediscovered by Pearl[3] and Reed and used by them with some success to fit the population growth of many countries. A typical fit is given in figure 1 where the solution of (2),

$$N(t) = \theta N(0)/\{N(0) + [\theta - N(0)] \exp(-kt)\} \quad (3)$$

is compared to US census data[4]. The curve is often called the logistic curve. Notice that the fit is remarkably good until the depression period when the population growth rate became lower than would be expected from (2). After World War II the increase

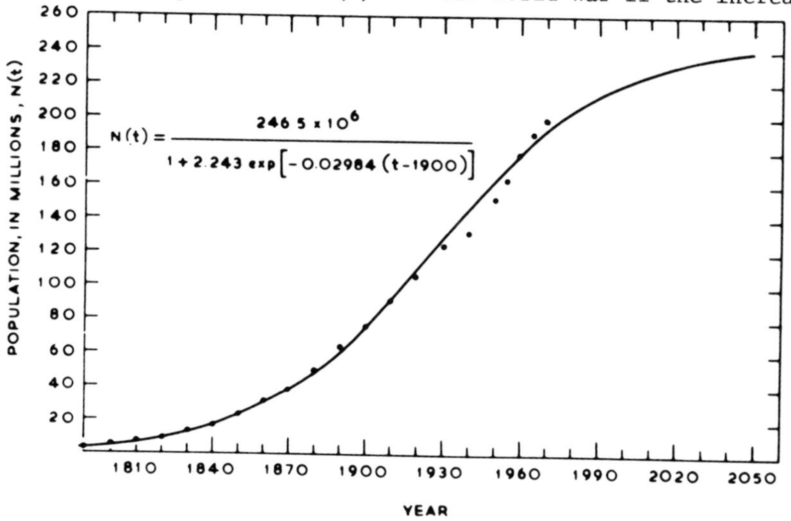

Fig. 1. Population of U.S. Logistic curve fitted so that observed points at 1840, 1900 and 1960 are exact. Points represent census data.

of birth rate more than compensated for the depression decline. Curves such as fig. 1 are common to many countries. They show that special economic forces, such as a depression, influence the birth rate as does the rebirth of public optimism after wars. More detailed discussion of these points will be given in section VI.

If one lets $f \equiv N/\theta$ be the fraction of the way to population saturation, then (2) has the form

$$\dot{f} = kf(1 - f) \tag{4}$$

which can be linearized by the transformation

$$g = f/(1 - f). \tag{5}$$

Then

$$\dot{g} = \dot{f}/(1 - f)^2 = kf/(1 - f) = kg \tag{6}$$

so that

$$\begin{aligned} \log g(t) \equiv {} & \log\{f(t)/[1 - f(t)]\} \\ & \log\{f(0)/[1 - f(0)]\} + kt \end{aligned} \tag{7}$$

This formula for $f(t)$ suggests that an appropriate way to decide whether the logistic equation fits a set of empirical data is to plot $f/(1 - f)$ on similog graph paper as a function of time and see if the points lie on a straight line.

J. Fisher and R. Pry[5] have made some remarkable plots of this sort in their investigation of the manner that new products or technologies replace older ones. If a new technology is introduced, clearly better in the production of some material, it will take over an increasingly larger fraction of the market until the full market is absorbed by it, unless in the interim a still better technology appears. Qualitatively one might expect the fraction of the market taken by the new method to follow a logistic type curve. The ratio $f(t)/[1 - f(t)]$ is just that of the fraction of the market captured by the new process to the fraction remaining to the old process.

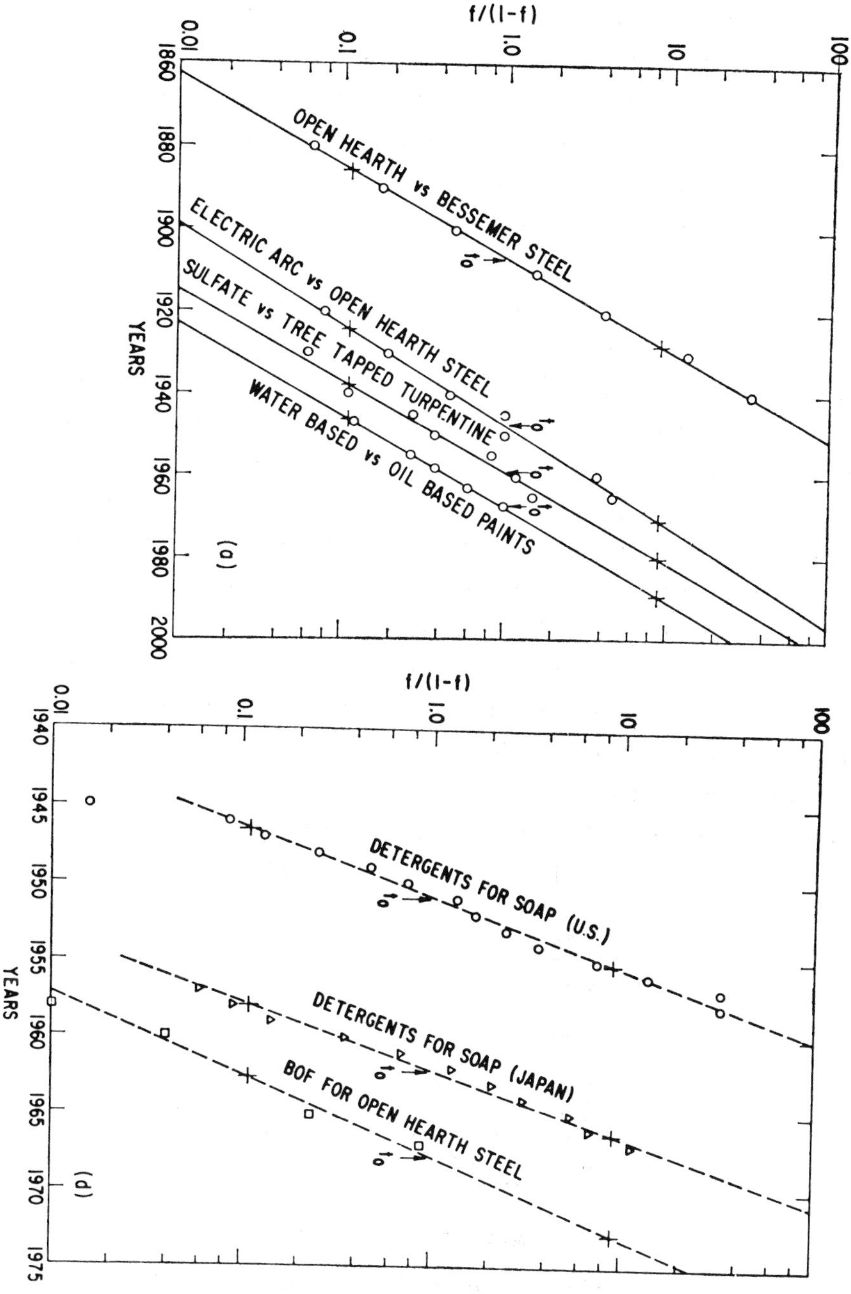

Figs. 2. Substitution data and fit to model for a number of products and processes. All data U.S. except detergents for soap as noted.

We have reproduced, in fig. 2, the Fisher-Pry replacement dynamics curves for a number of processes and materials ranging from replacement of the Bessemer Steel production method by the open hearth method to the replacement of soap by detergents.

It occurred to Robert Herman and the author[6] that the industrial revolution of the 19th century was an evolutionary process converting the labor force of a country from its agricultural nature to an industrially directed one in a manner that might be described by the logistic equation model. The fraction of the labor force doing agricultural work declined as the fraction involved in industry grew. Incidentally, the concept that there was an agricultural revolution in the 19th century is insufficiently emphasized; for if it were not for the fact that improvements in agricultural technology were developing rapidly there would not have been a surplus of farm laborers to migrate to the industrial centers, seeking employment.

Prior to 1840 the ratio of nonagricultural workers to agricultural workers in the labor force of the US remained fairly constant over many decades. The ratio of the fraction of nonagricultural workers to agricultural workers in the US is plotted as function of time in fig. 3 on similar graph paper (as is the

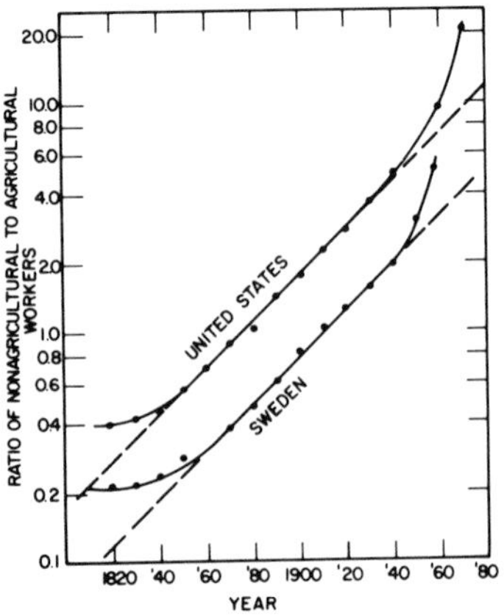

Fig. 3. Variation of the ratio of nonagricultural to agricultural, f/(1 - f), labor force in the U.S. and Sweden (4), 1820-1970.

corresponding fraction for Sweden). The US data was obtained from Statistical Abstracts of the United States. It is remarkable how well the data fits the straight line defined by the logistic equation for a period of about 100 years.

The rate at which agricultural workers left the farms in the early 1940's was somewhat greater than would have been expected on the basis of the logistic equation. The new force was generally a telegram from President Roosevelt which started with the word "Greetings... ." After World War II many of the young men who responded to the Greetings did not return to the farm. The curve for $f/(1 - f)$ (nonagricultural to agricultural worker ratio) for Sweden follows that of the US rather well. While the young Swedish farm hand did not get a greetings telegram, he was enticed to move to the city in response to the higher wages paid by companies which were selling factory products to pleading customers around the world.

It is said that we are now becoming a service-oriented country. Perhaps we should construct new curves which show the growth of the fraction of service workers to non service workers as a function of time. As an alternative to the above construction, we plot in section VI, curves of evolution which reflect the variation of four populations which give a finer characterization of the labor force.

We close this section with the introduction of alternative forcing functions which yield saturating population curves. A simple example having some advantages over (1) is

$$F\{N(t)\} = -k \log\{N(t)/\theta\} \tag{8}$$

This corresponds to the rate equation

$$d \log[N(t)/\theta]/dt = -k\{\log N(t)/\theta\} \tag{9}$$

Note that (1) and (8) are special cases of

$$F\{N\} = k\{1 - (N/\theta)^{\nu}\}/\nu, \tag{10}$$

the Verhulst equation (2) corresponding to $\nu = 1$ and (9), first discussed by Gompertz, corresponding to $\nu = 0$.

The attractive feature of (9) is that in terms of the variable

$$v = \log N/\theta, \tag{11}$$

(9) is linear:

$$dv/dt = -kv \tag{12}$$

so that

$$v(t) = v(0)\exp{-kt} \tag{13}$$

or in terms $N(t)$

$$N(t) = \theta[N(0)/\theta]^{\exp-kt} \tag{14}$$

The solution of[1]

$$dN/dt = kN\{1 - (N/\theta)^{\nu}\}/\nu \tag{15}$$

is plotted in fig. 4 for several values of ν to show the slight differences in the approach to saturation from below as ν is changed.

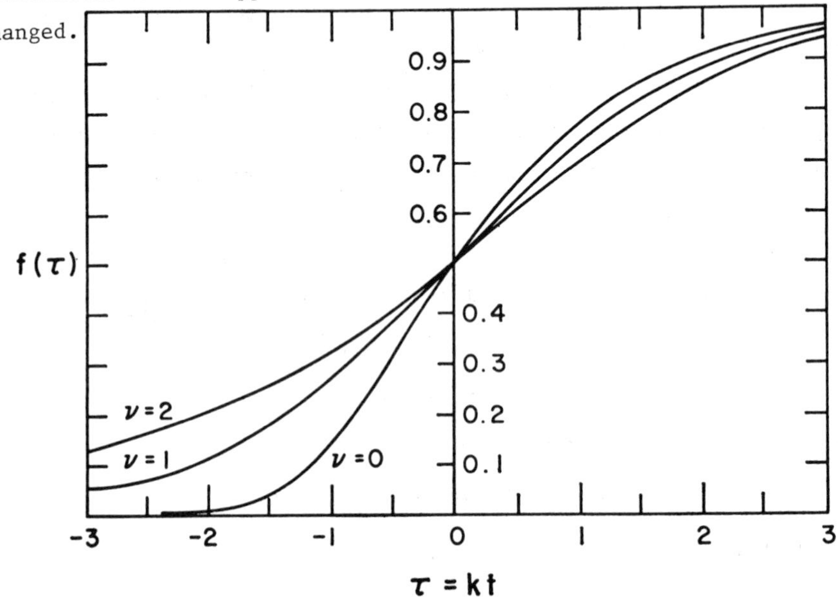

Fig. 4. A family of growth curves which saturate. The case $\nu = 0$ is the Gompertz growth curve, and the case $\nu = 1$ is the Verhultz growth curve.

We will sometimes introduce a general forcing function which leads monotonically to saturation, $k\ G(N/\theta)$, so that the generalization of (2) and (9) would be

$$d \log[N(t)/\theta]/dt = k\ G(N/\theta) \tag{16a}$$

with

$$G(1) = 0 \quad \text{and} \quad G'(x) > 0 \quad \text{if} \quad 0 < x < 1. \tag{16b}$$

IV. SOME REMARKS ON OBSERVED POPULATION FIGURES

Population growth curves such as that in fig. 1 are too gross to emphasize various short time changes in attitudes of the members of a population. The level of the population at a given time is a weighted average over the influence of motivations and catastrophes of the previous seventy years. Birth rate and death rate curves (as well as immigration and emigration curves) for a given country are a better image of the populations response to the problems of specified times. The birth rates and death rates (per 1000 persons) in the U.S. are plotted[4] in fig. 5 for the

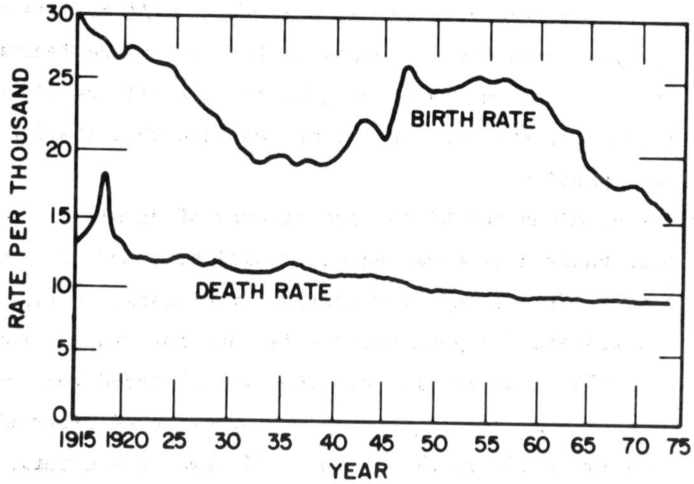

Fig. 5. Variation in birth and death rates per thousand per year in the U.S. during the period 1915-1970.

period 1915-1973.

Notice that in the period 1915-1933 the birth rate drop of the U.S. was essentially linear, as it also was in the period 1956-1973 (with about the same slope). The population literature of the 1920's decried the drop and urged people to have more children, while that of the late 1960's urged them to refrain from having children. Certain segments of the literature of the late 1960's were concerned with the observation that if one extrapolated along a straight line which connected the U.S. population of 1950 with that of 1960 (as given in fig. 3), the U.S. population of 2050 would exceed 400×10^6. Essentially none of the literature of the late 1960's observed that if the birth rate curve of fig. 5 were extrapolated linearly as indicated in the figure, the death rate would exceed the birth rate in the early 1980's so that the last years of the century would witness a decline in the U.S. population (in the absence of significant immigration).

The two contradictory observations made above indicate the dangers of pure extrapolation. A proper choice of curve permits one to predict whatever he wishes. Probably neither prediction is correct. However from certain patterns which will be presented for a variety of countries, I would be inclined to conjecture that the prediction based on a declining birth rate will be closer to the development of the next decade or two than that leading to a population explosion.

If one examines the birth rate records of developed countries for the past hundred or more years, it will be found that birth rates generally drop except for periods after wars. A typical curve for a western European country is that for France[7] for the period 1800-1970 shown in fig. 6. The overall trend for the period 1815-1935 is linear. A 10% rise was experienced after the Napoleonic Wars, a 20% rise over the 1914 level was experienced in 1920 (with the rate doubling over the 1916 wartime minimum of 10 per thousand), and a 25% rise over the 1935 level was achieved in 1946. While the Franco-Prussian War was much shorter than the big

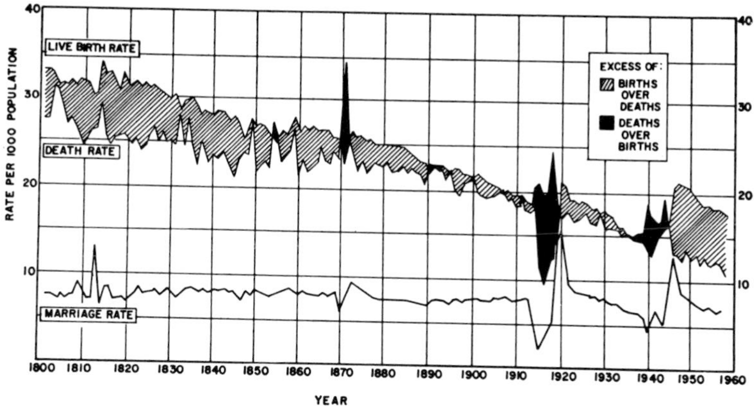

Fig. 6. France: Crude rates of birth, death, and marriage from 1801-1958. (Annuaire Statistique, 1951, pp. 35-37; Population, 1953, p. 754, and 1959, p. 106.) Overall graph from Marriage and the Family in France since the Revolution, W. D. Camp (Bookman Associates, N.Y., 1961.)

three wars, there was even a slight rise in birth rates after that one. Each of these accelerated birth rates followed large increases in the marriage rate as is evident in fig. 6. The U.S., the European countries, and Japan all experienced a baby boom after World War II. The birth rates of all of these countries has also been declining in the past ten years as is shown in fig. 7. The German record is quite dramatic, since in April 1973 the death rate exceeded the birth rate. With abortions becoming easily obtainable and with the accelerating trend for careers for women, it is hard to imagine a reversal of the declining birth rates in the countries listed above in the next few years. A cure for cancer and for cardiac disorders would induce a drop in the death rate, which has not significantly changed in the U.S. since 1920. The introduction of the antibiotics in the 1930's and 1940's caused only a minor decrease. In summary, we note that the most striking event in population statistics of the developed countries is the closing of the difference between birth and death rates.

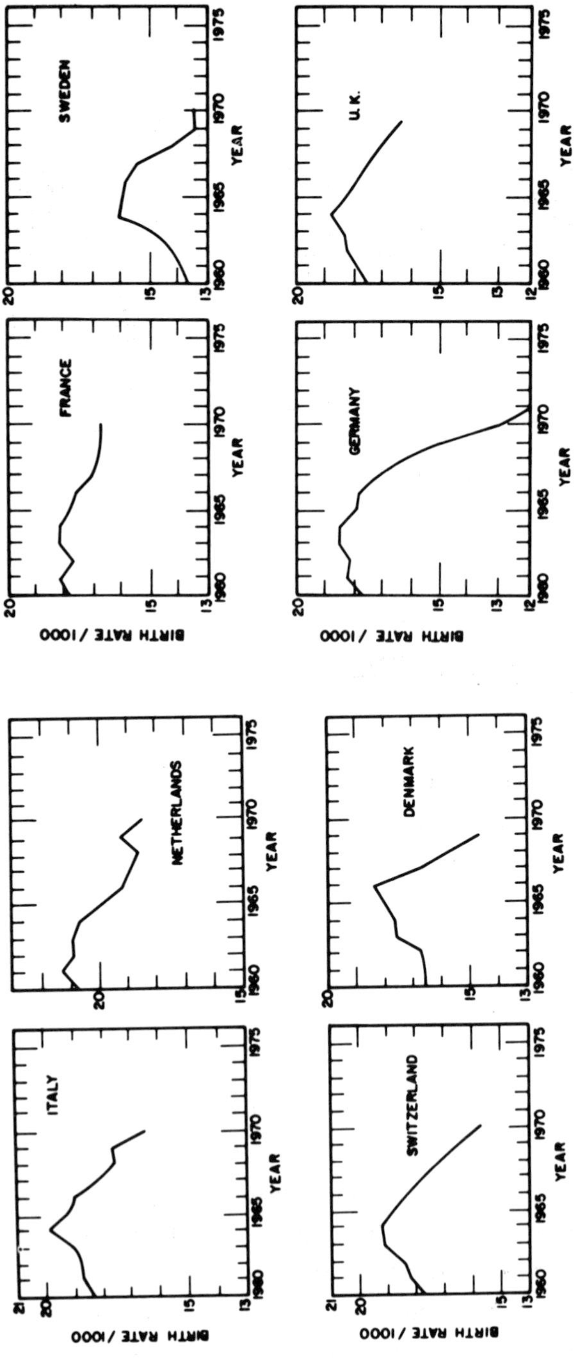

Fig. 7. Birth Rates per Thousand in Eight European Countries.

India, in recent years, has experienced the opposite population variation pattern. As is evident from fig. 8, the difference

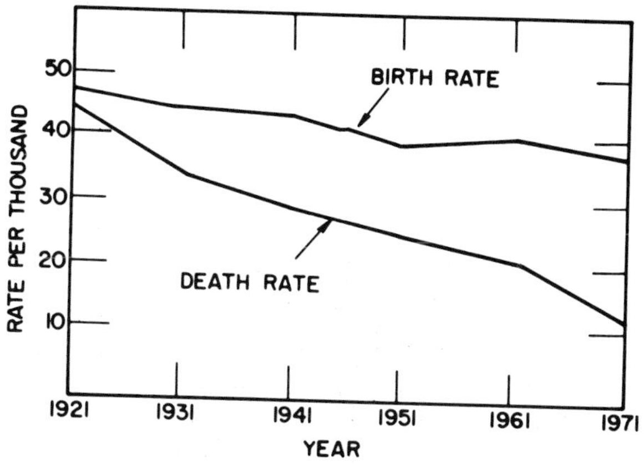

Fig. 8. Birth rates and death rates in India for period 1921-1971.

between birth and death rates has been growing rather than diminishing in the past twenty (and indeed in the past fifty years). The annual death rate has dropped from 48.6/1000 in 1921 to 14/1000 in 1971 while the birth rate has fallen only from 49.2/1000 to 39/1000 as exhibited in fig. 8. Prior to the 1920's, famines instituted a measure of population control; but in the past fifty years, foreign countries have rescued the Indians from that scourge. The improvement of public health standards and sanitation, while not comparable to that of the West, has still been sufficient to strongly contribute to the recorded drop in death rates and to allow an increase in life expectancy from 33 years in 1947 to 52 in 1972. It is interesting to note in fig. 9 that the death rate in New York City in the 1850's was only slightly lower than that in Calcutta in the 1920's; and that, indeed, the causes of death were rather similar.

Since it is doubtful that the death rate in India will drop significantly below 14/1000 in the next decade (that of New York City being about 10), the difference between birth and death rates will start to diminish with the two rates closing as in the case

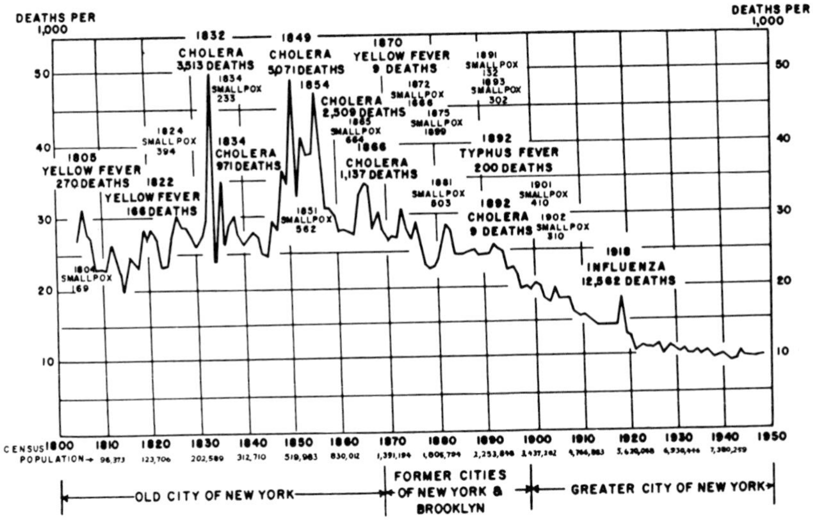

Fig. 9. Mortality and epidemics in New York City[8] *(1801-1948.*

in western countries. However unless two important traditions change in Indian social behavior, the closing will extend over a longer period than it has in the West or in Japan and China. As long as child brides are common, families will tend to remain large; since a girl married at 13 or 14 will have on the average at least three or four more children than one married at twenty or twenty-one. There are official pressures to deter child marriages but they are still common in the small villages and rural areas. Even if the average age of marriage is raised to 18, large families will remain the rule unless some form of national social security for the aged is developed. Traditionally, children care for their parents in their old age. Hence a thoughtful Indian father provides a safety factor for his security in the same manner that a good engineer introduces a safety factor in his designs. If four children would seem to be sufficient, eight would relieve most uncertainty. Without television the main evening entertainment for teenage couples in Indian villages contributes to population growth.

While death rates in Latin American countries have declined more rapidly than birth rates, the general population density in Latin America is small, so that there is more space for an expanding population than in India. The immediate Latin American problem is not one of a population explosion per se, but rather one of redistribution of population. The flow into the large cities, generally the capitals, has been fantastic in the past few decades; so that the inadequate facilities in those cities have gone progressively from bad to worse.

The population of Mexico City, for example, is increasing much more rapidly than that of most major cities of the world

1906	400,000	1950	3,050,000
1938	1,215,000	1960	4,870,000
1970	8,500,000		

While the country Mexico has doubled its population in thirty years, the capital, Mexico City, does so in about fifteen.

Incidentally, a ZPG or even a reduction in population will not necessarily solve the problems of the teeming cities since these problems are generally not the result of an overall population increase but are associated with a mass movement of people into the largest cities. The population of Ireland (see fig. 10)

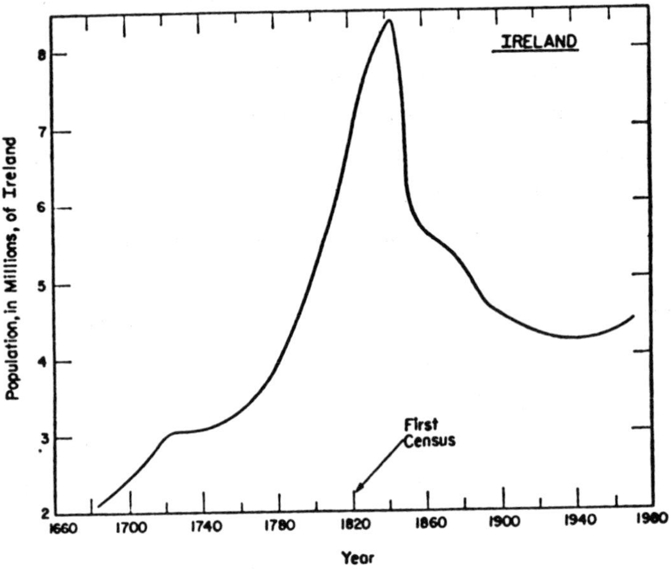

Fig. 10. Variation of population of Ireland since 1680.

dropped from a peak of 8.2×10^6 in 1841 just before the potato famine to approximately half that level, 4.7×10^6, in 1891. In the same 50 year period the population of the two largest cities[1], Belfast and Dublin, almost doubled in the jump from 3.4×10^5 to 6.0×10^5. A common experience everywhere is that during a national crisis there is a movement to the cities, especially the largest cities. One cause of this is the "cousin effect." If for some reason one wishes to leave a rural area, he generally moves to a place where he has a relative, perhaps a brother or a cousin. Since statistically, large cities have more brothers or cousins, they became a larger target for the relocater. It is also generally assumed that more opportunities exist in the largest cities.

The dangers of pure extrapolation of population growth curves were noted above. Demographers observe that such estimates are generally too high. While we conjectured that our extrapolation of birth rate and death rate curves might be better, a still better procedure for limited time predictions (10–20 years) would be based on age distribution and expected death rate and birth rate as a function of that distribution.

Any one alive at year t will either be one year older at year $t + 1$ or will have died by that time. Let $n_x(t)$ be the number of persons in their x-th year in a given country at year t. Then the total population at year t is

$$N(t) = \sum_{x=1}^{\infty} n_x(t)$$

Also let $p_x(t)$ be the probability of a person in his x-th year dying within the next year and let $b_x(t)$ be the probability that a woman in her x-th year will give birth to a baby in the next year. Then, if we make the approximation that the number of women in a given age group is the same as the number of men (an approximation which can be easily avoided at the expense of doing our bookkeeping on both sexes separately), the population in the year $(t + 1)$ will be

$$N(t+1) = \sum_{x=1}^{\infty} [1 + \tfrac{1}{2} b_x(t) - p_x(t)]\, n_x(t)$$

$$n_1(t+1) = \tfrac{1}{2} \sum b_x(t) n_x(t); \quad n_x(t+1) = [1 - p_x(t)] n_x(t), \quad x > 1.$$

The quantities $p_x(t)$ and $b_x(t)$ can be estimated by extrapolation from previous 10 years data. An error of 10% in these numbers yields an error of about 0.3% in $N(t + 1)$ since most people alive next year are alive this year.

V. ON THE INTERACTION BETWEEN TWO SPECIES, THE LOTKA[9]-VOLTERRA[10,11] MODEL AND CERTAIN EXTENSIONS OF IT.[12]

We now extend the ideas presented in section II to a discussion of the interaction between two species. We still postulate a linear form for the "force" which causes a violation of eq. 1a, but being concerned with two species we need a rate equation for the population of each and arrange the force term to be linear in the two populations N_1 and N_2 of our two species "1" and "2". Thus:

$$d \log N_1/dt = k_1 + c_{11}N_1 + c_{12}N_2 \tag{1}$$

$$d \log N_2/dt = k_2 + c_{21}N_1 + c_{22}N_2 \tag{2}$$

This might be considered to be the small population approximation to some general force which would be expressed as

$$d \log N_1/dt = F_1(N_1,N_2) \tag{3a}$$

$$d \log N_2/dt = F_2(N_1,N_2) \tag{3b}$$

By making various choices of k_1,k_2 and c_{ij}, especially with regard to signs of these terms, one obtains certain well known models of interacting populations.

The first choice we consider is

(A)

$$c_{11} = -k_1\gamma_1/\theta,\ c_{12} = -k_1\gamma_2/\theta \tag{4a}$$

$$c_{21} = -k_2\gamma_1/\theta,\ c_{22} = -k_2\gamma_2/\theta \tag{4b}$$

Then (1) and (2) become (with $k_1,k_2,\gamma_1,\gamma_2$ all positive)

$$d \log N_1/dt = k_1[1 - (\gamma_1 N_1 + \gamma_2 N_2)/\theta] \tag{5a}$$

$$d \log N_2/dt = k_2[1 - (\gamma_1 N_1 + \gamma_2 N_2)/\theta] \tag{5b}$$

This is a model suggested by Volterra for two species competing for the same resources. It is a generalization of Verhulst's model which leads to saturation and population stabilization when the combined population N_1 and N_2 reach a limit such that

$$\gamma_1 N_1 + \gamma_2 N_2 = \theta \tag{6}$$

Weight γ_i is given to species i in assessing its influence on the saturation level.

Equation (4) implies that there is a constant of the motion, c, with

$$N_2 = cN_1^{k_2/k_1}, \quad c = \text{constant} = N_2(0)/[N_1(0)]^{k_2/k_1} \tag{7}$$

The calculation of N_1 (and through 7, N_2) as a function of time can then be reduced to quadratures. We find, with $r = k_2/k_1$

$$k_1 t = \int_{N_1(0)}^{N_1(t)} dN_1/N_1[1 - (\gamma_1 N_1 + c\gamma_2 N_1^r)/\theta] \tag{8}$$

The saturation levels of N_1 and N_2 depend on the initial values $N_1(0)$ and $N_2(0)$ as well as on the various parameters. The equilibrium level $N_1(\infty)$ satisfies

$$N_1(\infty) + (\gamma_2/\gamma_1)N_2(0)\ [N_1(\infty)/N_1(0)]^r = (\theta/\gamma_1) \tag{9a}$$

while that of N_2 is

$$N_2(\infty) = (\theta/\gamma_2) - (\gamma_1/\gamma_2)N_1(\infty). \tag{9b}$$

A more interesting case is

$$\text{(B)} \quad \begin{array}{c} k_1 = \alpha_1 > 0, \quad k_2 = -\alpha_2 < 0, \quad c_{11} = c_{22} = 0 \\ -\lambda_1 \equiv c_{12} < 0 < c_{21} \equiv \lambda_2 \end{array} \tag{10}$$

This leads to the Lotka-Volterra preditor-prey model characterized by

$$\dot{N}_1 = \alpha_1 N_1 - \lambda_1 N_1 N_2 \tag{11a}$$

$$\dot{N}_2 = -\alpha_2 N_2 + \lambda_2 N_1 N_2 \tag{11b}$$

Species 2 is completely dependent on species 1 in this model. If $N_1 = 0$, N_2 decays exponentially to zero with time while N_1 grows in a Malthusian manner. When collisions occur between the preditor "2" and prey "1", the population of prey diminishes while the number of preditors increases.

An important feature of the model is that it leads to periodic oscillations in the population of both species.[10,12] Generally it is convenient to use the variables:

$$f_j = N_j/q_j \quad j = 1,\ 2 \tag{12}$$

with q_j being the steady state values of N_j as obtained from (11) by setting $\dot{N}_j = 0$:

$$q_1 = \alpha_2/\lambda_2, \quad q_2 = \alpha_1/\lambda_1. \tag{13}$$

The equations for f_1 and f_2 obtained from (11a) and (11b) imply, upon elimination of the time that

$$[f_1 \exp(-f_1)]^{1/\alpha_1} [f_2 \exp(-f_2)]^{1/\alpha_2} = \text{constant}. \tag{14}$$

It is shown in references 10 and 12 that this equation for the constant of the motion implies the periodicity of the solution of (11).

A natural extension of the two species Lotka-Volterra model would provide for a saturation level in the population of species 1 in the absence of species 2. In that case one would introduce a Verhulst type term in (1) as could be done by setting $c_{11} = \alpha_1/\theta_1$ so that (11) would be replaced by the set

$$\dot{N}_1 = \alpha_1 N_1 [1 - (N_1/\theta_1)] - \lambda_1 N_1 N_2 \tag{16a}$$

$$\dot{N}_2 = -\alpha_2 N_2 + \lambda_2 N_1 N_2 \tag{16b}$$

With this term the steady state population of species 1 remains

$$q_1 = \alpha_2/\lambda_2 \tag{17}$$

while that of species 2 becomes

$$q_2 = (\alpha_1/\lambda_1)\ \{1 - (\alpha_2/\lambda_2\theta_1)\} \tag{18}$$

which is somewhat smaller than the $\theta_1 \to \infty$ limit, α_1/λ_1. An important effect of the introduction of a saturation level is that the population actually achieves a steady state instead of oscillating about it.[12,13]

When $N_1 >> N_2$ one would hardly expect a linearized version of some general force as given in (3) to be appropriate. Consider the term $\lambda_2 N_1 N_2$ of eq. (11) when N_1 is large and N_2 is small. There is a limit to the rate at which the small number of preditors "2" can feast on the enormous number of prey N_1. Hence as N_2 becomes large we would expect the interaction term to become proportional only to N_2 independently of N_1. This is exactly the Malthusian exponention principle. Given an infinite food reservoir, the population of a species will grow exponentially. Watt[14] postulated a species interaction which reduces to the Lotka-Volterra model when N_1 is small and which yields the Mathus equation when N_1 is large. The term $N_1 N_2$ in eq. 16 would be replaced by

$$N_2 c^{-1}[1 - \exp(-cN_1)] \tag{19}$$

Then the Watt generalization of the Lotka-Volterra model with a Verhulst saturation term becomes[14]

$$\dot{N}_1 = \alpha_1 N_1 [1 - N_1/\theta_2] - \lambda_1 c^{-1}[1 - \exp(-cN_1)]\, N_2 \tag{20a}$$

$$\dot{N}_2 = -\alpha_2 N_2 + \lambda_2 N_2 [1 - \exp(-cN_1)] c^{-1} \tag{20b}$$

As $N_1 \to 0$, (20) reduces to (11).

More sophisticated models might take into account the age distribution of species "1" and "2" since very young and very old prey are more vulnerable to preditor attack, and very old and very young preditors are not as dangerous to the prey as middle-aged ones.[13]

The models presented above have been generalized to include more species and indeed classes of species. Theoretical approaches to the investigation of eco-systems employ such generalizations. Two reviews on this subject are given in references (12) and (15).

In conclusion we note that equations (11a) and (11b) with $\alpha_1 \equiv \alpha_2 = 0$ and $\lambda_1 > 0 > \lambda_2$ are a case of Lanchesters equations for "deadly combat", say for two sets of aerial "dogfighters".

VI. ON EVOLUTIONARY PATTERNS IN SYSTEMS OF SEVERAL VARIABLES[6]

One is sometimes required to exhibit evolutionary patterns in systems of several interacting variables. While he might attempt to construct a rate equation model for such systems, the nature of the nonlinearities inherent in the model and the values of the rate constants might be difficult to determine. The aim of this section is to describe a data-exhibiting style which in some cases can yield considerable insight into the evolutionary pattern without requiring an explicit form for the rate equations. We proceed by analyzing a specific example, the change in the structure of the labor force of countries as they have passed through the industrial revolution. It will be apparent that other systems can be investigated in a similar spirit.

At the end of Section I we discussed the manner in which non-agricultural workers replaced agricultural workers as the industrial revolution progressed. We now proceed to follow this process in more detail by characterizing a country in a given year by the fraction of its labor force occupied in each of four work categories; agriculture (actually we include all extractive activities such as mining and forestry in agriculture), trade, manufacturing, and service. In a cartesian coordinate system we represent the fraction of the labor force in agriculture by a point on the positive horizontal axis, the fraction in manufacturing by a point on the negative horizontal axis, the fraction in trade on the positive vertical axis and the fraction doing service work on the negative vertical axis. Hence the character of the labor force in a given year, with our rough categories is represented by four points, each on one ray. Only three of the points are independent since the sum of the four fractions must be unity. Three alternative variables which might be used to characterize the country are the mean of the four points along the rays and the radii which characterize the second moment ellipse of these four points.

The recent evolution of the U.S., the U.K., and Sweden is indicated by a trajectory formed by the motion of mean points with time is shown in fig. 11. This figure was taken from reference 6 where the basic data sources are listed as well as the details of the manner of construction of the curves. Notice that the U.K. moved into the manufacturing-service quadrant well over a hundred years ago, while the U.S. and Sweden are now just approaching that quadrant.

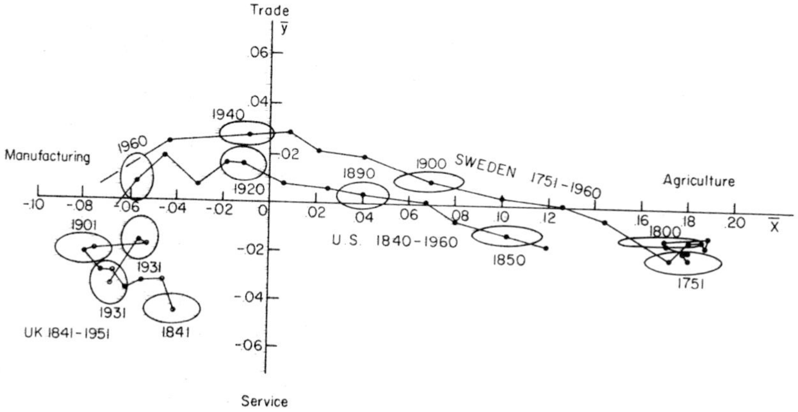

Fig. 11. Abstraction of the socio-economic history of the U.S., Sweden, and the United Kingdom as represented by the division of labor force on a four-variable phase plot.

We might ask if there exists a country which has progressed even beyond the state of the three listed. The small country of Monaco might be considered to be in the ultimate state. It has essentially no agricultural or industrial workers. Most of its labor force performs service work as they man the casinos and the hotels. The gambling spectrum of the U.S. and U.K. is somewhat broader, including the insurance companies, the parimutuals, the state lotteries, investment houses, and, to some degree, the Department of Defense. We produce the largest food crops in history with the order of five percent of our labor force, and if pressures continue from abroad we will expand our industrial automatization program so that few workers will be required to give us record industrial productivity. Most people will then be employed in service. In fig. 12 we have given a schematic trajectory of the development of a country from its agricultural beginnings, through industrial development into its service phase. A mathematical model of development would have to yield such a trajectory.

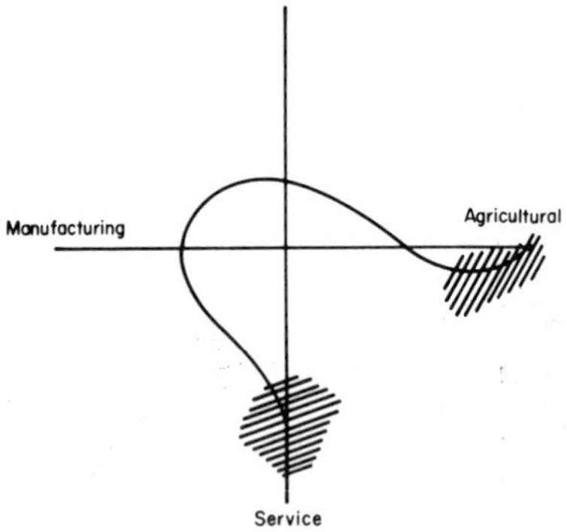

Fig. 12. Schematic phase point trajectory on a four-variable plot that starts in a primitive agricultural phase, develops into a manufacturing country, and finally evolves into the service phase.

It is easy to construct trajectories of other evolutionary curves in the same manner that fig. 11 was constructed. For example, the changes of cause of death of Americans is plotted in fig. 13. Notice that at the turn of the century, infectious and

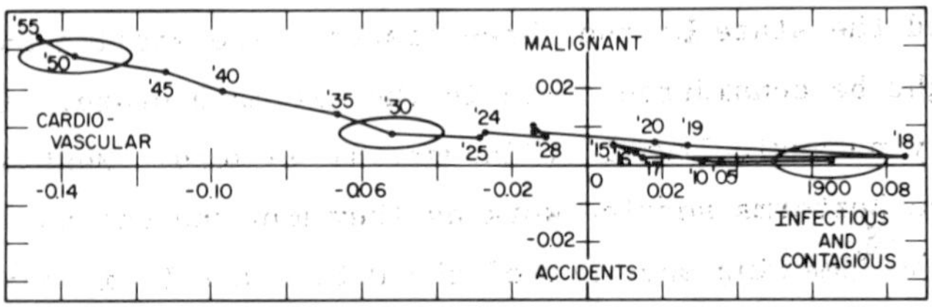

Fig. 13. Causes of death in the U.S. for period 1900-1955. Data plotted on a four-axis plot. The phase points and ellipses are constructed as described in Fig. 11. The influence of the 1918 influenza epidemic is exhibited through the backward motion of the points for 1916-1917 and 1918.

contageous diseases were the great killers,while by the mid 1900's this distinction was divided between the cardio-vascular diseases and malignancies. Notice also the precurser to the influenza epidemic of 1918. In 1916 the trajectory started on a retrograde motion which continued for several years.

The graphical representation of fig. 11 can be extended and refined to include more variables. For example, the labor force of a country might be divided into the six catagories,

1. Agriculture
2. Transportation
3. Commerce
4. Manufacturing
5. Service
6. Construction

Then, as indicated in fig. 14, a ray can be introduced for each of

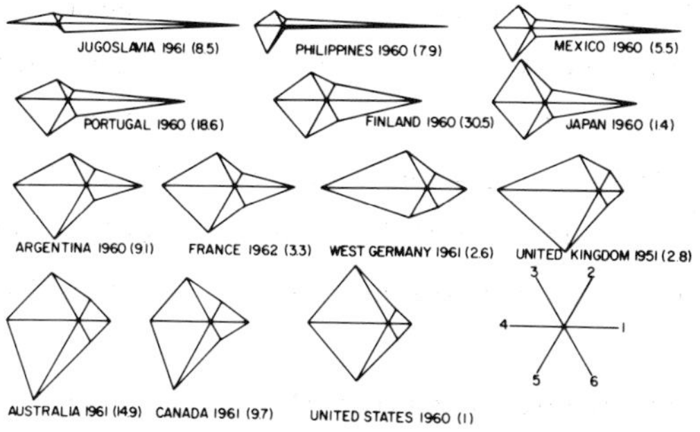

Fig. 14. Six-variable snowflakes for several countries for 1960. The axes, as numbered on the lower right figure, are as follows: 1. Agriculture; 2. transportation; 3. commerce; 4. manufacturing; 5. service; and, 6. construction. Numbers in parentheses are the ratio of the population of the U.S. to that of the subject country. Data from UNESCO statistics.

the six categories and for a given country in a given year a point can be placed on each ray to indicate the fraction of the labor force associated with the activity represented by that ray. For visual clarity it is convenient to connect the fraction points on each ray to construct a polygon which we call a snowflake. Fig. 14 contains a collection of polygons which R. Herman and the author[6] prepared from labor data of a number of countries in the early 1960's. Notice that the countries can be ordered so that it seems as though a natural evolution occurs in which a needle configuration slowly transforms into a square.

Ray patterns have also been used by cardiologists in the analysis of cardiac disorders.[16] Indeed the earliest application of rays to characterize and differentiate individuals, which we have noted in literature, was made by Roger Williams[17] in his book "Free and Unequal." As a physician, he was interested in characterizing patients and watching their changes. For that purpose, he used over 25 rays to represent the physiological state of an individual.

VII. INFLUENCE OF RANDOM FORCES

In section II we discussed the response of a population to various saturation inducing forces and in section IV we analyzed some interactions between two species. In generalizing those models for the investigation of the interaction of many species, we are confronted with the hopelessness of finding the enormous number of rate constants which would be required in the set of equations necessary to characterize the interaction of many species.

Statistical modeling avoids this difficulty. It has been used successfully in the modeling of physical systems of many molecules and is the basis of the branch of physics and chemistry known as statistical mechanics. The model presented in this section follow ideas originally used in the theory of Brownian motion of a large molecule in a fluid comprised of small molecules.

We start with eq. 16a which discribes the exposure of a population to a saturation inducing function and add to it a random driving force to mimic the role of many other not well characterized influences. This is to model the observation that as a population approaches what seems to be a saturation level, fluctuations start to appear. This is to be expected because there are generally many events which influence population growth and decline. In the case of wild animals there exist preditors to any given species, the population of available prey for nourishment fluctuates, epidemics suddenly appear, the weather changes, etc. In the case of man there are changes in the economy, changes in mores of society, famines and floods, wars, etc., which affect birth and death rates. Hence it is highly unlikely that as saturation is reached birth and death rates exactly balance each other. Two examples of fluctuations are shown in figures 15 and 16.

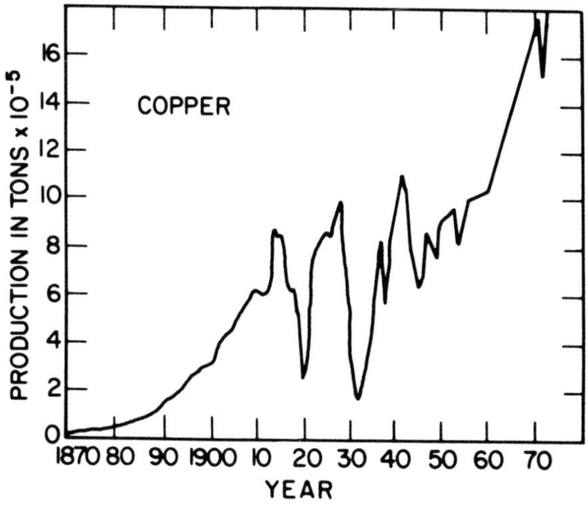

Fig. 15. One hundred years of copper production in the U.S.

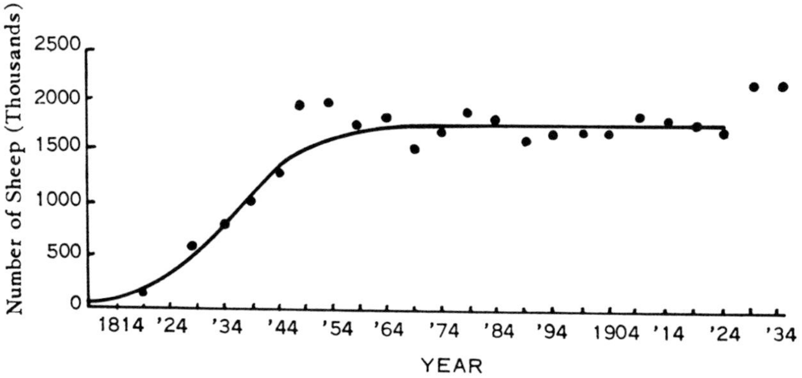

Fig. 16. Variation of sheep population in Tasmania (from J. Davidson, Transactions of the Royal Society of South Australia 62, 342 (1938)).

The most primitive model we can make for the influence of various forces of the type discussed above is to consider them to be random forces superimposed on the saturation-inducing force. A simple example of such a force in one with a short term memory and which is generated by a stationary Gaussian random process. Such a time dependent random force $F(t)$ would then be characterized by a single parameter σ and have the properties $\langle F \rangle_{av} = 0$ and

$$\langle F(t_1)F(t_2) \rangle_{av} = \sigma^2 \delta(t_2 - t_1) \tag{1}$$

The generalization of (II.16), when a saturation inducing force and a time dependent random force are applied together, is

$$d \log[N(t)/\theta]/dt = k\, G(N/\theta) + F(t). \tag{2}$$

The special case $G(x) = 1 - x$ was first considered by Leigh[18] and the general case by Goel, Maitra, and Montroll.[12]

When one has a process characterized by (1) and (2) he does not ask what the population will be at time t but rather, what is the probability $P(N,t)$ that the population will achieve a level N in time t. Equation (2) is similar to the Langevan equation for the Brownian motion of a large particle immersed in a fluid composed of small molecules. The term $\log(N/\theta)$ would be analogous to the velocity of the large particle, the term $k\, G(N/\theta)$ to an external force acting on the large particle, and $F(t)$ to the random force being applied by collisions between the large particle and many small molecules. Methods have been developed in the theory of Brownian motion to find an equation for the probability that the large particle achieve a velocity v at the end of time t. The basic differential equation for the discussion of the process is called the Fokker-Planck equation (cf. reference 19). The methods used in the derivation of the Fokker-Planck equation have been applied to (2). It is found that $P(N,t)$ satisfies[12]

$$\frac{\partial P}{\partial t} = -\frac{\partial}{\partial v} k\{PG(\exp v)\} + \frac{1}{2}\frac{\partial^2 P}{\partial v^2}\sigma^2 \tag{3}$$

where

$$v = \log(N/\theta) \tag{4}$$

The steady state solution with $\dot{P} = 0$ of this equation is easily verified to be

$$P(v,t) = P_0 \exp[2\sigma^{-2}\int_\theta^v G(\exp v)\, dv] \tag{5}$$

where P_0 is a normalization constant. In the Gomperz and Verhulst cases the steady state distribution functions are respectively

$$\text{Gompertz} \quad P = P_0 \exp(-kv^2/\sigma^2) \tag{6a}$$

$$\text{Verhulst} \quad P = P_0 \exp\{2k[v - \exp v]/\sigma^2\} \tag{6b}$$

when v is small the two distributions become indistinguishable. Eq. (6b) was first derived by Leigh[18] while eq. (5) and (6a) were first given in reference[12]. In terms of the population variable N, eq. 6b becomes the gamma distribution.

The time dependent distribution function $P(v,t)$ is perhaps better discussed in terms of a new function $\psi(v,t)$ which is related to it through the transformation

$$\psi(v,t) = P(v,t) \exp\{-k\sigma^{-2}\int_0^V G(e^v)dv\} \tag{8}$$

The differential equation for $\psi(v,t)$ has the form

$$(2/k)\,\psi_t = (\sigma^2/k)\psi_{vv} - \{\partial G(e^v)/\partial v + (k/\sigma^2)[G(e^v)]^2\}\psi. \tag{9}$$

This is to be compared with the Schrödinger equation

$$i\hbar\psi_t = (\hbar^2/2m)\psi_{xx} - U(x)\psi$$

and the Bloch equation in which $(-it/\hbar)$ is replaced by $\beta = 1/kT$ (which is used in statistical mechanics)

$$\psi_\beta = (\hbar^2/2m)\psi_{xx} - U(x)\psi$$

If in the Bloch equation, we choose the mass to be 1/2 and identify β with $\frac{1}{2}kt$, $\hbar^2$ with σ^2/k, and $U(x)$ with

$$W(v) \equiv k\sigma^{-2}[G(e^v)]^2 + \partial G(e^v)/\partial v$$

it has the same form as our basic eq. (9). Of course, there is no connection between the physical significance of the two equations. Since there is a great literature on the Bloch and Schrödinger equations this mathematically similarity is a great convenience.

In the Gompertz case $G(z) = -\log z$ so that if $z = e^v$, $G(e^v) = v$ and $W(v) = -1 + kv^2/\sigma^2$. Then (9) has the form of the Bloch equation for a harmonic oscillator potential

$$(2/k)\psi_t = (\sigma^2/k)\psi_{vv} - (-1 + kv^2/\sigma^2)\psi \tag{10}$$

Time dependent solutions of this and other examples of eq. 9 are discussed in references 1 and 12. We continue with another type of statistical modeling.

VIII. ON THE PRICES OF THINGS WITH SOME REMARKS ABOUT SEARS-ROEBUCK CATALOGUES

Many people feel that their wages should be higher and that prices should be lower. Even those who are normally satisfied with the balance become frightened during unusually inflationary periods as existed in the U.S. a few years ago. Such periods occur sporadically and occasionally they become so extreme as to cause a complete collapse of the economy.

One of my favorite examples of a public response to an inflationary instability concerns the popularity of counterfeit U.S. bills in Poland in 1947. During that highly inflationary period about 400 million dollars in such paper circulated in Poland.[20] To some, such a bill, having a stable questionable value, was preferable to a declining official zloty. If in January the probability that a counterfeit dollar could be passed as a real one were 1/10, it was agreed that that probability would have changed less by December than the value of a genuine zloty. The Hungarian inflation of 1946 produced some of the largest numbers recorded in economics, 5×10^{20} pengo having been the exchange value of one dollar U.S.

Considering the world history of economic crises, the U.S. economy has been remarkably stable. If it be any consolation to the worried reader, the amount of food purchasable by an average industrial worker for the wages of the work of one day increases almost every year. It grew tenfold in the century 1865-1965 as indicated in table I. The index numbers of wages and wholesale farm prices required for the construction of the table are available in reference 4. Thus improvement in purchasing power is a consequence of the improvement of agricultural technology to give greater yields per acre and per agricultural worker, and of the improved productivity of the industrial worker so that he commands higher wages.

Table I. Index of ratio of industrial daily wage to whole-sale Farm Price Index 1860-1965 (ratio 1865 is equal to 1)

1860	1.30	1920	2.55
1865	1.00	1925	3.41
1870	1.49	1930	4.35
1875	1.60	1935	4.83
1880	1.78	1940	6.78
1885	2.17	1945	5.53
1890	2.28	1950	5.96
1895	2.63	1955	8.13
1900	2.48	1960	10.00
1905	2.47	1965	11.53
1910	2.04	1970	12.62
1915	2.34	1975	10.50

In the transportation sector the rail fare for the New York City to Washington, D.C. run of \$13.75 is to be compared with the

1797 stage coach fares: New York to Philadelphia $6.00, Philadelphia to Baltimore $6.00, Baltimore to Georgetown $4.00, totaling $16.00 for the three and a half day trip. Colonial stage rates of 5-7 cents per mile are about the same as U.S. air fares of the period 1945-75.

A rich data base available for 20th century U.S. Prices is the set of Sears-Roebuck catalogues[21] since 1900. Robert Herman and one of the authors[22] have explored the possibility of using them to compare the state of various unrelated technologies. The data might also be used to record changes in cost of living as an alternative to the government bureau of labor statistics consumer price index.

As a data base for these purposes the SR catalogue has a number of attractive features:

1. Thousands of items are listed, so that any average taken over the prices is representative of averages over numerous technologies.

2. The listed prices are essentially in equilibrium with the market. If the price of an item is too high in the catalogue of a given year, the competition of other companies will not be met so that sales will be small. In the next catalogue that price would have to be reduced; or, if buying policies do not permit the reduction, it might be replaced in the catalogue by a more saleable item.

3. The items listed change somewhat from year to year and the number of lines, or of pages devoted to a given class of items changes, so that the catalogue tends to be at equilibrium with the market both pricewise and tastewise. As tastes change, generally becoming more sophisticated and expensive, the catalogue, so to speak, responds. The cost of living thus adjusts itself both to the variation of the value of money, and to the change in style of living.

We[22] have begun a statistical analysis of the prices in the numerous catalogues, a basic statistic extracted from the catalogue of a given year being the distribution of listed prices. Every eighth page was sampled from each catalogue; and on the pages chosen, the number of items in various price ranges were counted. The data from all the sampled pages was accumulated to give the distribution function of prices in the catalogue. Since the prices in the catalogues range from a few cents to thousands of dollars, the logarithm of the price was chosen to be the basic variable whose distribution was to be found. To a rather good approximation, the logarithm of the price has a normal distribution as exhibited in Fig. 17.

The histogram distributions for the years 1916, 1924, and 1974 plotted in fig. 17 are typical. A mild asymmetry is evident in the distributions of the figure in that they rise slightly faster than they fall. Our log normal distributions

$$F(\log_2 p) = (2\pi\sigma^2)^{-\frac{1}{2}} \exp[-(\log_2 p - \langle \log_2 p \rangle)^2/2\sigma^2] \qquad (1)$$

are not surprising since such distributions have also been observed for analogous variables; note for example the distribution of prices on a given day of common stock listed on the New York Stock Exchange and on the American Stock Exchange.[23] The range from the 5th to the 95th percentile of the distribution of incomes in the U.S. also can be fairly well fitted to the log normal distribution.

We have also observed that the dispersion in the distribution remains remarkably constant. The square of the dispersion is

$$\sigma^2 = \langle (\log_2 p_i - \langle \log_2 p_i \rangle)^2 \rangle \qquad (2)$$

where we use the notation

$$\langle \log_2 p_i \rangle = \frac{1}{N} \sum_{i=1}^{N} \log_2 p_i, \qquad (3)$$

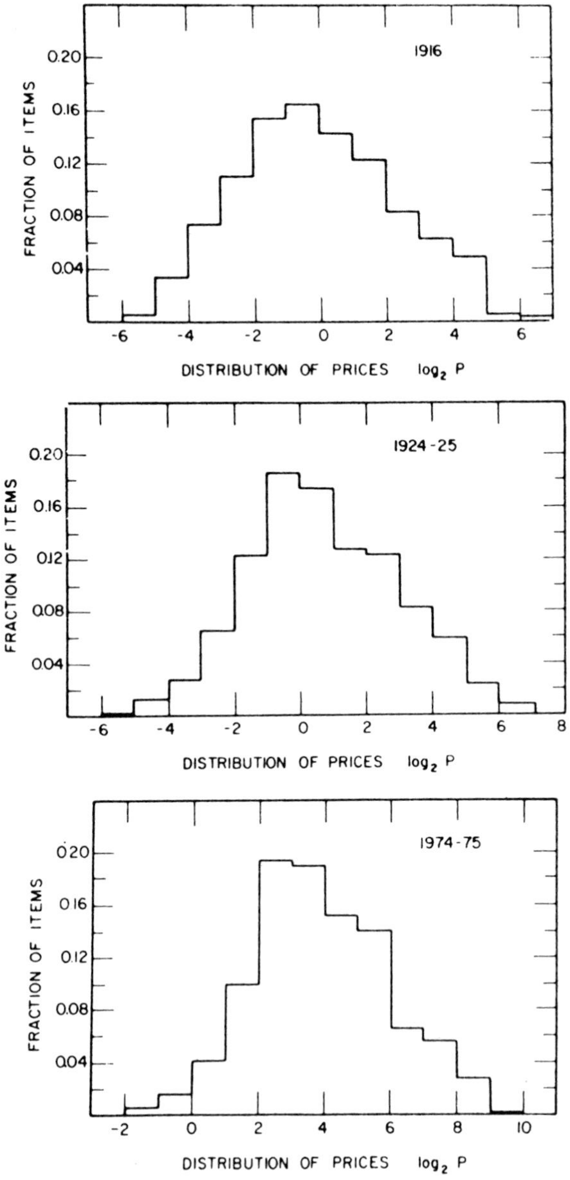

Fig. 17. Histogram of distribution of prices in SR catalogues for years 1916, 1924-25, and 1974-75. The fraction of items in each price range in each catalogue is plotted as a function of $\log_2 P$, *P being the price.*

N being the number of prices sampled. If one considers the following simple model for the construction of a catalogue for the year $(Y + 1)$ from that of the year Y, it becomes evident why σ should not change significantly from year to year.

Suppose the average inflation during a period $(Y, Y + 1)$ to be characterized by a factor α. Then most of the items in the catalogue will have their prices changed from p_i to αp_i. A few, due to changes in technology or other variables, may have a smaller price change than the average; while a few others will have a larger change. If we follow the simplest pattern of changing all prices from p_i to αp_i, then

$$\log p_i - \langle \log p_i \rangle = \log p_i - \frac{1}{N} \sum_{i=1}^{N} \log p_i \qquad (4)$$

becomes

$$\log \alpha p_i - \frac{1}{N} \sum_{i=1}^{N} \log \alpha p_i = \log p_i - \langle \log p_i \rangle \qquad (5)$$

so that σ^2 (defined by (2)) remains invariant in the process <u>independently of the inflation factor α</u>.

We have calculated σ for the years identified in the catalogue samples listed in reference 22. These σ's, which are listed in Table II are remarkably constant for a period of 75 years, deviating from the average value

$$\sigma = 2.26 \qquad (6)$$

with a standard deviation of only 0.17.

Let us now consider the time variation of the average of the logarithm of the prices found in annual catalogues since 1900. As was mentioned above, the graph of such a variable might be used as a price index chart. It might also be used to compare the progress of various technologies. If the price of the products of a given technology advance less rapidly than the SR average over many technologies, it might be considered to be developing well.

YEAR	$\langle \log_2 p \rangle$	$\sigma_{\log_2 p}$
1900	0.150	2.43
1902	0.212	2.34
1908	-0.228	2.29
1916	-0.068	2.38
1924-25	0.422	2.32
1929-30	0.998	2.26
1932-33	0.691	1.91
1934-35	0.673	2.22
1935-36	0.537	2.39
1939-40	0.627	2.62
1946-47	0.532	2.15
1948-49	1.336	2.37
1951-52	1.785	2.34
1962	2.403	2.24
1972-73	3.030	2.27
1973-74	3.322	2.05
1974-75	3.870	2.12
1975-76	4.060	2.03

$$\bar{\sigma} = 2.26; \quad \langle (\sigma - \bar{\sigma})^2 \rangle^{1/2} = 0.17$$

Table II. Standard deviation of $\log_2 p$ from the mean $\langle \log_2 p \rangle$, for various years in the period 1900-1976.

However, if prices rise more rapidly than the SR average, the technology should be examined to understand why it becomes less competitive with other products or services. The average value of the log of the SR prices is plotted in fig. 18 as a function of time since 1900. The straight line sections in the periods 1908-28 and 1940-70 correspond to an inflationary rate of about 4-1/2% ($\sim$ 15 year doubling time), a rate similar to savings bank interest rates in those periods. The low dip in the 1930's reflects the deflation of the depression years. It actually represents an over response since, to reduce cost, the widely distributed catalogues were thinner and separate catalogues of more expensive items were available on request. The fall in mean price from 1900 to 1910 is possibly a result of the SR learning process in the management of a mail order house rather than in a deflation on a national scale.

The employment of the SR curve in Fig. 18 as a guide to the evolution of a technology will now be presented in the case of automobile technology and poultry technology.

In Fig. 18 we have plotted the average factory sale price to dealers of an automobile in the U.S. from 1900-1975 as obtained by R. Herman and the author by analyzing data from Automobile Facts and Figures, 1967 and 1975 editions[24]. The tabulated listings found in this annual publication are the total number of passenger vehicles produced per year and the total factory value of all those vehicles. We give the following interpretation to the curve in the figure.

We identify the rise from 1904-1908 with the evolution of the horseless carriage into an object which had some reliability and which even slightly resembled a modern automobile. More people were willing to pay more money for the vastly improved product. The year 1909 marked a milestone in automobile technology. The Model T, dependable and primitively mass-produced, had arrived. The more modern production line was started in about 1911. For a while the Model T comprised a larger and larger percentage of the market, more by attracting new customers into it than by luring old ones away from other producers. By 1915, the average price of a car was $512, which is close enough to the Model T price to allow us to conclude that it and other cars like it dominated the market that year.

Since war years and the first few post-war years have a special effect on the economy which we do not wish to discuss here, we make no remarks about the period from 1918 to 1922 and we skip

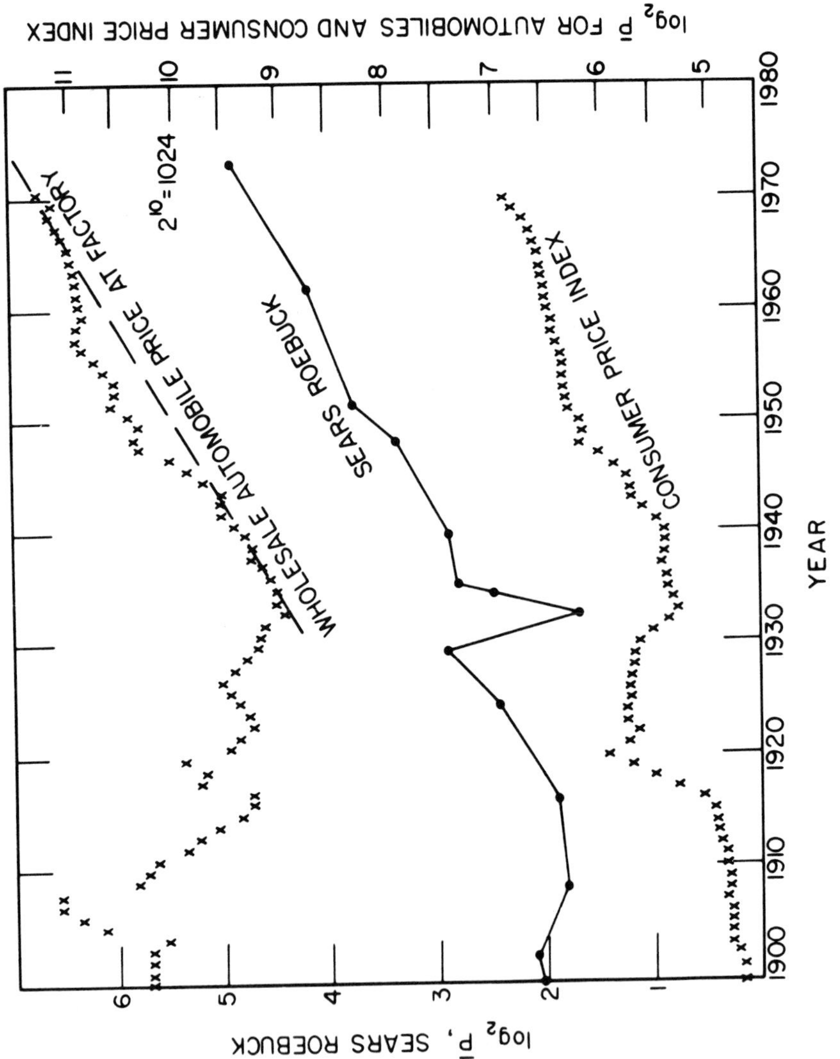

Fig. 18. Time variation of the logarithm to the base 2 of the average price in dollars of items found in general SR catalogues. Also time variation of the log of average wholesale automobile prices in dollars, and the log of Department of Labor Statistics Consumer Price Index for the period 1900-1975.

to the years 1923-27. In this period, the slope of the $\log_2 \overline{P}$ curve for automobile prices parallels that of the SR standard. The boom of the middle 1920's brought with it an increase in the price of the Model T. But more important, prosperity allowed more people to cultivate a taste for motor cars that were faster, roomier, and especially more elegant, so that the more modern vehicle was rapidly replacing the Model T, the low priced Chevrolet, and their equivalents.

Henry Ford's response to the competition was the introduction of the Model A which forced competitors to produce a similar good, inexpensive, stylish model so that a downward price trend started in 1927. This drop in car prices continued unabated through the inflationary period of the late 1920's into the deflationary mode of the great depression when the cost of cars decreased further and the cheapest models of any given line became the favorites of a dwindling market. The number of passenger cars sold per year fell from 4,455,178 in 1929 to 1,103,557 in 1932, the year of the depth of the depression. The turn-about of the mean auto price followed that of the SR curve rather closely into the World War II period. In the immediate post-war period, the tremendous demand for automobiles stimulated by an absence of civilian car production during the war years (a total of 749 passenger vehicles having left the factories in 1943-44) caused a rapid rise in the mean price, whose rate of increase exceeded that of the SR standard. Since 1950, the curve of mean auto price approximately follows the SR standard. This is not surprising since an automobile with its many components; steel, glass, battery, upholstery, electric wires, etc. is almost a small SR type listing of items.

Eggs (and poultry products generally) are most remarkable because their prices have changed very little in 70 years.[6] As can be seen in fig. 19 egg prices today are about the same as in 1920 and have less than doubled since 1910. This has been accomplished through a succession of technical advances in poultry genetics, poultry diet, space management and operations research

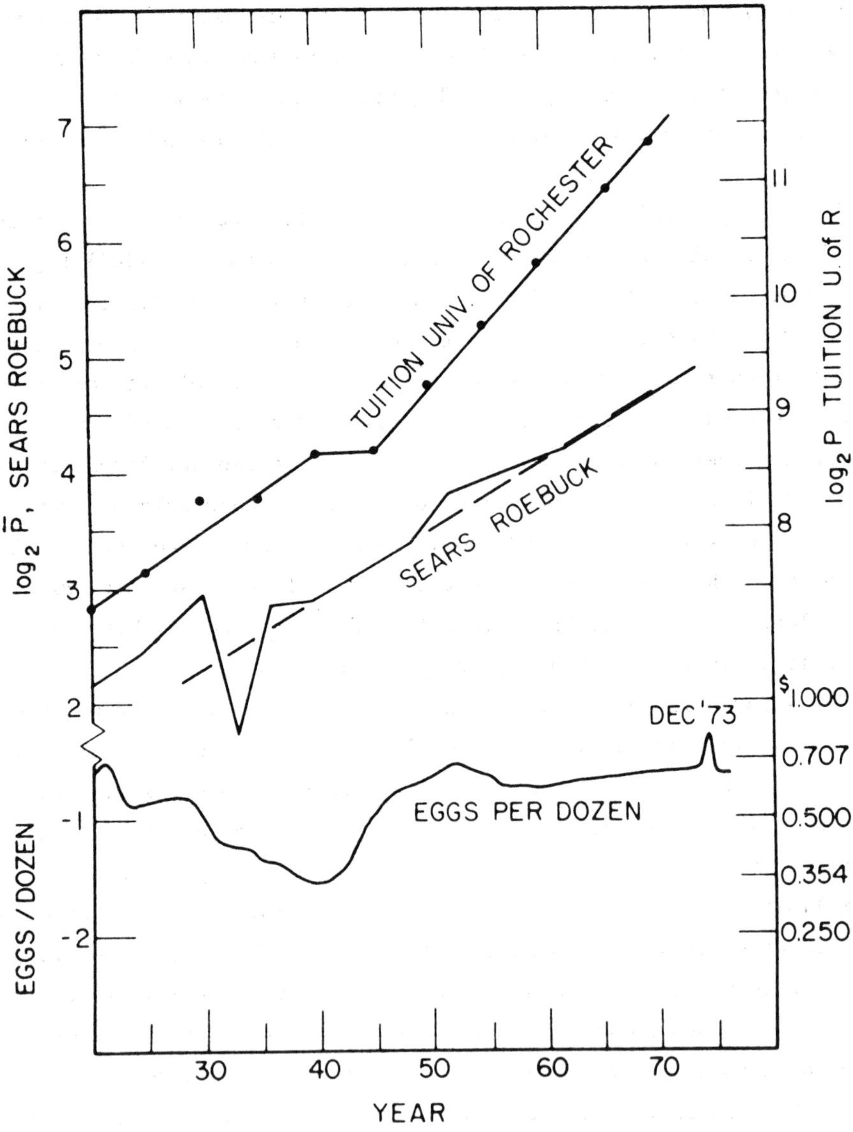

Fig. 19. Comparison of time variation of egg prices, university tuition, and SR price index.

on the product. While turkeys were a delicacy in the 1920's, they are now one of the cheapest of meats.

The computer industry with innovative integrated circuit technology is today's forefront technology, producing superior items with declining prices even in times of an abnormally high rate of inflation.

The prices of activities in the personal service sector, such as education and medical care, are rising considerably faster than the SR index. We have plotted the University of Rochester tuition variation in figure 19. The rent of a semi-private hospital room has been increasing even more rapidly than education. With Blue Cross, hospital insurance, and other advanced hospital payment plans, many hospital patients upon admission do not even inquire about the rental price of the room.

When the price of a commodity has been artificially regulated for a number of years and the regulation is removed, there seems to be a tendency for the price of that commodity to inflate to a level of the order of that which it would have achieved were it an average unregulated commodity. An interesting example is the recent change in the price of an ounce of gold. That change is plotted as a function of time in Fig. 20 in the period since the statutory $35 per ounce price set in 1935 was relaxed in 1971. The future course of gold prices will be determined by government policies concerning gold sales from their large stockpiles, from future plans on the monetary use of gold, and by the strength of important currencies such as U.S. dollars and German Marks.

The official U.S. Department of Labor Statistics consumer price index, exhibited at the bottom of fig. 18, deviated in certain periods from our SR index. Since the list items chosen by the government analysts are not adjusted annually to follow public taste, we feel that our SR index may better reflect the personal requirements of a population. During the period 1920-23 when public taste evolved through the appearance of a myriad of rather expensive automatic devices, increased agricultural productivity

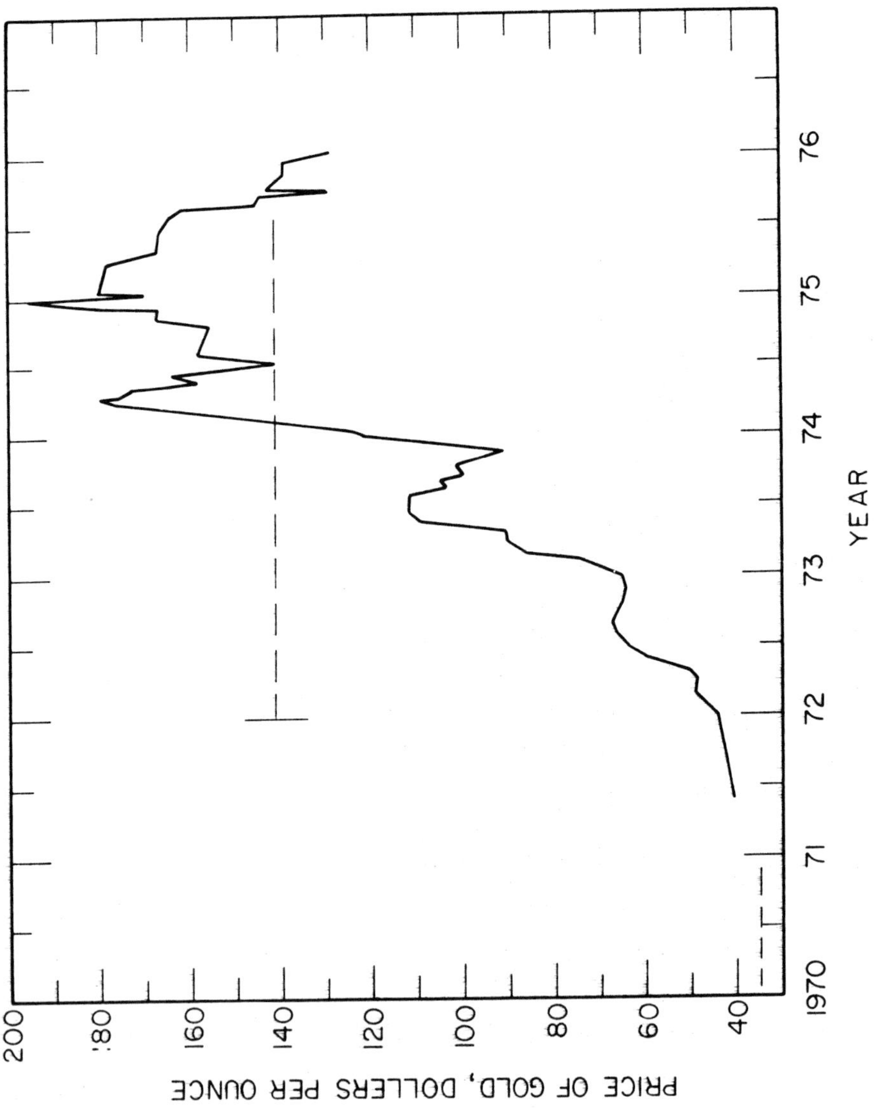

Fig. 20. The price of gold per ounce for the period 1971 to January 1976 after Government price regulations were relaxed.

leveled and even forced prices downward thus causing a smaller change in the official price index than ours based on the SR catalogue. Similarly in the "good life" period 1950-65, SR responded to the public demand for luxury items in a way not reflected in the Bureau of Labor Statistics figures.

We close this discussion of prices with a model of how the development costs (and time) of a new technology grows exponentially with the number of dimensionless constants required to determine the optimum operating ranges of the dimensionless constants. The standard procedure for investigating feasibility and the optimization of process variables in new complex technologies (especially those basically nonlinear) is by model testing. A crucial dimensionless constant for the design of airplanes is the Reynolds number; that for ship design is the Froude number, and for fission reactors the multiplication factor for neutron production with each fission reaction. Engineering studies in such systems generally required only a few years to establish feasibility and to produce a device. On the other hand, millions of dollars have been invested in confined fusion research for twenty years and we are still far from even a convincing feasibility determination.

Let N be the number of dimensionless constants required to characterize a process. Then an experimental program must sample $n_1 \times n_2 \times \ldots \times n_N$ points in the N dimensional space of characterization. The cost of the program P should be proportional to the number of sampling tests; i.e.,

$$P = kn_1 \times n_2 \times \ldots \times n_N = k \exp N\{\frac{1}{N} \sum_{j=1}^{N} \log n_j\}$$

Hence, if we define λ to be the average value of the logarithm of the number of observation for each dimensionless constant

$$P = k \exp N \lambda$$

as was suggested to be the case. Confined fusion testing involves some 8-10 dimensionless constants.

IX. THE PLANNER'S DILEMMA AND THE GUMBEL DISTRIBUTION OF EXTREME VALUES

As we approach the year 2000 an increasing concern with the future of the human race seems to be developing. Futurology becomes more fashionable and predictions, especially gloomy ones, become more common. Century end inventory-taking is not new and it is not surprising that a super round number like 2000 attracts more attention than did 1700, 1800 or 1900. Almost instant communication feeds the public with news of daily crises; and with increasing literacy, prophets find a wider audience than ever before. The beginning of the only other super round-numbered year, the year 1000, was expected by many to mark doomsday. As that year approached, a wave of hysteria spread through Europe. Great pilgrimages were taken to Jerusalem by those who wanted to be at the site of the Lord's Temple on the day of Judgement. Religious fanaticism absorbed many while the spirit of hedonism - enjoy life today for tomorrow we die, appealed to others. Some readers might find it entertaining to review the mood of those days as surveyed in writings such as chapter 10 of reference 27.

When one is worried about future events, whether he has become infected with the doomsday obsession, or whether he has merely found himself in a position that involves planning for the protection of his organization, or community, against extreme events: floods, earthquakes, financial reverses, etc., he is confronted with the planner's dilemma exhibited in fig. 21. While the probability of an extreme event diminishes rapidly as a function of its magnitude, the unpleasantness it might bring grows rapidly with its magnitude as does the cost of protection against it. The planner is presented with the need to spend increasingly larger amounts of money to prepare for events which become decreasingly likely. A strategy we might consider would be to (a) estimate the expected lifetime of the device to be protected (or of

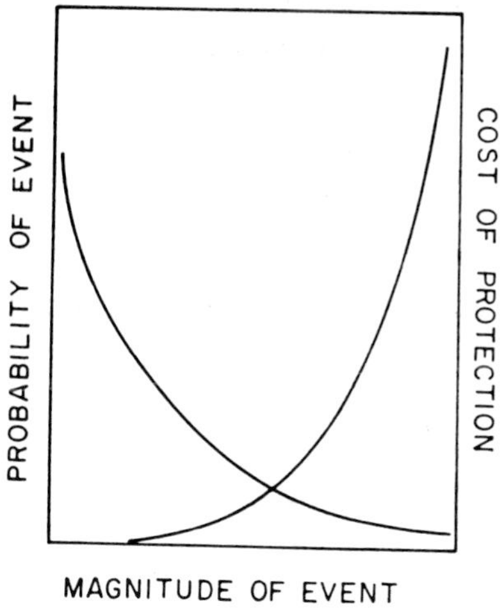

Fig. 21. The Planner's Dilemma.

the device affording the protection), (b) project the magnitude of the expected extreme event in that period, (c) introduce a safety factor $\alpha(>1)$, and prepare for the expected event of great magnitude in the period αt. Of course, this strategy is applicable only if one can reasonably estimate the magnitude of the expected extreme events. It is for this purpose that a statistical model pioneered by E. T. Gumbel[28,29] has been developed. Our discussion below will be given in terms of flood levels of rivers. The generalization to other situations will be obvious.

Every day the maximum river height at some location might be recorded. Over a number of years the mean daily height could be calculated as well as a distribution function of the deviation of the observed daily maxima from that mean. Now, let us suppose that river levels are generated by a stationary stochastic process whose character depends on winds, rainfall, temperature, melting rate of snow, etc., in a way which, while not understood, does not change from year to year.

The Gumbel theory of extreme values is concerned with the distribution of annual maxima, those levels which are the largest observed each year; i.e. the maximum value of daily maxima recorded in a single year. It can be shown that if (a) the basic stochastic process for the generation of levels is stationary, (b) the distribution of daily maxima levels is of exponential type (a class of distributions of which exponential, normal, log normal and Poisson are members), (c) correlations between fluctuations in daily maxima die out in a period much less than one year, then the cumulative distribution of annual maxima has the form

$$F(x) = \exp-(\exp-y) \tag{1a}$$

$$\text{with } y = \alpha(x - u) \tag{1b}$$

Here u is the mean annual extreme value and α is a parameter whose value can be deduced from empirical data.

While the general derivation of (1) is not complicated, it is somewhat lengthy (cf. references 28, 29, and section III.8 of reference 1). Since the derivation is very simple in the special case of an exponential daily cumulative distribution

$$F(x) = 1 - e^{-\alpha x} \tag{2}$$

we deduce (1) for this case. We divide our time scale of interest into many equal intervals such that many observations, n, are made in each interval. For example, let the intervals be one year and the many observations be made one a day for each year so that $n = 365$. Then we seek the extreme value distribution functions for one year intervals. If the observations are independent of each other, the probability that the largest value of X observed over a year is less than x is

$$\Phi_n(x) = [F(x)]^n \tag{3}$$

since for X to be less than x throughout the year it must be so every day. Then from (2)

$$\Phi_n(x) = (1 - e^{-\alpha x})^n \tag{4}$$

Now define a number u_n so that

$$n \exp(-\alpha u_n) = 1 \tag{5}$$

Hence

$$\Phi_n(x) = \{1 - \frac{1}{n} \exp{-\alpha(x - u_n)}\}^n \tag{6a}$$

$$\sim \exp(-e^{-y}) \quad \text{for large} \quad n \tag{6b}$$

where

$$y = \alpha(x - u_n). \tag{7}$$

The interpretation of u_n is determined by maximizing the probability density function

$$\phi_n(x) = d\Phi_n/dx \sim \alpha \exp-(y + e^{-y})$$

It is easy to show that $x = u_n$ is the value of x which is the most likely extreme value found in a set of n observations, since $\phi_n'(u_n) = 0$. In the general case the parameter α is precisely

$$\alpha \equiv \alpha_n \equiv nf(u_n) \tag{8}$$

where $f(x)$ is the probability density function of the daily distribution

$$f(x) \equiv dF/dx$$

Gumbel constructed a special graph paper such that a plot of cumulative distribution function of the form (6) appears as a straight line. The average number of years, $T(x)$, which one has to wait for a given value of the annual maximum level X to exceed a value x is

$$T(x) = 1/[1 - F(x)] \tag{9}$$

The Gumbel distribution of extreme values has been applied to river levels, snow levels and other weather phenomenon, to strength of materials and to certain aspects of vital statistics. To show how the theory can be used to make predictions, we consider death statistics records for Switzerland in the period 1879-1933. As shown in table III, in one year out of the 55 examined,

Table III. Oldest ages in Switzerland, 1879-1933, both sexes.

Oldest age	Number of individuals	Cumulative number, m	Plotting positions, $m/(N+1)$
97	1	1	0.018
98	3	2 to 4	0.036 to 0.071
99	6	5 to 10	.089 to .178
100	14	11 to 24	.196 to .428
101	12	25 to 36	.446 to .643
102	7	37 to 43	.660 to .768
103	7	44 to 50	.786 to .893
104	3	51 to 53	.911 to .946
105	1	54	0.9643
106	1	55	.9821
	$N = 55$		

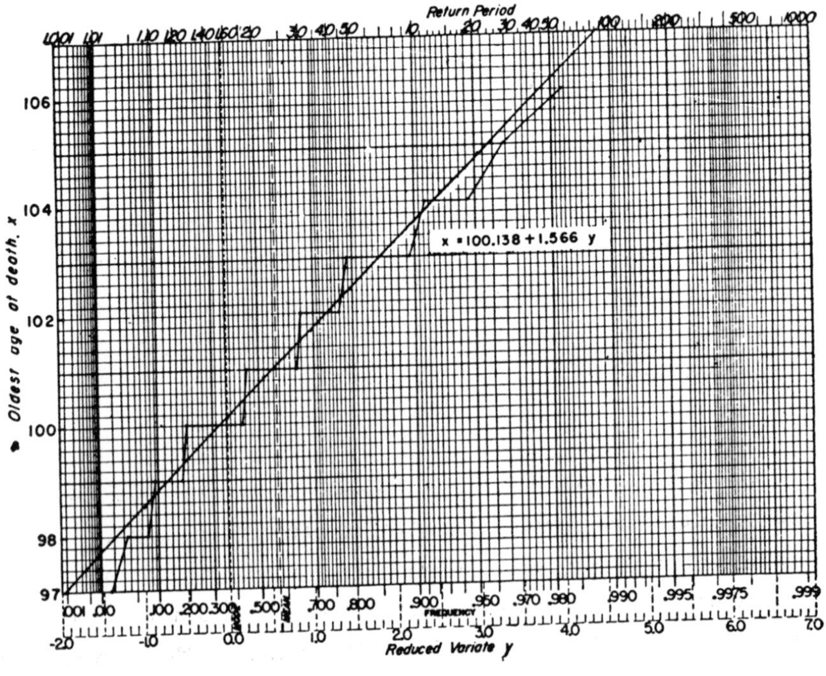

Fig. 22. Oldest ages at death, Switzerland, 1879-1933.

the oldest person who died was 97, in three out of the 55 the oldest who died was 98. The most common age at which the oldest person died in a given year was 100. For reasons discussed in reference (28) Gumbel recommends that the plotting position for the cumulative number n be at $m/(N + 1)$. On this basis, the cumulative distribution of the oldest ages at death in Switzerland is plotted on Gumbel graph paper. A given value of $F(x)$ listed as frequency at the bottom of the graph paper is associated with a particular "return period" listed at the top as obtained from (9). Thus, for example, if one asks how many years will be required on the average for the oldest person to die in a given year in Switzerland to be older than 105, the graph would indicate 20 years. The fact that a good straight line can be drawn through the recorded cumulative distribution function substantiates the idea that the data is consistent with the Gumbel distribution. Many similar analysis are given in references 28 and 29.

Now suppose that after a relatively few years the oldest age at death in Switzerland is 112 or greater. This could be interpreted in two ways; first, that a fantastically unlikely event occurred, since on the average one should have waited about 100 years for it to occur. Second, the stochastic processes which determines age at death for old people has changed its character. This might eventually be the case when new understanding of the aging process, or cancer causes, or of circulatory ailments is achieved and with it a new mode of caring for the aged which leads to increased life span for centegenarians. At least the appearance of points completely off the Gumbel trend line calls one's attention to the possibility of a basic change in the process and can act as an indicator. Thus Gumbel graphs might be used to monitor processes to judge when and if they change character.

Gumbel has noted that such unexpected points can also act as economic indicators when economic data is plotted. Variations in the Dow-Jones industrial averages are often used to exhibit trends in the stock market. These daily averages reflect both the stochastic, or random, component of the market and any systematic trends which might be developing. A trend line might merely reflect inflation and not necessarily a fundamental change in the relation of the public to the market. The theory of extreme values discussed above is applicable only to the random component of the market. Hence, if it is to be applied to boom-watching, the trend component must be subtracted from Dow-Jones averages.

The Dow-Jones central value characterizes the trend component. It is defined as the arithmetic mean of the median of daily highs and the median of the daily lows over a calendar year. An annual maxima can be defined as the difference between the largest daily high and the annual central value. These annual maxima for the years 1897-1947 have been plotted in fig. 23. Even without resorting to extreme value theory, it is clear that the stochastic process associated with the period 1928-1932 is quite unusual. The annual maxima were listed according to size and their cumulative distribution function and plotted by Gumbel on extreme value paper. The results are given in fig. 24. During the period 1897-1927 the straight line fitted the data quite well while the point representing 1928 is almost out of the picture, with its return period being over 2000 years. This strongly suggests that something changed in the basic mechanism in 1927.

With our great expertise in hindsight we can list a few of these changes:

(i) A tremendous growth of popular participation in stock speculation. Instead of it being a game of a few professionals, traders and wealthy families, it became an obsession of an enormous number of modest folk.

(ii) A relaxation of conditions required for margin purchases of stock. By 1929 all that was required by some brokers for the purchase of stock was a 25% down payment. This encouraged a speculator to buy four times as much stock as he could afford. The real money (and not promises

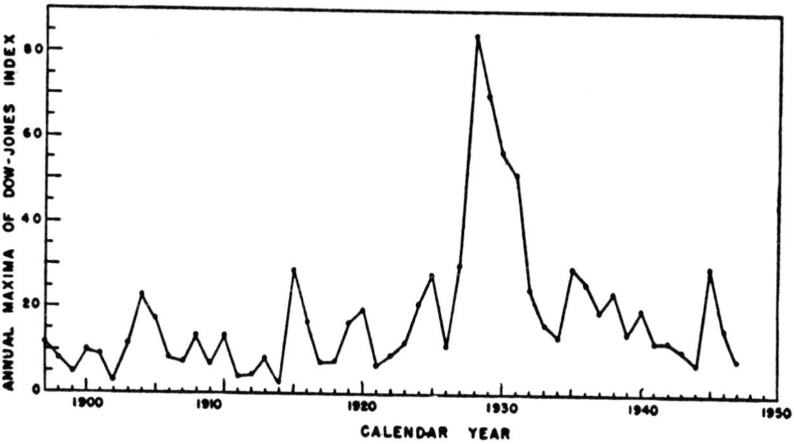

Fig. 23. Annual maxima, Dow-Jones Indes, 1897-1947.

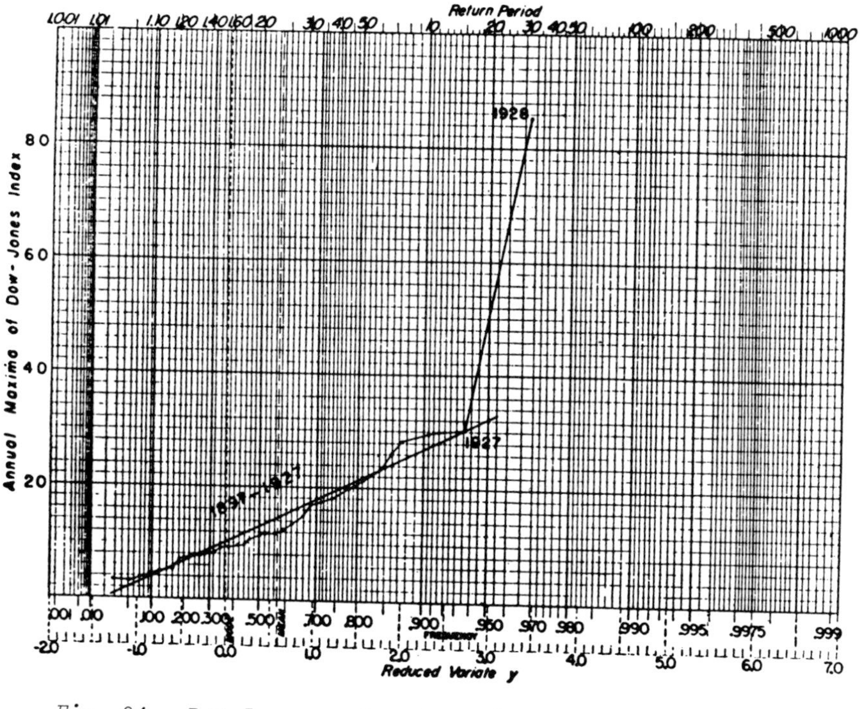

Fig. 24. Dow-Jones maxima plotted on probability paper, 1897-1928.

to pay) in the market was probably less than 35% of the listed market value.

(iii) The growing popularity of the investment trust, a company which produced nothing but merely held stock in other companies. The advertising of investment trusts claimed that with their wisdom, they could out-perform the novice in the trading of securities. The assets of investment trusts increased approximately eleven-fold in the period 1927-1929. The popularity of the trust rested on its amplification capability (or "leverage" as it was called then). The increase in value of the stock of the trust would generally grow more rapidly than the value of the stock of other companies that it held. A super-trust might specialize in stocks of other trusts increasing the amplification factor. A third trust might hold stock in the second one, yielding more leverage. Indeed the loop could be closed with the first trust owning stock of the third. The leverage was remarkable. Unfortunately, the amplification worked as well when the market dropped as it did on the upswing. Hence the stock in most investment trusts was practically worthless after the crash.

X. ACKNOWLEDGEMENTS

The preparation of this manuscript was partly supported by the General Electric Foundation. An earlier version of this report was the subject of the Sidney Chapman Memorial Lecture (Boulder, Colorado) of 1975.

XI. REFERENCES

[1] Montroll, E. W. and Badger, W. W., *Introduction to Quantitative Aspects of Social Phenomena*, Gordon and Breach (New York, 1974).

[2] Verhulst, P. F., Mem. Acad. Roy. Bruxelles, 28, 1 (1844).

[3] Pearl, Raymond, *Studies in Human Biology* (Baltimore, 1924).

[4] *Statistical Abstract of the United States* (U.S. Census Bureau, Washington, D.C.; Historical Statistics of the United States - Colonial Times to 1970, (U.S. Census Bureau, 1975).

[5] Fischer, J. C. and Pry, R. C., *Practical Applications of Technological Forcasting in Industries*, (Ed. M. J. Cetron, Wiley, 1971).

[6] Herman, R. and Montroll, E. W., Proc. Nat. Acad. Sci., 69, 3019 (1972).

[7] Camp, W. D., *Marriage and the Family in France Since the Revolution* (Bookman Assoc., N.Y., 1961).

[8] Woytinsky, W. S. and Woytinsky, E. S., *World Population and Production*, (20th Century Found., N.Y.).

[9] Lodka, A. J., *Elements of Mathematical Biology*, (Johns Hopkins, 1924; Dover Reprint, 1956).

[10] Volterra, V., *Lecons sur la Theorie Mathematique de la lutte pour la vie*, (Paris, 1931).

[11] D'Ancona, V., *The Struggle for Existence*, (Brill, Leiden, 1954).

[12] Goel, N., Maitra, S., and Montroll, E., Rev. Mod. Phys. 43, 231 (1971).

[13] Gazis, D., Montroll, E., and Ryniker, J., IBM J. of Res. and Dev., 17, 47 (1973).

[14] Watt, K. E. F., Canadian Entomology 91, 129 (1959).

[15] May, R. M., *Stability and Complexity in Model Ecosystems*, (Princeton, 1974).

[16] Siegel, J., Farrell, E., Goldwyne, R., and Friedman, H., Surgery 72, 126 (1972).

[17] Williams, R., *Free and Unequal*, (Univ. of Texas Press, Austin 1953).

[18] Leigh, E. G., *Some Math. Problems in Biology, 1*; Am. Math. Soc. (1969).

[19] Wang, M. C. and Uhlenbeck, G. E., Rev. of Mod. Phys. 43, 231 (1971).

[20] Fortune Magazine, April 1948, p. 79 (Vol. 37 No. 4).

[21] Sears and Roebuck Corp. Catalogues, 1900-1974.

[22] Herman, R. and Montroll, E. W., *Statistical Mechanics and Statistical Methods in Theory and Application*, p. 879 (Ed. U. Landman, Plenum, 1977).

[23] Osborn, M. F. M., *Operations Research*, 7, 145 (1959).

[24] *Automobile Facts and Figures*, American Automobile Manufacturers Assn., Annual Volumes.

[25] U.S. Labor Statistics Bulletin No. 1555 (1967), U.S. Dept. of Agric. Stat. Bulleting No. 525, (1973).

[26] Wall Street Journal lists gold prices daily.

[27] MacKay, C., *Extraordinary Popular Delusions and the Madness of Crowds*, (London, 1841; reprint Noonday Press, 1932).

[28] Gumbel, E. J., *Statistical Theory of Extreme Values and Some Practical Applications*, No. 33, National Bureau of Standards App. Math. Series, 1954.

[29] Gumbel, E. J., *Statistics of Extremes*, Columbia Univ. Press, 1958.

[30] Fisher, R. A. and Tippett, L. H. C., Proc. Comb. Phil. Soc. 24, 180 (1928).

GENERALIZED INVERSE MAPPING THEOREMS AND RELATED APPLICATIONS OF GENERALIZED INVERSES IN NONLINEAR ANALYSIS

M. Z. Nashed
University of Delaware

INTRODUCTION

The theory and approximation of generalized inverse operators have been developed thoroughly in recent years (see [2.1]-[2.4]). The methodology and results of generalized inverses of linear operators are playing an increasing role in various areas of nonlinear analysis and optimization. Uses of linear generalized inverses in nonlinear analysis can be roughly divided into two main streams: (1) local (or asymptotic) analysis, via the use of a generalized inverse of a derivative at a point x_0 (or an asymptotic derivative at ∞) or of a suitable approximating linear operator to a nonlinear operator, where the derivative or the linear operator is neither one-to-one nor onto; (2) nonlinear operator equations with a noninvertible linear part, either in an additive form, i.e. $F(x) = Lx - Nx = 0$, where L is linear and noninvertible, or in a multiplicative form, i.e. $F(x) = LG(x)$.

In this paper we develop several generalizations and an in-depth analysis of the inverse mapping theorem in the framework of outer, inner (and related generalized) inverses, and draw conclusions in the spirit of generalized inverse mapping theorems when the derivative $f'(x_0)$ (or an approximation to it) is neither one-to-one nor onto. The setting enables us, in particular, to draw natural conclusions when only a left or a right inverse to a linear approximation of the operator (at a point) is available.

ISBN 0-12-434160-8

The work follows and extends ideas of Graves [3.15] and Leach [3.22]. We obtain versions which include results of both [3.15] and [3.22] and streamline the proofs making heavy use of the geometry of outer, inner and generalized inverses, with particular reference to topological decompositions induced by the existence of such inverses. We also show that the set of all bounded linear operators with bounded outer inverses is open in the uniform operator topology. This establishes the stability of generalized inverse mapping theorems under small perturbations when outer inverses are used. This is in contrast to the fact that inner inverses and generalized inverses do not necessarily depend continuously on perturbations.

It also should be possible to apply generalized inverse theory to the case when the Gâteaux derivative of the nonlinear map is unbounded as a linear operator and has an unbounded generalized inverse, i.e. in the framework of the work of Nash and Moser on "hard" versions of the implicit function theorem. We comment on these directions in [3.28]. However, for completeness, we include in the bibliography of Section 3 extensive references on various contributions to inverse and implicit function theorems.

The generalized inverse mapping theorems of this paper are based on a smoothness condition in the form of strong differentiability (or a weaker but closely related approximate strong differentiability). This condition is stronger than the insufficient condition (to validate the classical inverse mapping theorem) of Fréchet differentiability at a point and weaker than Fréchet differentiability in a neighborhood of a point together with the continuity of f' at the point. Section 1 contains a review (with some new results) of certain aspects of strong differentiability in normed spaces. In view of the increasing role of strong differentiability in analysis and since there seems to be no comprehensive treatment of this topic in standard expositions on differential calculus in normed spaces (e.g. [1.2], [1.3], [1.10]), we provide a self-contained exposition which may be also of interest to some readers, independently of the rest of the paper. For this reason, we list the references by sections and include a comprehensive bibliography. Section 1 may be regarded as an addendum to our paper [1.10].

Section 2 provides a synoposis of the theory of bounded inner and outer inverses and generalized inverses of domain-decomposable operators, and includes some new observations. The material of Sections 1 and 2 will also serve as background for a sequel to this paper [3.28] which will provide a comprehensive treatment of the scope and limitations of applications of generalized inverses of linear operators in nonlinear analysis and optimization. An annotated bibliography on these applications as of 1975 is given in [2.11].

I. A. Strong Differentiability in Normed Spaces

Throughout this section we let X and Y be normed vector spaces and let F be an operator on an open subset U of X into Y. $\mathcal{L}(X,Y)$ will denote the space of all continuous linear operators on X into Y.

One of the objectives of calculus in normed space is to analyze locally a nonlinear operator F via an analysis of more tractable approximations to F. The mere existence of a linear operator $L = L(x_0)$ which approximates F in a neighborhood of x_0 in the sense that

$$(1.1)\quad F(x_0 + h) - F(x_0) = L(x_0)h + r(x_0;h) \text{ where } \lim_{h\to 0}\frac{||r(x_0;h)||}{||h||} = 0$$

is not in general strong enough to reflect local behavior of the operator F, or to validate in terms of conditions on L (such as invertability of L) some of the mapping theorems of analysis (e.g., the inverse mapping theorem). (1.1) implies immediately that F is continuous at x_0 if and only if $L \in \mathcal{L}(x,y)$. Fréchet differentiability (although itself is a strong notion in the hierarchy of much weaker concepts of differentiability at a point; see for example [1.10], [1.7] for a discussion of this hierarchy) is not strong enough to reject pathological situations in the behavior of F in a neighborhood of x_0. In other words, the existence of the Fréchet derivative at x_0 (which is an "infinitesimal" condition) does not imply a reasonable "smoothness" condition in a neighborhood of x_0 (which is a "local" property). For example, the function $f(x) = x^2 \sin \frac{1}{x}$ for $x \neq 0$, $f(0) = 0$ is Fréchet differentiable with $f'(0) = 0$. But in this case the limit of the slope of chords whose endpoints lie on the graph of f (as both endpoints approach the origin) does not exist. A notion of differentiability which eliminates such pathologies and provides a reasonable framework for local analysis is the notion of strong differentiability. This notion was introduced by Peano [1.13] in 1892 and has been rediscovered by several authors in recent years. The purpose of this section is to collect properties of this notion which are scattered in the literature and to develop other properties. We also supplement these by remarks which put the use of this notion in a historical perspective. Some of these properties will be used in Section 3.

Definition 1.1. An operator $F\colon U \to Y$ is said to have a strong (Fréchet) differential at $x_0 \in U$ if there exists $L(x_0) \in \mathcal{L}(X,Y)$ such that

$$F(x) - F(y) = L(x_0)(x - y) + \psi(x_0;x,y) \quad \text{where}$$

$$\lim_{(x,y)\to(x_0,x_0)} \frac{||\psi(x_0;x,y)||}{||x - y||} = 0.$$

That is, for each $\varepsilon > 0$, there exists a $\delta > 0$ such that

$$(1.2)\quad \begin{cases} ||F(x) - F(y) - L(x - y)|| \leq \varepsilon||x - y||, \quad \text{whenever} \\ \qquad ||x - x_0|| \leq \delta \quad \text{and} \quad ||y - x_0|| \leq \delta. \end{cases}$$

For a survey of various types of differentiability in normed spaces, see Nashed [1.10], where several properties of strong differentials are stated without proof. The following proposition gives some immediate consequences of the notion of strong differentiability. For part (c) recall that the Gâteaux variation of F at x with increment h is defined by

$$VF(x;h) = \lim_{t \to 0} \frac{F(x + th) - F(x)}{t} .$$

Proposition 1.2. (a) F is strongly differentiable at x_0 if and only if the Fréchet derivative $F'(x_0)$ exists and for each $\varepsilon > 0$ the map r defined by

$$r(x) := F(x) - F(x_0) - F'(x_0)(x - x_0)$$

is ε-Lipschitzian in x on some ball $\overline{S}(x_0,\delta) = \{x: ||x-x_0|| \leq \delta\}$, i.e., for each $\varepsilon > 0$, there exists $\overline{S}(x_0,\delta)$ on which the Lipschitz constant of the map r is ε.

(b) If F is strongly differentiable at x_0, then F satisfies a Lipschitz condition in a neighborhood of x_0.

(c) If F is strongly differentiable at x_0 and if F has a Gâteaux variation at $\overline{x} \in S(x_0,\delta)$, where $\delta = \delta(\varepsilon)$ satisfies (1.2), then $||VF(\overline{x};h) - Lh|| \leq \varepsilon||h||$ for $h \in X$. In particular, if F has a Gâteaux differential at $\overline{x}$ (so that $VF(\overline{x};h) = F'(\overline{x})h$, where $F'(\overline{x})$ is a linear operator), then $||F'(\overline{x}) - L|| \leq \varepsilon$.

Let D denote the subset of U on which F is Gâteaux differentiable and let D^* denote the subset of U on which F is strongly differentiable. Let $F'(x)$ and $F^*(x)$ denote, respectively, the Gâteaux and strong derivatives, with the corresponding differentials $dF(x;h)$ and $d^*F(x;h)$.

Proposition 1.3. If F is strongly differentiable at $a \in U$, then

$$\lim_{\substack{x \to a \\ x \in D^*}} d^*F(x;h) = \lim_{\substack{x \to a \\ x \in D}} dF(x;h) = d^*F(a;h)$$

uniformly in h on $\{h: ||h|| = 1\}$.

Proof. Let $\varepsilon > 0$. Since F is strongly differentiable at a, there exists a $\delta > 0$ such that $||x_1 - a|| \leq \delta$ and $||x_2 - a|| \leq \delta$, with $x_1 \neq x_2$ imply that $x_1,x_2 \in U$ and

(1.3) $||F(x_2) - F(x_1) - d^*F(a;x_2 - x_1)|| \leq \varepsilon||x_2 - x_1||.$

Let $x \in D$ with $||x - a|| \leq \frac{\delta}{2}$, and let h be a fixed nonzero element in X. If $|\tau| < \frac{\delta}{2||h||}$, then $||x + \tau h - a|| \leq ||x - a|| + ||\tau h|| \leq \delta$, and hence $x + \tau h$ is an admissible choice for x_2 in (1.3). Therefore if $||\tau h|| \leq \frac{\delta}{2}$, then

$$||F(x + \tau h) - F(x) - d^*F(a;\tau h)|| \leq \varepsilon||\tau h||,$$

from which it follows that

$$||dF(x;h) - d^*F(a;h)|| \leq \varepsilon||h||$$

if $x \in D'$ and $||x - a|| < \frac{\delta}{2}$. This proves that

$$\lim_{\substack{x\to a\\ x\in D}} F'(x) = F^*(a). \tag{1.4}$$

Now assuming that $\lim_{\substack{x\to a\\ x\in D^*}} dF(x;h)$ exists, we have

$$\lim_{\substack{x\to a\\ x\in D^*}} d^*F(x;h) = \lim_{\substack{x\to a\\ x\in D^*}} dF(x;h)$$

since F' and F^* are identical on D^*. But

$$\lim_{\substack{x\to a\\ x\in D^*}} d^*F(x;h) = \lim_{\substack{x\to a\\ x\in D}} dF(x;h) \tag{1.5}$$

since $D^* \subset D$. The combination of (1.4) and (1.5) completes the proof.

Proposition 1.3 implies that if $F\colon\ U \to Y$ is strongly differentiable at $a \in U$, and the Gâteaux derivative $F'(x)$ exists for all $x \in D \subset U$ and if a is an accumulation point of D, then the map $x \to F'(x)$ for $x \in D$ is continuous at a.

Proposition 1.4. Suppose F is Gâteaux differentiable in some neighborhood of $x_0 \in U$. Then F is strongly differentiable at x_0 if and only if F' is continuous at x_0.

Proof. If F is strongly differentiable at x_0 and is Gâteaux differentiable in some neighborhood of x_0, then it follows from the remark preceding the statement of Proposition 1.4 that F' is continuous at x_0.

Conversely, suppose F' is continuous at x_0, and let $r(x) = F(x) - F(x_0) - F'(x_0)(x - x_0)$. Then r is Gâteaux differentiable, $r'(x) = F'(x) - F'(x_0)$, and hence $\lim_{x\to x_0}||r'(x)|| = 0$. For any $\varepsilon > 0$, there exists a $\delta > 0$ such that $||r'(x)|| \leq \varepsilon$ for $||x - x_0||$. Using the mean-value theorem

$$||r(x_2) - r(x_1)|| \leq \sup_{0\leq t\leq 1}||r'(x_1 + t(x_2 - x_1)|| \; ||x_2 - x_2||.$$

This implies $||r(x_2) - r(x_1)|| \leq \varepsilon||x_2 - x_1||$ for $x_1, x_2 \in \overline{S}(x_0;\delta)$. Thus F is strongly differentiable at x_0.

We also give an elementary proof which does not use the mean-value theorem. Let F be Fréchet differentiable in some neighborhood N of x_0. Let $\varepsilon > 0$ be given and let $x_1 \in N$, $x_2 \in N$. Then

$$\text{(1.6)} \quad \begin{aligned} ||F(x_2) - F(x_1) - F'(x_0)(x_2 - x_1)|| &\leq ||F(x_2) - F(x_1) - F'(x_1)(x_2 - x_1)|| \\ &\quad + ||F'(x_1) - F'(x_0)|| \; ||x_2 - x_1||. \end{aligned}$$

Since $F'(x)$ is continuous at x_0, there exists a $\delta_1 > 0$ such that if $||x_1 - x_0|| \leq \delta_1$, then $x_1 \in N$ and $||F'(x_1) - F(x_0)|| \leq \frac{\varepsilon}{2}$. But since F is Fréchet differentiable at x_1, there exists $\delta_2 > 0$ such that $||x_2 - x_1|| \leq \delta_2$ implies that

$$||F(x_2) - F(x_1) - F'(x_1)(x_2 - x_1)|| \leq \frac{\varepsilon}{2}||x_2 - x_1||.$$

Now let $\delta = \frac{1}{2}\min\{\delta_1, \delta_2\}$. Let $||x_1 - x_0|| \leq \delta$ and $||x_2 - x_0|| \leq \delta$, so $||x_2 - x_1|| \leq 2\delta \leq \delta_2$. Then using the preceding two estimates in (1.6), we get

$$||F(x_2) - F(x_1) - F'(x_0)(x_2 - x_1)|| \leq \frac{\varepsilon}{2}||x_2 - x_1|| + \frac{\varepsilon}{2}||x_2 - x_1||,$$

which proves that F is strongly differentiable at x_0.

Proposition 1.4 implies that an operator F is strongly differentiable at every point of an open set $U \subset X$ is and only if

F is Fréchet differentiable on U and the map $F': U \to \mathcal{L}(X,Y)$ is continuous. However, a function which is strongly differentiable at x_0 is not necessarily Gâteaux differentiable in a neighborhood of x. Consider for example the real valued function of defined on $(-1,1)$ as follows: $f(0) = 0$, $f(\frac{1}{n}) = n^{-2}$ for integer $n \neq 0$, and f linear and continuous in the intervals $[\frac{1}{n}, \frac{1}{n+1}]$. Then f has a strong derivative at $x_0 = 0$ but is not differentiable in a neighborhood of 0, specifically $f'(\frac{1}{n})$, $n = \pm 1, 2, \ldots$, does not exist. In contrast, the function g with $g(0) = 0$, $g(x) = x^2 \sin \frac{1}{x}$, $x \neq 0$ is differentiable at 0, with $g'(0) = 0$. Also g' exists and is bounded (and hence absolutely continuous) on $[-1,1]$. But g' is discontinuous at 0 and therefore, by Proposition 1.4, g is not strongly differentiable at 0. With this function g, consider the function $f(x) = x + g(x)$. Then $f'(0) \neq 0$, yet f is not monotone in any neighborhood of 0. These pathologies cannot arise with strongly differentiable functions. For a real function f to be strongly differentiable at x_0 it is necessary and sufficient that (i) f is absolutely continuous in some closed interval $[a,b]$, $a < x_0 < b$, and (ii) $\lim_{\substack{x \to x_0 \\ x \in D}} f'(x)$ exists and and is finite, say L, where D is the set of points of differentiability of f (see [1.4]). The necessity follows from the fact that f satisfies a Lipschitz condition on $[a,b]$ (and hence f is absolutely continuous) and from Proposition 1.3. The sufficiency follows using $f(x) = f(a) + \int_a^x f'(t)dt$ (the integral being Lebesgue integral) with $a \leq x \leq b$. Let $\varepsilon > 0$ and let $\delta > 0$ be such that $[x_0 - \delta, x_0 + \delta] \subseteq [a,b]$, and such that $0 < |x - x_0| \leq \delta$, $x \in D$ imply $|f'(x) - L| \leq \varepsilon$. Then for $|x_1 - x_0| \leq \delta$ and $|x_2 - x_0| \leq \delta$,

$$|f(x_2) - f(x_1) - L(x_2 - x_1)| = \left| \int_{x_1}^{x_2} \{f(t) - L\}dt \right| \leq \varepsilon |x_2 - x_1|.$$

(An excellent exposition on differentiability properties of functions of locally bounded variation is given in Asplund and Bungart [1.1]).

B. Strong Partial Differentiability

Definition 1.5. Let X and Y be normed linear spaces, W a topological space, and U an open subset of X. A function $f\colon U \times W \to Y$ is called strongly partially differentiable with respect to X at $(x_0,w_0) \in U \times W$ if there is a continuous linear operator $A = A(x_0,w_0) \in \mathcal{L}(X,Y)$ such that for every $\varepsilon > 0$ there exist a $\delta > 0$ and a neighborhood $N(w_0)$ of $w_0 \in W$ such that

$$(1.7)\quad \|f(x_2,w) - f(x_1,w) - A(x_2 - x_1)\| \leq \varepsilon\|x_2 - x_1\|$$

for all $x_1,x_2 \in \overline{S}(x_0,\delta)$ and $x \in N(w_0)$. In this case A is denoted by $\partial_x f(x_0,w_0)$ and is called the strong partial derivative of f with respect to X at (x_0,w_0).

Clearly if f is strongly partially differentiable, then the partial derivative $f_x(x_0,w_0)$ exists and the map $x \to f(x,w_0)$ is strongly differentiable at x_0. The converse is not necessarily true since in (1.7) it is allowed that $w \neq w_0$, whereas in the definition of $f_x(x_0,w_0)$, w is fixed at w_0 throughout.

For a function of several real variables, and more generally for a function on a Cartesian product of n normed spaces into another normed space, (i) separate continuity does not necessarily imply joint continuity; (ii) the existence of partial derivatives does not imply continuity of the function; (iii) the continuity of all, except possibly one, of the partial derivatives is sufficient (but not necessary) to imply (Fréchet) differentiability; (iv) the continuity of all the partial derivatives is a necessary and sufficient condition for the continuity of the Fréchet derivative. In the framework of strong differentiability these properties take a simpler and tighter form, (as the next theorem shows), attesting to the elegance of the notions of strong differentiability and partial strong differentiability; these are indeed the natural concepts in this context. Furthermore, the proofs are simple.

Theorem 1.6. (i) If f has a strong partial derivative $\partial_x f(x_0,w_0)$ and if the map $w \to f(x_0,w)$ is continuous at w_0, then f is continuous at (x_0,w_0).

(ii) Let $f\colon U \to Z$, where U is an open subset of $X \times Y$, and X, Y and Z are normed spaces, be strongly partially differentiable with respect to both X and Y at $(x_0,y_0) \in U$, with strong partial derivatives $\partial_1 f(x_0,y_0)$ and $\partial_2 f(x_0,y_0)$ respectively. Then f is strongly differentiable at (x_0,y_0) and

$$f'(x_0,y_0)(h_1,h_2) = \partial_1 f(x_0,y_0)h_1 + \partial_2 f(x_0,y_0)h_2.$$

(iii) Let $f\colon U \subset X \times Y \to Z$ be strongly partially differentiable with respect to X and partially (Fréchet) differentiable with respect to Y at (x_0,y_0). Then f is Fréchet differentiable at (x_0,y_0). More generally, if $f\colon U \subset X_1 \times X_2 \times \ldots \times X_n \to Z$ has partial Fréchet derivatives at $x = (x_1, x_2, \ldots, x_n) \in U$ and if all, except possibly one, of these partial derivatives are strong, then f has a Fréchet derivative at x and $f'(x)h = \sum_{i=1}^{n} \partial_i f(x)h_i$, $h = (h_1, \ldots, h_n)$.

(iv) Let $f\colon U \subset X \times Y \to Z$. If the partial derivative f_x exists in a neighborhood of $(x_0,y_0) \subset U$ and is continuous at (x_0,y_0), then f has a strong partial derivative (with respect to X) at (x_0,y_0).

Proof. (i) follows immediately using the inequality

$$\begin{aligned}||f(x,w) - f(x_0,w_0)|| &\le ||f(x,w) - f(x_0,w) - A(x - x_0)|| \\ &+ ||f(x_0,w) - f(x_0,w_0)|| + ||A||\,||x - x_0||,\end{aligned}$$

where $A = \partial f_x(x_0,y_0)$.

(ii) Let $\varepsilon > 0$, and let $A_1 := \partial_1 f(x_0,y_0)$, $A_2 := \partial_2 f(x_0,y_0)$. Then there exists a $\delta > 0$ such that

$$||x_1 - x_0|| \le \delta, \quad ||x_2 - x_0|| \le \delta, \quad ||y - y_0|| \le \delta$$

together imply

$$||f(x_2,y) - f(x_1,y) - A_1(x_2 - x_1)|| \leq \varepsilon||x_2 - x_1||,$$

and

$$||x - x_0|| \leq \delta, \quad ||y_1 - y_0|| \leq \delta, \quad ||y_2 - y_0|| \leq \delta$$

together imply

$$||f(x,y_2) - f(x,y_1) - A_2(y_2 - y_1)|| \leq \varepsilon||y_2 - y_1||.$$

Then for all x_1, x_2, y_1, y_2 for which

$$||x_i - x_0|| \leq \delta \text{ and } ||y_i - y_0|| \leq \delta, \quad i = 1, 2,$$

we have

$$\begin{aligned} &||f(x_2,y_2) - f(x_1,y_1) - A_1(x_2 - x_1) - A_2(y_2 - y_1)|| \\ &\quad \leq ||f(x_2,y_2) - f(x_1,y_2) - A_1(x_2 - x_1)|| \\ &\quad + ||f(x_1,y_2) - f(x_1,y_1) - A_2(y_2 - y_1)|| \\ &\quad \leq \varepsilon(||x_2 - x_1|| + ||y_2 - y_1||), \end{aligned}$$

which proves that f is strongly differentiable at (x_0,y_0).

(iii) This follows from the proof of (ii), by setting in the above inequality $x_1 = x_0$ and $y_1 = y_0$. The general case is treated similarly.

(iv) For given $\varepsilon > 0$, choose $\delta > 0$ so that $f_x(x,y)$ exists for all $x \in S_1 := \bar{S}(x_0,\delta_1)$ and $y \in S_2 := \bar{S}(y_0,\delta)$, and such that

$$||f_x(x,y) - f_x(x_0,y_0)|| \leq \varepsilon \text{ for all } x \in S_1, \quad y \in S_2.$$

For $y \in S_2$, let $r(x) := f(x,y) - f_x(x_0,y_0)x$. Then $||r'(x)|| = ||f_x(x,y) - f_x(x_0,y_0)|| \leq \varepsilon$ for all $x \in S_1$. Now as before we apply the mean-value theorem for any $x,x' \in S_1$:

$$\begin{aligned} ||r(x) - r(x')|| &= ||f(x,y) - f(x',y) - f_x(x_0,y_0)(x - x')|| \\ &\leq \sup_{0 \leq t \leq 1} ||r'(x + t(x' - x))|| \; ||x - x'|| \leq \varepsilon||x - x'||. \end{aligned}$$

One final observation on the role and significance of the strong differential is in order. Dieudonné [1.3, p. 142] regards the mean-value theorem as "probably the most useful theorem in Analysis." The search for suitable forms of the mean-value theorems is a key to a variety of existence proofs in nonlinear analysis, and to inference of "local" properties of a function from "infinitesimal" conditions (e.g., properties of say a "derivative" at a point). A careful phrasing of the mean-value theorem strengthens and simplifies the proof of inverse and implicit function theorems. The observation that we like to make is that strong differentiability itself is a form of the mean-value theorem. To justify this claim, we reexamine a classical result in the light of strong differentiability.

Let F be a continuously Fréchet differentiable function on a convex open subset Ω of a normed space X into a Banach space Y. Then for each $x_1, x_2 \in \Omega$,

$$(1.8)\quad F(x_2) - F(x_1) = \int_0^1 dF(x_1 + t(x_2 - x_1); x_2 - x_1)dt.$$

This is the "fundamental theorem of calculus" relating a differential to its primitive, stated here under stronger hypotheses than necessary for the purpose to be used (see, e.g. [1.10, pp. 171–173] for a proof under weaker hypotheses). Now let $x_0 \in \Omega$ and let $||x_1 - x_0|| \leq \delta$ and $||x_2 - x_0|| \leq \delta$. Then $||x_1 + t(x_2 - x_1) - x_0|| \leq t||x_2 - x_0|| + (1 - t)||x_1 - x_0|| \leq \delta$. Thus for any $\varepsilon > 0$, there exists $\delta > 0$ such that $||F'(x_1 + t(x_2 - x_1)) - F'(x_0)|| \leq \varepsilon$ for $||x_1 - x_0|| \leq \delta$ and $||x_2 - x_0|| \leq \delta$, and hence, using the inequality in (1.8),

$$(1.9)\quad ||F(x_2) - F(x_1) - F'(x_0)(x_2 - x_1)|| \leq \varepsilon||x_2 - x_1||.$$

This provides another proof of the proposition: If F is continuously differentiable on a neighborhood S of x_0, then F is strongly differentiable on S.

The above argument appears frequently in analysis. The inequality (1.9) is crucial to some of the results of L. M. Graves (see, e.g., [3.15], [3.17], [3.18]) on calculus in normed spaces and mapping theorems, e.g., the inverse mapping theorem. Most of these results are based on the validity of (1.9), i.e., it suffices to assume strong differentiability at x_0 in these contexts. (1.8) is a substitute for the mean-value theorem. (1.9) is even a sharper substitute than (1.8) for the mean-value theorem since we can require (1.9) to hold at a point in the absence of continuity of the derivatives.

Notes and Remarks. 1. The notion of the strong derivative of a function of a real variable seems to have been introduced first in 1892 by Peano [1.13] who felt "it portrayed the concept of the derivative used in the physical sciences more closely than does the usual derivative." One can hardly disagree with this statement. Fréchet differentiable functions may exhibit pathologies which have no counterpart in functions arising from physical sciences; this does not appear to be the case with strongly differentiable functions. It is perhaps unfortunate that this notion has not found its way into elementary textbooks. Gleason [1.5] expressed support for the use of this notion in calculus. Esser and Shisha [1.4] define strong differentiability of a function of a real variable and prove Propositions 1.3 and 1.4 in this case. Leach [1.6] defines strong differentiability of an operator on a normed space and uses the strong derivative to prove an inverse function theorem, but he does not develop properties of the derivative. The notion appears also in the books by Dieudonné [1.3] and Cartan [1.2]. However, none of these five references mention any earlier appearance of the notion. (Fortunately, the term "strong differentiability" is used in all of them.) Some properties of strong differentiability are given in Nashed [1.9], [1.10], Wiener [1.15], Ortega and Rheinholdt [1.12] and in the more recent paper of Nijenhuis [1.11]. Strong differentiability with respect to a subspace and higher order derivatives are introduced in [1.10], [1.8], [1.12] and [1.11]. The exposition of this section includes and extends the main properties of strong differentiability mentioned in the references cited above.

2. The set of all functions which are strongly differentiable at a point is obviously closed under addition, scalar multiplication, and composition. However, the composition $g \circ f$ of a strongly differentiable function f and a Gâteaux differentiable function does not necessarily obey the chain rule. For example, let $f\colon \mathbb{R} \to \mathbb{R}^2$ be defined by $f(t) = (t,t^2)$ and $g\colon \mathbb{R}^2 \to \mathbb{R}$ be defined by $g(x,y) = x$ if $y = x^2$ and 0 otherwise. Then $(g \circ f)x = x$, f is strongly differentiable at 0, g is Gâteaux differentiable at $(0,0)$, with $g'(0,0) = 0$, but $(g \circ f)'(0) = 1$. The composition $f \circ g$ of a strongly differentiable function f and a Gâteaux differentiable function g is Gâteaux differentiable, but is not necessarily Fréchet differentiable.

3. An operator F is said to have a <u>locally uniform</u> Fréchet differential $dF(x;h)$ on an open set $U \subset X$ if F has a Fréchet differential on U and the remainder $r(x) := F(x + h) - F(x) - dF(x;h)$ is locally uniformly bounded, i.e., for each $\varepsilon > 0$ and each $x_0 \in U$ these exist $\delta(x_0;\varepsilon)$ and $\eta(x_0;\varepsilon)$ such that

$$||h|| \leq \delta \text{ and } ||x - x_0|| \leq \eta \text{ imply } ||r(x;h)|| \leq \varepsilon||h||.$$

The Fréchet differential is said to be <u>locally bounded</u> on U if every $x_0 \in U$ has a neighborhood in which $||F'(x)||$ is bounded. Then F is strongly differentiable on an open convex set Ω if and only if F has in Ω a locally uniform Fréchet differential and F' is locally bounded on Ω. This characterization follows from the fact that these conditions characterize continuously differentiable functions on Ω (see, e.g., [1.14]).

It would be interesting to find characterizations of strong differentiability <u>at a point</u> in normed spaces and for mappings on function spaces of analysis.

II. A. <u>Generalized Inverses of Domain-Decomposable Linear Operators in Banach Spaces</u>.

Let X and Y be Banach spaces and let A be a linear operator with domain $D(A)$ in X and range $R(A)$ in Y. We say that A is <u>domain-decomposable</u> if there exists a projector (i.e. a continuous linear idemponent) P on X such that (i) $N(A) \subset R(P)$, where $N(A)$ is the null space of A, (ii) $Px \in N(A)$ for all $x \in D(A)$, and (iii) $D(A) \cap N(P)$ is dense in $N(P)$. In this case we call $C_P := D(A) \cap N(P)$ the <u>carrier</u> of A with respect to P.

A projector P on X induces a decomposition of X into two topological complements PX and $(I - P)X$. A subspace M of X is said to have a topological complement if and only if there exists a subspace N such that $X = M \oplus N$ (topological direct sum). A subspace M has a topological complement if and only if there exists a projector P of X onto M.

Examples: 1. Suppose $D(A)$ is dense in X, and that $N(A)$ has a topological complement M. Let P be the projector of X onto $N(A)$ along M, so $X = R(P) \oplus N(P) = N(A) \oplus M$, $N(P) = M$, $R(P) = N(A)$, and $X = N(A) \oplus \overline{M \cap D(A)}$. Thus A is <u>domain-decomposable</u> with respect to P.

2. If H is a Hilbert space and if A is closed linear operator with $\overline{D(A)} = H$, or if A is a bounded linear operator on H, then $N(A)$ has a topological complement (in particular an orthogonal complement) and hence A is domain-decomposable. However, this is not necessarily the case in Banach spaces since an infinite-dimensional closed subspace does not necessarily have a topological complement.

3. If $N(A)$ is finite dimensional, then A is domain-decomposable.

Theorem 2.1. (See [2.3]). Let X and Y be Banach spaces and let $A\colon D(A) \subset X \to Y$ be a linear operator which is domain-decomposable with respect to a projector P. Suppose also that $\overline{R(A)}$, the closure of the range of A, has a topological complement S; and let Q be the projector of Y onto $\overline{R(A)}$ along S. Then there exists a unique linear operator $A^\dagger = A^\dagger_{P,Q}$ which satisfies the following relations:

$$(2.1)\qquad \begin{cases} D(A^{\dagger}) = R(A) + N(Q) \\ R(A^{\dagger}) = C_P(A) \\ A^{\dagger}A = I - P \quad \text{on} \quad D(A) \\ AA^{\dagger} = Q \quad \text{on} \quad D(A^{\dagger}) \end{cases}$$

The operator $A^{\dagger}_{P,Q}$ is called the generalized inverse of A with respect to the projectors P and Q. From (2.1) it follows immediately that $AA^{\dagger}A = A$ and $A^{\dagger}AA^{\dagger} = A^{\dagger}$ on $D(A)$ and $D(A^{\dagger})$ respectively, and that $N(A^{\dagger}) = N(Q)$. The generalized inverse $A^{\dagger}$ is also characterized as the unique solution of the equations (i) $BABx = Bx$ for $x \in R(A) + N(Q) =: D(B)$, (ii) $BA = I - P$ on $D(A)$, and (iii) $AB = Q$ on $D(B)$.

The dependence of $A^{\dagger}$ on A is discontinuous in general, even in finite-dimensional spaces. Let A_n and A be $m \times n$ matrices and let $A_k \to A$. A necessary and sufficient condition for $A_k^{\dagger}$ to converge to $A^{\dagger}$ is that rank A_k = rank A for all sufficiently large k. On the other hand, $A^{\dagger}$ depends "continuously" on the choice of the topological complement, or equivalently on the projectors, regardless of the dimension of spaces. The precise nature of this continuous dependence is discussed in Nashed [2.9]. For other aspects of perturbation and approximation theory of generalized inverses of linear operators see Nashed [2.1], [2.4], [2.6] and the bibliography [2.11].

We now consider the important case of a bounded linear operator $A: X \to Y$ where $N(A)$ and $R(A)$ are assumed to be complemented in the Banach spaces X and Y respectively. That is, $X = N(A) \oplus M$ and $Y = R(A) \oplus S$. The projector P of X onto $N(A)$ along M and the projector Q of Y onto $R(A)$ along S are both continuous. In this case, by the open mapping theorem, the generalized inverse $A^{\dagger}_{P,Q}$ is bounded and defined on all of Y.

For a detailed theory of generalized inverses of closed or bounded operators in Banach spaces, with emphasis on the algebraic, projectional and proximinal properties, see [2.10] and [2.3].

Generalized inverses of linear operators in Hilbert spaces are discussed in [2.1], [2.3], [2.4], [2.7], and [2.8]; other references are cited in [2.11].

B. Bound Inner and Outer Inverses

Let A be a bounded linear operator on X into Y. An inner inverse for A is a linear operator $B: Y \to X$ such that $ABA = A$. While every linear operator has an inner inverse, a bounded linear operator need not necessarily have any bounded inner inverse. $A \in \mathcal{L}(X,Y)$ has a bounded inner inverse if and only if $N(A)$ and $R(A)$ are closed complemented subspaces of X and Y, respectively.

An outer inverse for A is a linear operator $B: Y \to X$ such that $BAB = B$. Every operator has a nonzero outer inverse. If B is an outer inverse of A, then BA and AB are idempotent, $R(BA) = R(B)$, $R(AB) \subset R(A)$, $N(B) = N(AB)$, $N(A) \subset N(BA)$, and B induces the algebraic direct sum decompositions:

$$X = R(B) \dotplus N(BA)$$

$$Y = N(B) \dotplus R(AB).$$

Clearly B is an outer inverse for A if and only if A is an inner inverse for B. Thus we immediately have the following result:

Proposition 2.2. Let $A \in \mathcal{L}(X,Y)$, where X and Y are Banach spaces. A necessary and sufficient condition for $B \in \mathcal{L}(Y,X)$, $B \neq 0$, to be an outer inverse for A is that $R(B)$ and $N(B)$ be closed complemented subspaces of X and Y respectively. If B is a bounded outer inverser for A, $B \neq 0$, then BA is a (continuous) projector of X onto $R(B)$ and $I - AB$ is a (continuous) projector of Y on $N(B)$, and

$$X = R(B) \oplus N(BA)$$

$$Y = N(B) \oplus R(AB).$$

Note that if B is an inner inverse for A, then BAB is both an inner and an outer inverse for A. Hence, the existence of a bounded inner inverse for $A \in \mathcal{L}(X,Y)$ implies the existence of a bounded outer inverse. Thus, if $N(A)$ and $R(A)$ are closed complemented subspaces of X and Y, respectively, then A has both a bounded inner inverse and a bounded outer inverse.

If A is surjective, the inner inverses of A are precisely the right inverses of A. If A is injective, the inner inverses of A are precisely the left inverses of A. If A is neither onto nor one-to-one, then we cannot talk about either a right or a left inverse; however (algebraic) inner and outer inverses of A always exist, even though bounded inner (outer) inverses for A may not. In fact, if every bounded linear operator $A: X \to Y$ has a bounded inner inverse, then X and Y are finite dimensional. Examples of classes of operators which have bounded inner inverses are given in [2.3], [2.5], [2.10]. Structural properties of the affine space $I(A)$ of all inner inverses of A and the set $O(A)$ of all outer inverses are developed in [2.10]. Finally, we remark that the relations $ABA = A$ and $BAB = B$ are symmetric, properties of $I(A)$ and $O(A)$ are markedly different. Also, the set of all bounded operators which have bounded outer inverses in open in $\mathcal{L}(X,Y)$ (see Lemma 3.11), whereas the set of all bounded operators which have bounded inner inverses is not open.

III. INVERSE MAPPING THEOREMS IN THE FRAMEWORK OF GENERALIZED INVERSES

A. Generalized Inverse Mapping Theorems

Theorem 3.1. Let X and Y be Banach spaces and let $f: U \to Y$, where U is a neighborhood of $x_0 \in X$. Suppose that there exist a bounded linear operator A (depending on x_0) and positive numbers ε and δ such that the sphere $\overline{S}(x_0,\delta) = \{x: ||x - x_0|| \leq \delta\}$ is in U, and

$$(3.1)\quad \begin{cases} ||f(x_2) - f(x_1) - A(x_2 - x_1)|| \leq \varepsilon||x_1 - x_2|| \\ \quad\text{whenever } x_1 \text{ and } x_2 \text{ are in } \overline{S}(x_0,\delta). \end{cases}$$

Suppose that A has a (nonzero) bounded outer inverse B such that $k := \varepsilon\|B\| < 1$. Let

(3.2) $$N_2 := \{y \in Y: \|y - y_0\| < \frac{\delta}{\|B\|}(1 - k)\}$$

where y_0 is such that $y_0 - f(x_0) \in N(B)$.

(i) Define the modified Newton-Raphson sequence

(3.3) $$x_{n+1} = x_n - B(f(x_n) - y)$$

associated with the equation $f(x) = y$. Then for any $y \in N_2$, the sequence (3.3) converges to $x^* \in \overline{S}(x_0, \delta) \cap \{x_0 + R(B)\}$ and

(3.4) $$f(x^*) - y \in N(B).$$

In particular, if $x_0 \in R(B)$, equivalently $BAx_0 = x_0$, then $x_n \in R(B)$ for all n and $x^* \in \overline{S}(x_0, \delta) \cap R(B)$.

(ii) There exist a neighborhood N_1 of $x_0 \in U$, a neighborhood of $y_0 \in Y$ and a unique function $g\colon N_2 \to N_1$, g is continuous with $g(y_0) = x_0$, such that for all $y \in N_2$

(3.5) $$g(y) \in x_0 + R(B) = x_0 + N(I - BA)$$

(3.6) $$f(g(y)) - y \in N(B)$$

Furthermore g satisfies for some $\gamma > 0$ the smoothness condition

(3.7) $\|g(y_1) - g(y_2) - B(y_1 - y_2)\| \leq \gamma\|y_1 - y_2\|$ for $y \in N_2$,

(γ depends on ε and $\|B\|$), and on N_2,

(3.8) $$gfB = B$$

(3.9) $$gAB = g \quad \text{and} \quad gfg = g.$$

Proof: We prove inductively that

(3.10) $$\|x_n - x_0\| < \delta \quad \text{and} \quad \|x_{n+1} - x_n\| < k^n(1 - k)\delta$$

for any $y \in N_2$. By definition of x_1,

$$\|x_1 - x_0\| = \|B(f(x_0) - y)\| \leq \|B\|\,\|y - y_0\| < \delta$$

so that $(3.10)_n$ is valid for $n = 0$. Suppose $(3.10)_n$ is valid for

$n \leq m - 1$. Then

$$\|x_m - x_0\| \leq \sum_{i=1}^{m} \|x_i - x_{i-1}\| \leq \sum_{i=1}^{m} (\varepsilon\|B\|)^{i-1}\delta(1 - k)$$
$$\leq \delta(1 - k^m) < \delta.$$

Also,

$$\begin{aligned} x_{m+1} - x_m &= B(y - f(x_m)) \\ &= B(y - f(x_{m-1})) + B(f(x_{m-1}) - f(x_m)) \\ &= BAB(y - f(x_{m-1})) + B(f(x_{m-1}) - f(x_m)) \\ &= BA(x_m - x_{m-1}) + B(f(x_{m-1}) - f(x_m)). \end{aligned}$$

Thus,

$$\begin{aligned} \|x_{m+1} - x_m\| &\leq \|B\|\ \|f(x_m) - f(x_{m-1}) - A(x_m - x_{m-1})\| \\ &\leq \|B\|\ \varepsilon\ \|x_m - x_{m-1}\| \end{aligned}$$

by (3.1) since $\|x_{m-1} - x_0\| \leq \delta$ and $\|x_m - x_0\| \leq \delta$. That is,

(3.11) $$\|x_{m+1} - x_m\| \leq k\|x_m - x_{m-1}\|, \quad k < 1,$$

and $\|x_{m+1} - x_m\| < k^m(1 - k)\delta$, so that $(3.10)_n$ is valid for all n. From this it follows that $\{x_n\}$ converges to $x^* \in \bar{S}(x_0,\delta)$ and (3.3) implies that $f(x^*) - y \in N(B)$ since f is continuous. Now we show that $x^* \in x_0 + R(B)$. Note that $x_{n+1} - x_n \in R(B)$ and that BA projects X on $R(B)$ along a complement S. So $BA(x_{n+1} - x_n) = x_{n+1} - x_n$ and

$$x_n = x_0 + \sum_{m=1}^{n} BA(x_m - x_{m-1}).$$

Thus for all n,

(3.12) $$x_n \in x_0 + R(B).$$

and $x^* \in x_0 + R(B)$.

The vector x_0 has a unique decomposition $x_0 = x_B + x_S$, where $x_B \in R(B)$ and $x_S \in S$. Hence, $BAx_0 = x_B = x_0 - x_S$ and $BA(x_1 - x_0) = BAx_1 - x_B = x_1 - x_0$, so $BAx_1 = x_1 - x_S$ and for all n

$$BAx_n = x_n - x_S.$$

That is, for all n the component of x_n along the subspace S is the same as the component of x_0 along S. In particular, if $x_0 \in R(B)$, then $x_S = 0$ and $x_n \in R(B)$ for all n, and so $x^* \in R(B)$ since $R(B)$ is closed.

(ii) Let ε and δ be related as in (3.1) and suppose $k := \varepsilon||B|| < 1$. By (i) the Newton-Raphson sequence (3.3) converges for any $y \in N_2$. Let $\lim_{n\to\infty} x_n = g(y)$ for $y \in N_2$. Then by (i) $f(g(y)) - y \in N(B)$ and $g(y) \in x_0 + R(B)$ for $y \in N_2$.

We next prove that g satisfies a Lipschitz condition on N_2. Let y_1 and y_2 be in N_2. Then $||g(y_1) - x_0|| \leq \delta$ and $||g(y_2) - x_0|| \leq \delta$ and thus by (3.1),

$$||f(g(y_1)) - f(g(y_2)) - A(g(y_1) - g(y_2))|| \leq \varepsilon||g(y_1) - g(y_2)||$$

$$||B(fg(y_1))-B(f(g(y_2))-BA(g(y_1)-g(y_2))||\leq\varepsilon||B|| \; ||g(y_1)-g(y_2)||.$$

Since $g(y_1) - g(y_2) \in R(B)$, this simplifies using (3.6) to

(3.13) $$||By_1 - By_2 - (g(y_1) - g(y_2)|| \leq \varepsilon||B|| \; ||g(y_1) - g(y_2)||.$$

From this relation and the triangle inequality we obtain

$$||B(y_1 - y_2)|| \geq ||g(y_1) - g(y_2)|| - \varepsilon||B|| \; ||g(y_1) - g(y_2)||, \text{ or}$$

(3.14) $$||g(y_1) - g(y_2)|| \leq \frac{||B||}{1 - \varepsilon||B||} ||y_1 - y_2||, \text{ for } y_1, y_2 \in N_2.$$

Substituting (3.14) into (3.13) we get

$$||g(y_1) - g(y_2) - B(y_1 - y_2)|| \leq \frac{||B||^2\varepsilon}{1 - \varepsilon||B||} ||y_1 - y_2||.$$

which proves (3.7). In particular, this shows that g satisfies a smoothness condition of the same type that f satisfies, although f is not assumed to be differentiable.

We now prove (3.8). For convenience and without loss of generality, we now take $x_0 = 0$ and $f(0) = 0$. Since both f and g are continuous at zero, $||g(f(By))||$ and $||By||$ are less than δ for sufficiently small y. Hence

$$||f(g(f(By))) - f(By) - A(g(f(By) - By)|| \leq \varepsilon||g(f(By)) - By||.$$

and

$$||Bf(g(f(By)))-Bf(By))-BA(g(f(By)))-By|| \leq \varepsilon ||B|| \; ||g(f(By))-By||.$$

Now using (3.6), the first two terms on the left side cancel out, and the inequality simplifies to

$$||g(f(By)) - By|| \leq \varepsilon ||B|| \; ||g(f(By)) - By||$$

Since $\varepsilon||B|| < 1$, this implies that $g(f(By)) = By$.

To prove uniqueness (i.e., local uniqueness) of g with properties (3.5) and (3.6), suppose that h is another function which has properties (3.5) and (3.6), and let y be such that $||g(y) - x_0|| \leq \delta$ and $||h(y) - x_0|| \leq \delta$. Then using the same argument as before, we obtain

$$||Bfg(y) - Bfh(y) - BA(g(y) - h(y))|| \leq \varepsilon ||g(y) - h(y)||.$$

Now the first two terms cancel and the inequality simplifies to

$$||g(y) - h(y)|| \leq \varepsilon ||B|| \; ||g(y) - h(y)||.$$

This inequality contradicts $\varepsilon||B|| < 1$ unless $g(y) = h(y)$ for all $y \in N_2$.

Finally, the proof of (3.9) follows easily invoking uniqueness of g and using similar manipulations.

Corollary 3.2. Let X and Y be Banach spaces and let $f\colon U \to Y$, where U is a neighborhood of $x_0 \in X$. Suppose that f has a strong derivative $A = f'(x_0)$ at x_0 and that A has a bounded outer inverse, say B. Then the conclusions of Theorem 3.1 hold. Furthermore, the function g (which depends on B) is strongly differentiable at y_0 and $g'(y_0) = B$.

Proof. Since f has a strong differential at x_0, for every $\varepsilon > 0$ there exists a $\delta > 0$ such that with $A = f'(x_0)$,

$$(3.15) \quad \begin{cases} ||f(x_2) - f(x_1) - A(x_2 - x_1)|| \leq \varepsilon ||x_2 - x_1|| \\ \text{whenever } ||x_1 - x_0|| \leq \delta \text{ and } ||x_2 - x_0|| \leq \delta. \end{cases}$$

Let $\varepsilon < 1/||B||$ and let δ be related to ε by (3.15). Then the assumptions of Theorem 3.1 are satisfied.

Now to prove that g is strongly differentiable at y_0, we note that (3.13) holds for every $\varepsilon > 0$ such that $\varepsilon||B|| < 1$; thus as in the proof of Theorem 3.1,

$$||g(y_1) - g(y_2) - B(y_1 - y_2)|| \le \frac{||B||^2\varepsilon}{1 - \varepsilon||B||} ||y_1 - y_2||.$$

Now for each $\gamma > 0$, let $k := \frac{\gamma}{\gamma + ||B||}$ and $\varepsilon := \frac{k}{||B||}$. Then $\frac{||B||^2\varepsilon}{1-\varepsilon||B||} = \frac{k||B||}{1 - k} = \gamma$. Thus for each $\gamma > 0$, there exists a $\delta > 0$ such that

$$||g(y_1) - g(y_2) - B(y_1 - y_2)|| \le \gamma||y_1 - y_2||$$

for $y_1, y_2 \in \overline{S}(y_0, \delta)$. This proves that B is the strong differential of g at y_0.

An important special case of Corollary 3.2 is when the null space and the range of the strong derivative $A\colon X \to Y$ are complemented in the spaces X and Y, respectively. As noted in Section 2, the existence of such complements is a necessary and sufficient condition for $A \in \mathcal{L}(X,Y)$ to have a <u>bounded inner</u> inverse. The existence of a bounded inner inverse to A implies the existence of a (nonzero) bounded outer inverse to A, but not conversely (see, e.g., [2.3] for an example). It would be interesting to obtain useful characterizations for the existence of bounded outer inverses.

Corollary 3.3. Let X and Y be Banach spaces and let $f\colon U \to Y$, where U is a neighborhood of $x_0 \in U$. Suppose that f has a strong derivative $A = f'(x_0)$ at x_0 and that X and Y admit topological direct sum decompositions relative to $N(A)$ and $R(A)$, respectively, i.e.,

$$X = N(A) \oplus M \quad \text{and} \quad Y = R(A) \oplus S.$$

Then A has a bounded outer inverse and the conclusion of Corollary 3.2 holds. In particular, if B_1 is any bounded inner inverse of A, then $B := B_1AB_1$ is a bounded outer inverse.

The conclusions of Theorem 3.1 and Corollaries 3.2 and 3.3

are in one respect weaker than the classical inverse mapping theorem, obtained in the latter case under the assumption of invertibility of $f'(x_0)$, since one can only conclude in our setting that $f(g(y)) - y \in N(B)$, rather than $f(g(y)) = y$ for all sufficiently small y. However, if A is <u>onto</u> then we can choose B such that $N(B) = \{0\}$ and obtain the conclusion $f(g(y)) = y$ for sufficiently small y, and it becomes clear that the weaker assumptions of Theorem 3.1 and Corollaries 3.2 - 3.4 are indeed more natural. This situation often arises with mapping on infinite-dimensional Banach spaces.

Corollary 3.4. Let $f\colon U \to Y$ and suppose that the strong derivative of f at $x_0 \in U$ exists and that $A = f'(x_0)$ maps X onto Y. Then A has a bounded outer inverse B with $N(B) = \{0\}$ and for each such choice of B there exists a (unique) function $g\colon N_2 \to N_1$ (as in Theorem 3.1, where g depends on B) such that the conclusion of Corollary 3.2 holds with $f(g(y)) = y$ for $y \in N_2$. If in addition $N(A)$ is complemented in X, then any right inverse of A is a bounded outer inverse with $N(B) = \{0\}$.

Remark 3.5. The case when $f'(x_0)$ is <u>not</u> invertible has been considered by Graves [3,15] who proved that if $f\colon U \to Y$ has a continuous Fréchet differential on a neighborhood of $x_0 \in U$ such that $f'(x_0)$ maps X <u>onto</u> Y, then f is an <u>open mapping</u> for x near x_0, i.e. for y near $y_0 = f(x_0)$, the equation $f(x_0 + h) = y$ has a solution for small $||h||$. (One cannot expect to drop the hypothesis that $f'(x_0)$ be onto and get an open mapping - e.g. $f(x) = x^2$, $x_0 = 0$. But the hypothesis is not necessary either - e.g. $f(x) = x^3$, $x_0 = 0$, in which case $f'(0)$ does not map onto $\mathbb{R}$, but f is open.) Under the assumption that $f'(x_0)$ is onto, one can only conlcude that f is open; obviously f need not be one-to-one. The point of departure between this result of Graves and the generalized inverse approach to inverse mapping theorems is that in the latter approach <u>each choice of a (bounded) outer inverse</u> B <u>induces a unique selection</u>

of an "inverse" mapping g (in the sense of (3.6)) even when f is not open. This approaches provides a unique generalized inverse mapping (for each choice of B) satisfying (3.5), (3.6), (3.8) and (3.9).

In the remainder of this section, we recast and relate several results on inverse mapping theorem in the framework of the preceding exposition. For convenience, we take $x_0 = 0$ and $f(0) = 0$. The inverse mapping theorem in Banach spaces was established fifty years ago by Hildebrandt and Graves [3.18]: if f is of class C' and $f'(0)$ is invertible, then f is open and uniquely invertible near the origin. This obviously follows from Corollary 3.2 in view of Proposition 1.4. In 1950, Graves considered the case when $f'(0)$ is not invertible. The following lemma, which follows immediately from the open mapping theorem, was used by Graves, to prove Theorem 3.7 below.

Lemma 3.6. Let X and Y be Banach spaces and let T be a continuous linear transformation of X onto Y. Then there exists a constant M such that for each $y \in Y$, there is at least one $x \in X$ with $Tx = y$ and $||x|| \leq M||y||$.

Theorem 3.7. (Graves [3.18; Theorem 1]). Let $f\colon U \to Y$ and $f(0) = 0$. Let $\{x\colon ||x|| < \delta\} \subset U$. Let T be a bounded linear transformation of X onto Y such that

$$(3.16) \qquad ||f(x_1) - f(x_2) - T(x_1 - x_2)|| \leq \varepsilon||x_1 - x_2||$$

whenever $||x_1|| < \delta$ and $||x_2|| < \delta$, where $M\varepsilon < 1$ and M is the constant defined in Lemma 3.6. Then the equation $f(x) = y$ has at least one solution with $||x|| < \delta$, whenever $||y|| < \rho$, where $\rho = \frac{\delta}{M}(1 - M\varepsilon)$.

Graves' proof is based on successive approximations by alternate application of the formulas

$$y_{n-1} = T(x_n - x_{n-1}),$$

$$y_n = T(x_n - x_{n-1}) - f(x_n) - f(x_{n-1}),$$

where $x_0 = 0$ and $y_0 = y$.

As a corollary to Theorem 3.7, Graves also proved:

Theorem 3.8. (Graves [3.18; Theorem 2]). Let f have a continuous Fréchet derivative for $||x|| < \delta$ with values in Y, and $f(0) = 0$. Suppose that the derivative $f'(0)$ maps X onto Y. Then there exists a constant $\rho > 0$ such that whenever $||y|| < \delta$, the equation $f(x) = y$ has a solution x with $||x|| < \delta$, i.e., f is an open mapping for x near x_0.

Proof. Let $T := f'(0)$. By Proposition 1.4, f has a strong derivative T at zero; hence (3.16) would be satisfied if δ is chosen sufficiently small to correspond to ε satisfying $M\varepsilon < 1$. Then the desried result follows from Theorem 3.7.

It should be noted that Theorem 3.7 is stated in terms of an approximate strong derivative and that Graves' proof of Theorem 3.8 is tantamount to showing that a continuous Fréchet derivative is a strong derivative, a result which has since been rediscovered by many authors (see Section 1). It is clear that the use of strong derivative prevails in the Graves' papers on inverse and implicit functi^n theorem, although he did not frame the concept as a definition!

In order to see that Theorem 3.7 is a special case of Theorem 3.1, it suffices to relate the constant M to the norm of a generalized inverse of T. Let $J(T)$ be the infimum of all positive numbers M with the property $||x|| \leq M||y||$ as in Lemma 3.6. Then

$$J(T) = \sup_{||y||=1} [\{ \inf ||x|| : Tx = y\}]. \tag{3.17}$$

Since T is onto Y, for each $y \in Y$ and each (algebraic) inner inverse S of T, $x = Sy$ is a solution of $Tx = y$ and all so solutions are obtained as S ranges over the set $I(T)$ of all inner inverses of T (see [2.3]). It follows that there exists $S \in I(T)$ such that $||Sy|| \leq M||y||$ for each $y \in Y$. Hence T has a bounded inner inverse and therefore also a bounded outer inverse. Thus, $J(T)$ is the infimum of the norms of all bounded inner inverses of T. (In Hilbert spaces, $J(T) = ||T^{\dagger}||$, where $T^{\dagger}$ is the (Moore-Penrose) operator generalized inverse of T.) This discussion also provides a proof for the first part of Corollary 3.4.

In 1955, R. Nevalinna [3.30] showed that, in the case of Hilbert space, unique inversion is possible when f' is continuous only at $x = 0$ and $f'(0)$ is invertible. In 1957, F. Nevalinna [3.29] proved that, in the case of Hilbert space, the assumption that f' is continuous can be relaxed somewhat and the inversion of f remains possible. In 1958, Bartle [3.5] showed that the theorem of Graves (Theorem 3.7 above) can be used to generalize the results of F. Nevanlinna in two ways: he extended the results to Banach spaces and assumed that $f'(0)$ maps onto Y instead of the hypothesis that $f'(0)$ be one-to-one and onto Y. The conclusion in this case is that f is open.

Theorem 3.9. (Bartle [3.5]). Let $f\colon U \to Y$, with $f(0) = 0$. Suppose f is Fréchet differentiable for $||x|| < r$ and that $T := f'(0)$ maps X onto Y. Let $\phi(\rho)$ be the monotone function defined for $0 \le \rho < r$ by

$$\phi(\rho) := \sup \{||f'(x) - f'(0)||: ||x|| \le \rho\}.$$

Assume in addition that

$$\phi'(0^+) := \limsup_{x \to 0} ||f'(x) - f'(0)|| < \frac{1}{J(T)}, \tag{3.18}$$

where $J(T)$ is defined by (3.17). Let δ be the supremum of all numbers ρ such that $J(T)\phi(\rho) < 1$. Then f is an open mapping near $x = 0$. More specifically, the image under f of any open set contained in $|x| < \delta$ is an open set.

We note that (3.18) involves the modulus of the operator $f'(x)$ and is obviously satisfied in the case that the map $x \mapsto f'(x)$ is continuous at $x = 0$.

From the assumption of Theorem 3.9, it follows easily that $f'(x)$ maps X onto Y for $||x|| < \delta$ (see Lemma 3.10 below) and that f has an approximate strong derivative, i.e. it satisfies a condition of the form (3.1). Then Theorem 3.9 follows from Theorem 3.1, noting that $f'(0)$ serves as an approximate strong derivative.

In his review of [3.5] in Mathematical Reviews, D. G. Bourgin remarks that "throughout this and antecedent papers in the literature heavy restrictions are placed on the Frẽchet derivative, especially at θ. Presumably, the use of differences instead of derivatives would allow weakening of the hypotheses of the existence of T_θ and perhaps of T_x also. Indeed, many other results such as the Hildebrandt-Graves implicit function theorem can possibly be generalized in this way." We feel that the generalized inverse mapping theorems presented in this section, in the framework of the geometry of generalized inverses and using the concepts of strong differentiability and approximate strong differentiability, achieve this purpose. As remarked in Section 1, these notions of differentiability are strong forms of a mean-value theorem (using differences).

Finally we remark that generalized implicit function theorems analogous to these generalized inverse mapping theorems can be formulated and easily deduced using the notion of strong partial differentiality and approximate variant thereof, extending results in [3.8], [3.31], [3.32] as well as the Hildebrandt-Graves implicit function theorem [3.18], [3.19], [3.23].

B. Stability of Generalized Inverse Mapping Theorems

The classical inverse mapping theorem assumes the invertibility of the linear operator $A = f'(x_0)$. It is well known that the set of all invertible operators in $\mathcal{L}(X,Y)$ is an open subset of $\mathcal{L}(X,Y)$. In particular, let A and B be in $\mathcal{L}(X,Y)$ and assume $A^{-1} \in \mathcal{L}(Y,X)$ exists and $||A^{-1}||\ ||A - B|| < 1$. Then B^{-1} exists and is bounded. Thus a "close" approximation to $F'(x_0)$ is necessarily invertible if $F'(x_0)$ is invertible. Two questions then arrise: (1) How do such approximations to $F'(x_0)$

(or equivalently the error in $F'(x_0)$) propagate into the conclusion of the inverse mapping theorem? (2) What can be said about the stability of the generalized mapping theorems under perturbations of the approximating operator (or the strong derivative) A, assuming that A is not invertible. We address these questions in this section.

Lemma 3.10. The set of all bounded linear operators of X onto Y is open in the space $\mathcal{L}(X,Y)$.

This lemma appears essentially as Theorem 3 in Graves [3.18] and is an immediate consequence to Theorem 1 of [3.18]. Let T be a continuous linear transformation of X onto Y and let L be another linear mapping such that $||Lx - Tx|| \leq \delta ||x||$ for all x, where $\delta M < 1$ and M is the constant of Lemma 3.6. Then L maps X onto Y. The lemma also appeared earlier in a paper of Dieudonné [3.12], where it shown that the set of all right (left) invertible operators in $\mathcal{L}(X,Y)$ is an open set in $\mathcal{L}(X,Y)$.

Lemma 3.11. Let B be a bounded outer inverse of $A \in \mathcal{L}(X,Y)$. Let $\tilde{A} \in \mathcal{L}(X,Y)$ be such that $||B(\tilde{A} - A)|| < 1$. Then $\tilde{B} := [I + B(\tilde{A} - A)]^{-1}B$ is a bounded outer inverse to $\tilde{A}$ with $N(\tilde{B}) = N(B)$ and $R(\tilde{B}) = R(B)$. Furthermore,

$$||\tilde{B} - B|| \leq ||B|| \; ||\tilde{B}|| \; ||\tilde{A} - A||. \tag{3.19}$$

Proof. Let $C := I + B(\tilde{A} - A)$. Then by Banach's lemma, C has a bounded inverse. Since $BAB = B$, we have $BAC = BAB\tilde{A} = B\tilde{A}$. Thus

$$BA = B\tilde{A}C^{-1}. \tag{3.20}$$

Let $\tilde{B} := C^{-1}B$. Then $\tilde{B}\tilde{A}\tilde{B} = C^{-1}B\tilde{A}C^{-1}B = C^{-1}BAB = C^{-1}B = \tilde{B}$, using (3.20) in the second term. Clearly $\tilde{B}$ is bounded and $N(\tilde{B}) = N(B)$.

Let $x \in R(B)$, i.e., $BAx = x$. Then $\tilde{B}\tilde{A}x = C^{-1}B\tilde{A}x = C^{-1}B\tilde{A}BAx$. Now $CB = [I + B(\tilde{A} - A)]B = B\tilde{A}B$, so $B = C^{-1}B\tilde{A}B$. Thus $\tilde{B}\tilde{A}x = C^{-1}B\tilde{A}BAx = BAx = x$. This proves that $x \in R(\tilde{B})$. The reverse inclusion $R(\tilde{B}) \subset R(B)$ is established by symmetry. Let

$\tilde{C} = I + \tilde{B}(A - \tilde{A})$. Then $C\tilde{C} = C[I + C^{-1}B(A - \tilde{A})] = C + B(A - \tilde{A})$ $= I + B(\tilde{A} - A) + B(A - \tilde{A}) = I$. Now since C^{-1} exists, it follows that $\tilde{C} = C^{-1}$ and $\tilde{C}^{-1} = C$. This shows that if $\tilde{B}\tilde{A}\tilde{B} = \tilde{B}$, then $BAB = B$ and $R(\tilde{B}) \subset R(B)$. Finally, (3.19) follows from $\tilde{B} = C^{-1}B = \tilde{C}B = [I + \tilde{B}(A - \tilde{A})]B$.

The condition $||B(\tilde{A} - A)|| < 1$ in Lemma 3.11 can be replaced by less stringent conditions involving a polynomial function and a corresponding $\tilde{B}$ can be constructed to allow a wider class of perturbations in A. These are modelled after the general approximation theory developed in Moore and Nashed [2.6]. Further results on perturbation theory are given in [2.4] and [2.9].

Using Lemmas 3.10 and 3.11, we can easily show that the various inverse mapping theorems of this section are stable under "small" perturbations. For example, in Theorem 3.7, if T is replaced by $L \in \mathcal{L}(X,Y)$ such that $||Lx - Tx|| \leq \delta||x||$ for all x, where $M\delta < 1$, then $||f(x_2) - f(x_1) - L(x_2 - x_1)||$ $\leq (\varepsilon + \delta)||x_2 - x_1||$ and $||x|| \leq \frac{M}{1-M\delta}||Lx||$.

With reference to Corollary 3.2, Lemma 3.10 may be used to obtain the following result on stability. Suppose instead of using $A := f'(x_0)$, we take a linear approximation to A by $\tilde{A}$ with bounded outer inverse $\tilde{B}$ such that $I + \tilde{B}(A - \tilde{A})$ has a bounded inverse. Then the conclusion of Theorem 3.1 holds with $g(y) \in x_0 + R(B)$ and $fg(y) - y \in N(B)$. In contrast, if instead of bounded outer inverses, we use bounded inner inverses, then stability does not necessarily follow.

IV. REFERENCES

[1.1] Asplund, E. and Bungart, L., *A First Course in Integration*, Holt, Rinenhart and Winston, New York, 1966.

[1.2] Cartan, H., *Calcul différentiel*, Herman, Paris, 1967. English Edition: *Differential Calculus*, Houghton Mifflin, Boston, 1971.

[1.3] Dieudonné, J., *Treatise on Analysis I*, (Foundations of Modern Analysis), Academic Press, New York and London, 1969.

[1.4] Esser, M., and Shisha, O., *A modified differentiation*, Amer. Math. Monthly 71 (1964), pp. 904-906.

[1.5] Gleason, A. M., *The geometric content of advanced calculus*, Proceedings of CUPM Geometry Conference, Part II: Geometry in other subjects, L. K. Durst, Ed., Math. Assoc. of America, 1967.

[1.6] Leach, E. B., *A note on inverse function theorems*, Proc. Amer. Math. Soc. 12 (1961), pp. 694-697.

[1.7] Nashed, M. Z., *Some remarks on variations and differentials*, Amer. Math. Monthly 73 (1966), Slaught Memorial Papers, pp. 63-76.

[1.8] Nashed, M. Z., *Higher order differentiability of non-linear operators on normed spaces. I, II*, Comment. Math. Univ. Carolinae 10 (1969), pp. 509-533, 535-557.

[1.9] Nashed, M. Z., *On strong and uniform differentials*, unpublished notes (presented at a short on nonlinear analysis at UCLA), 1966.

[1.10] Nashed, M. Z., *Differentiability and related properties of nonlinear operators: Some aspects of the role of differentials in nonlinear functional analysis*, in Nonlinear Functional Analysis and Applications, L. B. Rall, ed., Academic Press, New York, 1971, pp. 109-309.

[1.11] Nijenhuis, A., *Strong derivatives and inverse mappings*, Amer. Math. Monthly 81 (1974), pp. 969-981. Addendum, ibid., 83 (1976), p. 22.

[1.12] Ortega, J. M. and Rheinboldt, W. C., *Iterative Solution of Nonlinear Equations in Several Variables*, Academic Press, New York, 1970.

[1.13] Peano, G., *Sur la definition de la dérivée*, Mathesis (2) 2 (1892), pp. 12-14. = Opere Scelte, V. 1, Edisioni Cremonsense, Rome, 1957, pp. 210-212.

[1.14] Vainberg, M. M., *Variational Methods for the Study of Nonlinear Operators*, Holden-Day, San Francisco, 1964.

[1.15] Wiener, S. M., *Differentiability in general analysis*, M.S. Thesis, Georgia Institute of Technology, Atlanta, 1966, 97 pp.

[2.1] Nashed, M. Z., *Generalized inverses, normal solvability, and iteration for singular operator equations*, Nonlinear Functional Analysis and Applications, L. B. Rall, ed., Academic Press, New York (1971), pp. 311-359.

[2.2] Nashed, M. Z. and Votruba, G. F., *A unified approach to generalized inverses of linear operators: Algebraic, topological and projectional properties*, Bull. Amer. Math. Soc., 80 (1974), pp. 825-830.

[2.3] Nashed, M. Z. and Votruba, G. F., *A unified operator theory of generalized inverses*, Generalized Inverses and Applications, M. Z. Nashed, ed., Academic Press, New York, 1976, pp. 1-109.

[2.4] Nashed, M. Z., *Perturbations and approximations for generalized inverses and linear operator equations*, Generalized Inverses and Applications, M. Z. Nashed, ed., Academic Press, New York, 1976, pp. 325-396.

[2.5] Nashed, M. Z., *Aspects of generalized inverses in analysis and regularization*, Generalized Inverses and Applications, M. Z. Nashed, ed., Academic Press, New York (1976) pp. 193-244.

[2.6] Moore, R. H. and Nashed, M. Z., *Approximations to generalized inverses of linear operations*, SIAM J. Appl. Math., 27 (1974), pp. 1-16.

[2.7] Ben-Israel, A. and Greville, T. N. E., *Generalized Inverses: Theory and Applications*, Wiley-Interscience, New York, 1974.

[2.8] Groetsch, C. W., *Generalized Inverses of Linear Operators*, Dekker, New York, 1977.

[2.9] Nashed, M. Z., *Perturbation theory of generalized inverse operators*, Functional Analysis Methods in Numerical Analysis, M. Z. Nashed, ed., Spring-Verlag, 1978, to appear.

[2.10] Nashed, M. Z., ed., *Recent Advances in Generalized Inverses: Theory and Applications*, Pitman, London-San Francisco, 1978, to appear.

[2.11] Nashed, M. Z. and Rall, L. B., *Annotated bibliography on generalized inverses and applications*, Generalized Inverses and Applications, M. Z. Nashed, ed., Academic Press, New York, (1976), pp. 771-1041.

[3.1] Altman, M. A., *Inverse differentiability, contractors and equations in Banach spaces*, Studia Math. 46 (1973), pp. 1-15.

[3.2] Altman, M. A., *A generalization of Newton's method*, Bull. Acad. Polon. Sci. cl. III 3 (1955), pp. 189-193.

[3.3] Altman, M. A., *On the generalization of Newton's method*, Bull. Acad. Polon. Sci. cl. III 5 (1957), pp. 789-795.

[3.4] Altman, M. A., *Contractors and Contractor Directions: Theory and Applications*, Marcel Dekker, New York, 1977.

[3.5] Bartle, R. G., *On the openness and inversion of differentiable mappings*, Ann. Acad. Scient. Fennicae A.I. 257, 1958, 8 pp.

[3.6] Bartle, R. G., *Newton's method in Banach spaces*, Proc. Amer. Math. Soc. 6 (1955), pp. 827-831

[3.7] Ben-Israel, A., *A Newton-Raphson method for the solution of systems of equations*, J. Math. Anal. Appl. 15 (1966), pp. 243-252.

[3.8] Ben-Israel, A., *On applications of generalized inverses in nonlinear analysis*, Theory and Applications of Generalized Inverses of Matrices, T. L. Boullion and P. L. Odell, eds., Texas Tech Math. Series, No. 4, (1968), pp. 183-202.

[3.9] Ben-Israel, A., and Greville, T. N. E., *Generalized Inverses: Theory and Applications*, Wiley-Interscience, New York, 1974.

[3.10] Berger, M. S., *Nonlinearity and Functional Analysis*, Academic Press, New York, 1977.

[3.11] Cesari, L., *The implicit function theorem in functional analysis*, Duke Math. J. 33 (1966), pp. 417-440.

[3.12] Dieudonné, J., *Sur les homomorphismes d'espaces normés*, Bull. Sci. Math., Sér. 2, 67 (1943), pp. 72-84.

[3.13] Dieudonné, J., *Foundations of Modern Analysis*, Academic Press, New York, 1960.

[3.14] Ehrmann, H. H., *On implicit function theorems and the existence of solutions of nonlinear equations*, Enseignement Math. 9 (1963), pp. 129-176.

[3.15] Graves, L. M., *Some mapping theorems*, Duke Math. J. 17 (1950), pp. 111-114.

[3.16] Graves, L. M., *Implicit functions and differential equations in general analysis*, Trans. Amer. Math. Soc. 29 (1927), pp. 514-552.

[3.17] Graves, L. M., *Nonlinear mappings between Banach spaces*, Studies in Real and Complex Analysis, I. I. Hirschman, Jr. ed., Math. Assoc. of America, 1965, pp. 34-54.

[3.18] Hildebrandt, T. H. and Graves, L. M., *Implicit functions and their differentials in general analysis*, Trans. Amer. Math. Soc. 29 (1927), pp. 127-153.

[3.19] Kantorovich, L. V. and Akilov, G. P., *Functional Analysis in Normed Spaces*, Pergamon Press, New York, 1964.

[3.20] Krasnoselskii, M. A., Vainikko, G. M., Zabreiko, P. P., Rutickii, Ja. B., and Stecenko, V. Ja., *Approximate Solutions of Operator Equations*, "Nauka," Moscos, 1969; Engl. Transl., Wolters-Noordhoff, Groningen, 1972.

[3.21] Lang, S., *Differentiable Manifolds*, Addison-Wesley, Reading, Mass., 1972.

[3.22] Leach, E. B., *A note on inverse function theorems*, Proc. Amer. Math. Soc. 12 (1961), pp. 694-697.

[3.23] Liusternik, L. A. and Sobolev, V. J., *Elements of Functional Analysis*, Ungar, New York, 1961.

[3.24] Moser, J., *A technique for the construction of solutions of nonlinear differential equations*, Proc. National Academy of Sciences 47 (1961), pp. 1824-1831.

[3.25] Moser, J., *A rapidly convergent iteration method and nonlinear partial differential equations, I.* Ann. Scuola Norm. Sup. Pisa 20 (1966), pp. 226-315; *II.*, ibid, pp. 449-535.

[3.26] Nash, J., *The imbedding problem for Riemannian manifolds*, Annals of Mathematics, 63 (1956), pp. 20-63.

[3.27] Nashed, M. Z., *Aspects of generalized inverses in analysis and regularization*, Generalized Inverses and Applications, M. Z. Nashed, ed., Academic Press, New York, (1976), pp. 193-244.

[3.28] Nashed, M. Z., *The scope and limitations of generalized inverse applications in nonlinear analysis and optimization*, J. Nonlinear Analysis, to appear.

[3.29] Nevalinna, F., *Über die Umkehrung differenzienzierbarer Abbildungen*, Ann. Acad. Scient. Fennicae A I 245 (1957), 14 pp.

[3.30] Nevalinna, R., *Über die Umkehrung differenzierbarer Abbildungen*, Ann. Acad. Scient. Fennicae A I 185 (1955), 12 pp.

[3.31] Nijenhuis, A., *Strong derivatives and inverse mappings*, Amer. Math. Monthly 81 (1974), pp. 969-981.

[3.32] Ortega, J. M. and Rheinboldt, W. C., *Iterative Solution of Nonlinear Equations in Several Variables*, Academic Press, New York, 1970.

[3.33] Rheinboldt, W. C., *A unified convergence theory for a class of iterative processes*, SIAM J. Numer. Anal. 5 (1968), pp. 42-63.

[3.34] Rheinboldt, W. C., *Local mapping relations and global implicit function theorems*, Trans. Amer. Math. Soc. 138 (1969), pp. 183-198.

[3.35] Sard, A., *The measure of critical values of differentiable maps*, Bull. Amer. Math. Soc. 48 (1942), pp. 883-890.

[3.36] Schwartz, J., *On Nash's implicit function theorem*, Comm. Pure Appl. Math. 13 (1960), pp. 509-530.

[3.37] Schwartz, J., *Nonlinear Functional Analysis*, Gordon and Breach, New York, 1969.

[3.38] Smale, S., *An infinite dimensional version of Sard's theorem*, Amer. J. Math. 87 (1965), pp. 861-867.

[3.39] Zehnder, E. J., *A remark on Newton's method*, Comm. Pure Appl. Math. 27 (1974), pp. 361-366.

ITERATION FOR SYSTEMS OF NONLINEAR PARTIAL DIFFERENTIAL EQUATIONS

J. W. Neuberger
North Texas State University

I. STATEMENT OF MAIN LEMMA

Denote by H a real Hilbert space and by L a strongly continuous function on H so that $L(x)$ is an orthogonal projection on H for all $x \in H$.

Lemma. Suppose $0 < c \leqslant 1$, $W \in H$, P is an orthogonal projection on H, $Q_0 = P$, $Q_{n+1} = Q_n - cQ_n L(Q_n W) Q_n$, $n = 0, 1, \ldots$. Then $[Q_n W]_{n=0}^{\infty}$ converges to $Z \in H$ so that $PZ = Z$ and $L(Z)Z = 0$.

This lemma is used to obtain solutions to a variety of nonlinear problems. The choice of $c = 1$ seems to give most rapid convergence but other choices seem of interest too.

The next section is intended to illustrate how this lemma may be applied to a simple class of partial differential equations. More extensive applications are in section 3. A related numerical method is given in section 4.

This work represents a considerable improvement over [6] in that square roots of operators are not used. This is especially beneficial in numerical computations.

II. A SIMPLE APPLICATION

Suppose Ω is an open subset of R^2. Denote by H_0 the space of triples (f,g,h) so that f,g,h are real-valued continuous functions on Ω so that $\|(f,g,h)\| \equiv (\int_\Omega (f^2 + g^2 + h^2))^{1/2}$

ISBN 0-12-434160-8

exists. Denote by H_0' the subspace of H_0 consisting of all elements of H_0 of the form (u,u_1,u_2) where u_1,u_2 represent partial derivatives of u in the first and second places respectively. Denote by H a completion of H_0 relative to the given norm and denote by H' the closure of H_0' in H. This is just a way to introduce L_2 - generalized derivatives $(cf[1])$. Denote by P the orthogonal projection of H onto H'. A simple variational argument shows that if $g_1,h_2 \in L_2(\Omega)$ and Ω is bounded with a smooth boundary $\partial\Omega$, then $P(f,g,h) = (u,u_1,u_2)$ so that $u - \Delta u = f - g_1 - h_2$ on Ω, $du/dn = \langle (g,h), n \rangle$ on $\partial\Omega$ where $n(p)$ is the outward normal at p for all $p \in \partial\Omega$.

Consider now the problem of finding u on Ω so that

$$(*) \quad r(u,u_1,u_2)u_1 + s(u,u_1,u_2)u_2 = 0$$

where r,s are continuous real-valued functions on R^3 so that $r(z)^2 + s(z)^2 > 0$ for all $z \in R^3$. Define for each $F = (f,g,h) \in H$, $A(F)\colon H \to L_2(\Omega)$ so that

$$A(f,g,h)(a,b,c) = r(f,g,h)b + s(f,g,h)c \quad \text{for all} \quad (a,b,c) \in H.$$

Then $A(f,g,h)^*z = (0,r(f,g,h)z, s(f,g,h)z)$ for all $z \in L_2(\Omega)$. It may be seen that $A(F)A(F)^*\colon L_2(\Omega) \to L_2(\Omega)$ is invertible for all $F \in H$ and if $L(F) = A(F)^*(A(F)A(F)^*)^{-1}A(F)$ for all $F \in H$, then $L(F)$ is the orthogonal projection of H onto the orthogonal complement of the null space of $A(F)$. In particular, if $F,G \in H$, then $L(F)G = 0$ if and only if $A(F)G = 0$.

Theorem 1. With P and L defined as above, the hypothesis of the lemma is satisfied.

Hence given any $W = (f,g,h) \in H$, an application of the lemma yields $Z \in H$ so that $PZ = Z$ and $L(Z)Z = 0$. But $PZ = Z$ implies that Z is of the form (u,u_1,u_2) for some $u \in L_2(\Omega)$ and so $L(Z)Z = 0$ implies $A(Z)Z = 0$, i.e.,

$$r(u,u_1,u_2)u_1 + s(u,u_1,u_2)u_2 = 0.$$

A variation on this development is achieved by taking $H = L_{2,\alpha}(\Omega) \times L_{2,\beta}(\Omega) \times L_{2,\gamma}(\Omega)$ where α,β,γ are measures on Ω

chosen absolutely continuous relative to Lebesgue measure so that $\delta(E) > 0$ if E is an open subset of Ω, $\delta = \alpha,\beta,\gamma$. Such a change gives a different projection P and also alters the definition of $A(F)^*$, $F \in H$, and hence the resulting definition of L. Computational evidence suggests, for example, if $\beta = \gamma =$ (Lebesgue measure) and α is heavily weighted on an arc Φ designated as a 'boundary' then the values of f in the 'initial estimate' $W = (f,g,h)$ 'tend' to be preserved under the iteration provided there is a solution u to $(*)$ which agrees with f on Φ. Some of the nature of this numerical evidence is perhaps suggested by section 4.

III. APPLICATION TO CONSERVATION SYSTEMS

Here it is pointed out that the lemma may be applied to a variety of conservation systems. For simplicity we consider two equations in three dimensions (time and two space dimensions). Some references which give an idea of the importance of conservation equations are [2], [3], [4], [5].

Take H_0 to be the space of all continuous F from Ω to $R^2 \times R^6$ which are square integrable on Ω, an open subset of R^3. Take H to be a completion of H_0 under the L_2 norm. Take H_0' to be all $F \in H_0$ of the form (u,u') where $u\colon \Omega \to R^2$ and u' is the derivative of u (here $u'(x)$ is considered to be a vector in R^6, $x \in \Omega$).

For each of $\alpha,\beta,\gamma,\delta$ a real-valued $C^{(1)}$ function on all of R^2, consider the problem of finding real-valued functions v,w on Ω so that

$$v_1 + D_2(\alpha(v,w)) + D_3(\beta(v,w)) = 0$$

$$w_1 + D_2(\gamma(v,w)) + D_3(\delta(v,w)) = 0$$

where the subscript i denoted differentiation in the i-th place, $i = 1, 2, 3$.

After differentiation, one has

$$(v_1 + \alpha_1(v,w)v_2 + \beta_1(v,w)v_3) + (\alpha_2(v,w)w_2 + \beta_2(v,w)w_3) = 0$$

$$(\gamma_1(v,w)v_2 + \delta_1(v,w)v_3) + (w_1 + \gamma_2(v,w)w_2 + \delta_2(v,w)w_3) = 0.$$

Write $\binom{v}{w}$ as u and for $(t,x,y) \in \Omega$, $u'(t,x,y)$ is the member of $L(R^3,R^2) \sim R^6$ represented by the matrix

$$\begin{pmatrix} v_1(t,x,y) & v_2(t,x,y) & v_3(t,x,y) \\ w_1(t,x,y) & w_2(t,x,y) & w_3(t,x,y) \end{pmatrix}.$$

Define $B\colon\ R^2 \times R^6 \to L(R^2 \times L(R^3,R^2), R^2)$ so that if $(p,q) \in R^2$, $r,s,t \in R^2$,

$$B(p,q,r,s,t)\left(\begin{pmatrix} (m,n) \\ \begin{pmatrix} a & b & c \\ d & e & f \end{pmatrix} \end{pmatrix}\right) = \begin{pmatrix} a + \alpha_1(p,q)b + \beta_1(p,q)c + \alpha_2(p,q)e + \beta_2(p,q)f \\ \gamma_1(p,q)b + \delta_1(p,q)c + d + \gamma_2(p,q)e + \delta_2(p,q)f \end{pmatrix}.$$

Hence the system may be written

$$B(u,u')\binom{u}{u'} = 0 \quad .$$

Examination shows that

$$B(p,q,r,s,t)B(p,q,r,s,t)^* = I + CC^* \qquad \text{where}$$

$$C = \begin{pmatrix} \alpha_1(p,q) & \beta_1(p,q) & \alpha_2(p,q) & \beta_2(p,q) \\ \gamma_1(p,q) & \delta_1(p,q) & \gamma_2(p,q) & \delta_2(p,q) \end{pmatrix}$$

Clearly $I + CC^*$ is invertible. Now define A with domain H so that $(A(z))(x) = (B(z(x))B(z(x))^*)^{-1/2}B(z(x))$, $z \in H$, $x \in \Omega$. The system may now be written

$$A(u,u')\binom{u}{u'} = 0$$

and A satisfies $A(z)A(z)^* = I$ for all $z \in H$.

For $L(z) = A(z)^*A(z)$, $z \in H$ and P the orthogonal projection of H onto H', it is asserted that the hypothesis of the lemma is satisfied. A proof is similar to one for Theorem 1.

Observe that much more general systems may be dealt with using the lemma together with ideas from this section.

IV. NUMERICAL APPROXIMATIONS

Suppose each of m and k is a positive integer, Ω is a bounded open subset of R^m and G_0 is a finite rectangular grid with uniform spacing δ between adjacent points. Assume G_0 intersects Ω in such a way that if $p \in G \equiv G_0 \cap \Omega$, then for $i \in [1, \ldots, m]$, either $p + \delta e_i$ or $p - \delta e_i$ is in G where $e_1, \ldots, e_m$ denotes the standard basis for R^m.

Denote by K the vector space of all R^k - valued functions on G. If $u \in K$ and $i \in [1, \ldots, m]$, define

$$(D_i u)(p) = \begin{cases} (1/2\delta)(u(p+\delta e_i) - u(p-\delta e_i)) & \text{if } p+\delta e_i,\ p-\delta e_i \in G \\ (1/\delta)(u(p+\delta e_i) - u(p)) & \text{if } p+\delta e_i \in G,\ p-\delta e_i \notin G \\ (1/\delta)(u(p) - u(p-\delta e_i)) & \text{if } p-\delta e_i\ \ g,\ p+\delta e_i \notin G. \end{cases}$$

Take $D_0 u = u$, $u \in K$. Define $H = K^{m+1}$. For $u \in K$ and $Z = (z_0, z_1, \ldots, z_m) \in H$, define $\|u\| = (\Sigma_{r\in G}\|u(r)\|^2)^{1/2}$ and $\|Z\| = (\Sigma_{i=0}^m \|z_i\|^2)^{1/2}$.

For $i \in [1, \ldots, k]$, denote by G_i a subset of G. For $u \in K$, Du denotes $(D_0 u, D_1 u, \ldots, D_m u) \in H$.

For $i \in [1, \ldots, k]$, define $P_i : K \to K$ so that for $u \in K$ and $r \in G$, $(P_i u)(r) = 0 \in R^k$ if $r \notin G_i$, $(P_i u)(r) = (y_1, \ldots, y_k) \in R^k$ if $r \in G_i$ where $y_s = 0$, $s \neq i$, $y_s = (u(r))_s$, $s = i$, for $s = 1, \ldots, k$. Denote $\Sigma_{i=1}^k P_i$ by T and, if $(x_0, x_1, \ldots, x_m) \in H$, take $M(x_0, x_1, \ldots, x_m) = (Tx_0, 0, \ldots, 0) \in H$. Note that each of $P_1, \ldots, P_k, T, M$ is an orthogonal projection.

Suppose E is a finite dimensional inner product space and A is a continuous function from H to $L(H,E)$. We seek solutions $u \in K$ of

$$(**) \quad \begin{cases} A(Du)Du = 0 \\ Tu = Tw \end{cases}$$

where w is some given element of K.

The equation $A(Du)Du = 0$ is intended to represent a finite difference approximation to a system of nonlinear partial differential equations (or more generally a system of functional partial differential equations - see [7] for some linear functional equations which fit the present theory if the right sides are set equal to zero). The expression $Tu = Tw$ represents Dirichlet boundary conditions for u - it says that certain components of u must match corresponding components of w at certain points of G.

We now describe a sufficient condition for an iteration process based upon the lemma to converge to a solution u of (**) given $w \in K$. For each positive integer j denote by H_j the inner product space whose points are those of H but with $\|Z\|_j \equiv (\|Z\|^2 + (j - 1)\|MZ\|^2)^{1/2}$ for $Z \in H_j$. Note that if $Z, X \in H_j$, then $< Z.X >_j = < Z,X > + (j - 1) < MZ,X >$. Denote by $P^{(j)j}$ the orthogonal projection (relative to $\| \|_j$) of H_j onto $R(D)$. In the proof of Theorem 3 an expression for $P^{(j)j}$ is given.

For j a positive integer and $F \in H_j$, denote by $A(F)^{*j}$ the adjoint of $A(F)$ regarded as a transformation from H_j to K:

$$< A(F)X,y > = < X,A(F)^{*j}y >_j \quad \text{for all} \quad X \in H_j,\ y \in K \quad \text{where}$$

$< , >$ here denotes inner product in K and $< , >_j$ denotes inner product in H_j.

Denote by π the transformation on H to K (and on H_j, $j = 1, 2, \ldots$) so that $\pi(x_0,x_1, \ldots, x_m) = x_0$ for all $(x_0,x_1, \ldots, x_m) \in H$. For each $F \in H$, denote by $A_0(F)$ the restriction of $A(F)$ to the kernel of π. Condition B is to denote the proposition that

$$(A_0(F)A_0(F)^*)^{-1} \text{ exists for all } F \in H.$$

Note that condition B implies that $(A(F)A(F)^{*j})^{-1}$ exists for all $F \in H$, $j = 1, 2, \ldots$ since $A(F)A(F)^{*j} \geqslant A(F)A(F)^{*(j+1)} \geqslant A_0(F)A_0(F)^*$, $j = 1, 2, \ldots$ Define $A_1(F)$ to be the restriction of $A(F)$ to the range of $I - M$. Then

$A(F)A(F)^{*j} \geqslant A(F)A(F)^{*}$ and $\lim_{j\to\infty}(A(F)A(F)^{*j})^{-1} = (A(F)A(F)^{*})^{-1}$.

The following is an immediate consequence of the lemma:

Theorem 2. Suppose condition B holds, $0 < c \leqslant 1$, $W \in H$, j is a positive integer, $Q_{0,j} \equiv P^{(j)}$ and

$L_j(F) \equiv A(F)^{*j}(A(F)A(F)^{*j})^{-1}A(F)$ for all $F \in H_j$. Suppose also that $Q_{n+1,j} \equiv Q_{n,j} - cQ_{n,j}L(Q_{n,j}W)Q_{n,j}$, $n = 0, 1, \ldots$. Then $[Q_{n,j}W]_{n=0}^{\infty}$ converges to $Z_j \in H_j$ such that $A(Z_j)Z_j = 0$, $P^{(j)}Z_j = Z_j$.

Note that since H_j is a finite dimensional space it follows that $\lim_{n\to\infty}Q_{n,j}$ exists, $j = 1, 2, \ldots$.

Fix $w \in K$. Take $W \equiv Dw$. For this $W \in H$ choose $[[Q_{n,j}]_{n=0}^{\infty}]_{j=1}^{\infty}$ as in Theorem 2. Denote $\pi Q_{n,j}D$ by $R_{n,j}$, $n = 0, 1, \ldots, j = 1, 2, \ldots$.

Theorem 3. $[R_{n,j}w]_{n=0}$ converges, $j = 1, 2, \ldots$. If this convergence is uniform for $j = 1, 2, \ldots$, and z_j denotes $\lim_{n\to\infty}R_{n,j}w$, $j = 1, 2, \ldots$, then $[z_j]_{j=1}^{\infty}$ converges to $z \in K$ such that

$$A(Dz)Dz = 0 \quad \text{and} \quad Tz = Tw.$$

V. PROOFS

Proof of Lemma. Clearly $0 \leqslant Q_0 \leqslant I$ and Q_0 is symmetric. By induction one has that each of $Q_1, Q_2, \ldots$ is also symmetric. Moreover $\langle Q_{n+1}x,x\rangle = \langle Q_nx,x\rangle - c\langle Q_nL(Q_nW)Q_nx,x\rangle = \langle Q_nx,x\rangle - c\|L(Q_nW)Q_nx\|^2$ for all $x \in H$ since $L(Q_nW)$ is an orthogonal projection. Therefore $Q_{n+1} \leqslant Q_n$, $n = 1, 2, \ldots$. In particular $Q_n \leqslant Q_0 = P \leqslant I$, $n = 0, 1, \ldots$. But

$$\langle Q_nx,x\rangle - c\langle L(Q_nW)Q_nx,Q_nx\rangle \geqslant \langle Q_nx,x\rangle - \langle Q_nx,Q_nx\rangle \geqslant$$

$$\langle Q_nx,x\rangle - \langle Q_nx,x\rangle \geqslant 0, \quad \text{since } Q_n \leqslant I,\ n = 0, 1, \ldots .$$

Hence by induction $Q_n \geqslant 0$, $n = 0, 1, \ldots$. Therefore $[Q_n]_{n=0}^{\infty}$

converges strongly on H to some symmetric, nonnegative continuous linear transformation Q. In particular, $[Q_n W]_{n=0}^{\infty}$ converges to $Z \equiv QW$. So $< QZ, Z > = \lim_{n\to\infty} < Q_{n+1}W, W > = \lim_{n\to\infty} < Q_n W, W >$ $- c\lim_{n\to\infty} < L(Q_n W)Q_n W, Q_n W > = < QZ, Z > - c < L(Z)Z, Z >$ since L is strongly continuous. Hence $L(Z)Z = 0$ since $L(Z)$ is an orthogonal projection. Note that $PQ_0 = Q_0$. By induction, $PQ_n = Q_n$, $n = 0, 1, \ldots$ and so $PZ = P(QW) = \lim_{n\to\infty} PQ_n W$ $= \lim_{n\to\infty} Q_n W = Z$. Hence $PZ = Z$ and the proof is complete.

Indication of proof of Theorem 1. P is an orthogonal projection on H. For each $F \in H$, $L(F)$ is an orthogonal projection on H since $L(F)$ is both symmetric and idempotent. That L is strongly continuous follows from a vector-valued version of the following:

Suppose S is an open subset of R, $y \in L_2(S)$, and g is a bounded continuous function from R to R. If $f_1, f_2, \ldots$ is a sequence convergent in $L_2(S)$ to $f \in L_2(S)$, then

$$\lim_{n\to\infty} \int_S y^2 \cdot (g(f_n) - g(f))^2 = 0.$$

Proof of Theorem 3. That $\lim_{n\to\infty} R_{n,j}$ exists follows from the fact that $\lim_{n\to\infty} Q_{n,j}$ exists and that $R_{n,j} = \pi Q_{n,j} D$, $n = 0, 1, \ldots, j = 1, 2, \ldots$. Now for $x \in K$, $Y \in H$, $< D^{*j} Y, x > = < Y, Dx >_j = < Y, Dx > + (j - 1) < MY, Dx >$ $= < D^* Y, x > + (j - 1) < D^* MY, x >$. Therefore $D^{*j} = D^* + (j - 1)D^* M$ and so $D^{*j} D = D^* D + (j - 1)D^* MD$. Therefore $(D^{*j} D)^{-1}$ exists since $D^{*j} D \geqslant D^* D = I + D^* D + \ldots + D_m^* D_m$ and so $D^{*j} D > 0$. It follows that

$$P^{(j)} = D(D^{*j} D)^{-1} D^{*j}$$

since this last expression is symmetric (relative to the inner product for $\| \|_j$), is idempotent, is fixed on elements of the form Dz, $z \in K$, and $R(D(D^{*j} D)^{-1} D^{*j}) \subset R(P^{(j)})$.

Using a slight variation of Theorem 1 of [6].

$$\lim_{j\to\infty} (D^{*j} D)^{-1} = J$$

so that $TJ = JT = 0$.

Note that $R_{0,j} = \pi Q_{0,j} D = I$, $j = 1, 2, \ldots$ so that $\lim_{j\to\infty} R_{0,j}$ exists. Suppose n is a nonnegative integer and $R_n \equiv \lim_{j\to\infty} R_{n,j}$ exists. Define $B_j(Z) = (A(Z)Z(Z)^{*j})^{-1/2}A(Z)$, $Z \in H_j$. Now $Q_{n+1,j} = Q_{n,j}$

$$- cQ_{n,j}A(Q_{n,j}W)^{*j}(A(Q_{n,j}W)A(Q_{n,j}W)^{*j})^{-1}(Q_{n,j}W)Q_{n,j} \quad \text{and}$$

$Q_{n,j} = P^{(j)}Q_{n,j}$ so $R_{n+1,j} = \pi Q_{n+1,j}D = R_{n,j}$

$$- c\pi D(D^{*j}D)^{-1}D^{*j}Q_{n,j}A(Q_{n,j}W)^{*j}(A(Q_{n,j}W)A(Q_{n,j}W)^{*j})^{-1}A(Q_{n,j}W)$$

$$Q_{n,j}D = R_{n,j} - c(D^{*j}D)^{-1}(B_j(DR_{n,j}w)DR_{n,j})^*(B_j(DR_{n,j}w)DR_{n,j})$$

using the fact that $DR_{n,j} = Q_{n,j}D$, $j = 1, 2, \ldots$.

Hence for $S_n \equiv (A_1(DR_n w)A_1(DR_n w)^*)^{-1/2}$,

$R_{n+1,j} \to R_n - cJ(S_n A(DR_n w)DR_n)^*(S_n A(DR_n w)DR_n)$ as $j \to \infty$. Hence $\lim_{j\to\infty} R_{n+1,j}$ exists.

Since $TJ = 0$, it follows by induction that $TR_{n+1} = TR_n$, $n = 0, 1, \ldots$ and so $TR_n w = TR_0 w = Tw$, $n = 0, 1, \ldots$.

Denote $R_n w$ by z_n, $n = 0, 1, \ldots$. Since $[R_{n,j}w]_{n=0}^{\infty}$ converges uniformly for $j = 1, 2, \ldots$, it follows that $z_0, z_1, \ldots$ converges to some $z \in K$. Since, as noted above, $Tz_n = TR_n w = Tw$, $n = 0, 1, \ldots$, it follows that $Tz = Tw$. This is the second part of the conclusion to the theorem.

Denote $\lim_{n\to\infty} R_{n,j}w$ by y_j, $j = 1, 2, \ldots$. It follows that $y_1, y_2, \ldots$ converges to z. Since $A(Dy_j)Dy_j = 0$, $j = 1, 2, \ldots$, it must be that

$$A(Dz)Dz = 0.$$

This completes an argument for the theorem.

VI. CONCLUSION

The procedure outlined in section 4 has been used for a number of computer runs although many details regarding the relationship of the discrete process of section 4 and the continuous processes of sections 2 and 3 have yet to be explored. The following example sheds some light on the relationship, however.

Consider the linear problem of finding y on $[0,1]$ so that $y' = y$. Take $A\binom{f}{g} = 2^{-1/2}(g - f)$, $f,g \in L_2([0,1])$, and take $P\binom{f}{g}$ to be the nearest (in $L_2([0,1]) \times L_2([0,1])$) element of the closure of the set of all elements of the form $\binom{f}{f'}$ where $f \in C^{(1)}([0,1])$. Take $Lz = A^*Az$, $z \in L_2([0,1]$. Calculation yields that if W in the lemma is of the form $\binom{f}{f'}$, f' continuous, then the limit Z is given by

$$Z(x) = ((ef(1) - f(0))/(e^2 - 1))\binom{e^x}{e^x},\ x \in [0,1].$$

Compare this now with a numerical procedure based upon section 4. Take n a positive integer, K the vector space of all real-valued functions on $[0/n, 1/n, \ldots, n/n]$, $H_d = K \times K$, $\delta_n = 1/n$. For $u \in K$,

$$(Du)(p) = \begin{cases} (1/2\delta_n)(u(p+\delta_n) - u(p-\delta_n)), & p = 1/n, \ldots, (n-1)/n \\ (1/\delta_n)(u(1) - u(1-\delta_n)), & p = 1 \\ (1/\delta_n)(u(1/n) - u(0)), & p = 0 \end{cases}.$$

Take $A_d\binom{r}{s} = 2^{-1/2}(s - r)$, $r,s \in K$, $L \equiv A_d^*A_d$, $P \equiv D(D^*D)^{-1}D^*$. For this choice of P and L, the hypothesis of the lemma is satisfied. For $W_d \equiv \binom{r}{s}$, the lemma yields a limit $Z_d = \binom{u}{u}$ so that $Du = u$. Philip Walker (unpublished) has discovered that u must by 0 since the equation $Du = u$ has only the zero element of K for a solution! However, for $n = 100$, say, the above iteration seems to work nicely on a computer. Specifically, suppose $f \in C^{(1)}([0,1])$ is chosen, $r \in K$ is defined so that $r(i/n) = f(i/n)$, $i = 0, 1, \ldots, n$, $W \equiv \binom{f}{f'}$, $W_d \equiv \binom{r}{Dr}$ and Z, Z_d denote the respective limits for the continuous and discrete applications of the lemma. Denote by $(r_j, s_j) \in K \times K$ the j-th iterate of the discrete process and denote by k the first component of Z. Then for j in the range $10 - 20$, $r_j(i/n)$ and $k(i/n)$, $i = 0, 1, \ldots, n$ seem in good agreement for a reasonable choice of f in the first place. However, for j large enough, a perfect computer 'must' yield r_j close to 0 since the first component of the limit Z_d is 0. One 'explanation' is that the subspaces $[\binom{r}{r}: r \in K]$ and $[\binom{r}{Dr}: r \in K]$, even though they intersect only in the zero element of K, are sufficiently close that they appear to intersect in a one-dimensional space (to a computer). Try it.

This writer has tentatively concluded that it is not so much the *limits* of the discrete process which are of interest but rather some appropriate iterate. Such an appropriate stopping place has so far been easily 'recognized' from printouts of computations but a careful investigation seems to be in order regarding precise relationships between the continuous and discrete processes.

VII. REFERENCES

[1] Agmon, S., "*Lectures on Elliptic Boundary Value Problems*", D van Nostrand Co., 1965.

[2] Crandall, M. G., *The semigroup approach to first order quasi-linear equations in several space variables*, Israel J. Math 12 (1972), pp. 108-132.

[3] Dafermos, C. M., *The entropy rate admissibility criterion for solutions of hyperbolic conservation laws*, J. Differential Equations 14 (1973), pp. 202-212.

[4] Friedrichs, K. O., *On the laws of reativistic electro-magneto -fluid dynamics*, Comm. Pure Appl. Math. 27 (1974), pp. 749- .

[5] Lax, P., *Hyperbolic systems of conservations laws, II*, Comm. Pure Appl. Math. 10 (1957), pp. 537-566.

[6] Neuberger, J. W., *Boundary value problems for systems of nonlinear partial differential equations*, to appear, Springer Lecture Notes.

[7] ________, *Square integrable solutions to linear inhomogeneous systems*, to appear, J. Differential Equations.

EXISTENCE THEOREMS AND APPROXIMATIONS IN NONLINEAR ELASTICITY

J. T. Oden*
Texas Institute for Computational Mechanics
The University of Texas at Austin

I. INTRODUCTION

The mechanical foundations of the theory of elasticity began to be pieced together in the early nineteenth century. They emerged from the writings of Cauchy and Green during the period 1829-1843, and were further developed in the twentieth century by Rivlin, Truesdell, and others. The book by Wang and Truesdell [1] contains a readable account of the history and current status of the subject.

Despite its long and rich history, the mathematical theory of nonlinear elastostatics is very incomplete. There is, in fact, still disagreement as to what conditions should be imposed in order that one can formulate boundary-value problems which have physically reasonable solutions. Over twenty years ago Truesdell [2] referred to this situation as the "main unsolved problem" in the theory of elasticity, and a multitude of inequalities have been proposed as solutions to this problem (for a summary, see [1]). However, none of these has escaped criticism on either physical or mathematical grounds, so that the "main unsolved problem" seems to be still open.

In this paper, I will summarize some recent results on

* The support of this work by the National Science Foundation under Grant NSF-ENG-75-07846 is gratefully acknowledged.

ISBN 0-12-434160-8

existence theorems for a class of problems in elastostatics. Details can be found in [3, 4].

II. BOUNDARY-VALUE PROBLEMS IN ELASTOSTATICS

We consider the motion of a bounded material body $\bar{\Omega} \subset R^3$ relative to a fixed reference configuration. The position x of a particle $X \subset \Omega$ at time t is given by

$$x = \chi(X, t) \tag{2.1}$$

The function χ is differentiable with respect to X and t and the tensor

$$F = \nabla \chi \tag{2.2}$$

where ∇ is the material gradient, is called the deformation gradient tensor. We generally require that (2.1) be locally invertible and orientation preserving, which is guaranteed by the condition

$$\det F > 0 \tag{2.3}$$

We also use the notations

$$u = x - X = u(X, t) \quad ; \quad \nabla u = H = F - 1$$

$$G = 1 + H^T + H + H^T H \tag{2.4}$$

$$I_1 = tr\, G = I_1(H) \quad ; \quad I_2 = \frac{1}{2}(tr\, G)^2 - \frac{1}{2}\, tr\, G^2 = I_2(H)$$

$$I_3 = \det G = I_3(H)$$

Here u is the displacement vector, H the displacement gradient tensor, G the Green deformation tensor, and I_1, I_2, I_3 the principal invariants of G.

A theory of elasticity is obtained through the assumption that there exists a differentiable, frame-indifferent, function σ of H called the strain energy function, for which

$$T_R = \frac{\partial \sigma}{\partial H} \equiv Q(H, X) \tag{2.5}$$

where T_R is the first Piola Kirchhoff stress tensor. For isotropic, homogeneous materials (which we exclusively consider here), σ can be assumed to be given as a function of I_1, I_2, and I_3. Then the mechanical constitution of the material is described by

$$Q(H) = \frac{\partial\sigma(H)}{\partial I_1} \cdot \frac{\partial I_1}{\partial H} + \frac{\partial\sigma(H)}{\partial I_2} \cdot \frac{\partial I_2}{\partial H} + \frac{\partial\sigma(H)}{\partial I_3} \cdot \frac{\partial I_3}{\partial H} \tag{2.6}$$

The energy σ should also exhibit the singular behavior

$$\sigma \to \infty \quad \text{as} \quad \det F \to 0 \quad \forall\; X \in \Omega, \quad t \geqslant 0 \tag{2.7}$$

which physically means that an infinite energy results when a finite volume is compressed to zero.

The boundary value problem of place in elastostatics consists of seeking the displacement vector u such that

$$\begin{aligned} -\operatorname{Div} Q(\nabla u) &= \rho_0 f \quad &\text{in} \quad \Omega \\ u &= g \quad &\text{on} \quad \partial\Omega \end{aligned} \tag{2.8}$$

Here ρ_0 is the initial mass density, f the body force vector, and g the prescribed displacement on the boundary $\partial\Omega$ of Ω. We choose to replace (2.8) by the equivalent variational problem: find $u \in U$ such that

$$\int_\Omega Q(\nabla u) : \nabla v \, dX = \int_\Omega \rho_0 f \cdot v \, dX \quad \forall\; v \in U \tag{2.9}$$

where U is an appropriate space of admissible displacement vectors (here we take $u = 0$ on $\partial\Omega$).

Abstractly, we will have a reflexive Banach space U and (2.9) defines a formal operator A from U into the dual U' of U; i.e., if $\langle \cdot,\cdot \rangle$ denotes duality pairing on $U' \times U$, then

$$\langle A(u), v \rangle = \int_\Omega Q(\nabla u) : \nabla v \, dX; \quad \langle f, v \rangle = \int_\Omega \rho_0 f \cdot v \, dX \tag{2.10}$$

and (1.9) can be written

$$\langle A(u),v\rangle = \langle f,v\rangle \quad \forall \quad v \in U \tag{2.11}$$

We have here $U \hookrightarrow H = H' \hookrightarrow U'$, where H is a Hilbert (pivot) space.

III. AN EXISTENCE THEOREM

The following theorem is proved in [3].

Theorem 1. Let U and V be reflexive separable Banach spaces with $U \hookrightarrow V$ and the injection of U into V compact. Let A be an operator from U into U' such that

(i) A is bounded

(ii) A is hemicontinuous

(iii) $\exists$ a non-negative, continuous, real-valued function G with the property

$$\lim_{\theta\to 0} \frac{1}{\theta} G(x,\theta y) = 0 \qquad \forall \quad x,y \in \mathbb{R}^+ \tag{3.1}$$

such that $\forall\, u,v \in B_\mu(0)$ $\big(B_\mu(0)$ being the ball of radius $\mu > 0$ in U centered at the origin$\big)$,

$$\langle A(u) - A(v),\ u - v\rangle \geq -G(\mu,\ \|u - v\|_V) \tag{3.2}$$

Then A is pseudo-monotone (in the sense of Brezis [5] and/or Lions [6]).

Moreover, if, in addition to conditions (i), (ii), and (iii),

(iv) A is coercive, i.e., we have

$$\lim_{\|u\|_U\to\infty} \frac{\langle A(u),u\rangle}{\|u\|_U} = +\infty \tag{3.3}$$

Then A is surjective, i.e., $\forall\ f \in U'$ $\exists$ at least one $u \in U$ such that

$$A(u) = f \tag{3.4}$$

■

Conditions (ii) and (iv) are satisfied by operators encountered in elastostatics, and (i) holds for many important cases. Inequality (iii) appears to hold in several nontrivial cases. In general, the satisfaction of (iii) will depend on the existence of a nonlinear Gårding inequality of the form

$$\langle A(u) - A(v), u - v \rangle \geqslant F(\|u - v\|_U) - G(\mu, \|u - v\|_V) \tag{3.5}$$

where G satisfies the conditions of (iii) (particularly (3.1)) and

$$F(x) \geqslant 0 \qquad x \in \mathbb{R}^+ \tag{3.6}$$

We prepose conditions (3.3) and (3.6) as partial answers to "the main unsolved problem"; the answer is not complete because we have not adequately handled the constraint (2.3) or the singular behavior (2.7).

IV. A MODEL PROBLEM

As a test problem, we consider the case $\Omega \subset \mathbb{R}^2$,

$$\sigma = E_1(I_1 - 3) + E_2(I_1 - 3)^2 + E_3(I_2 - 3) + E_4(\sqrt{I_3} - 1) \tag{4.1}$$

where E_i = constants. We ignore (2.3) and do not include the singular behavior (2.7). Here we seek a vector field

$$u = \left\{ (u_1, u_2) \in \underset{\sim}{W}_0^{1,4}(\Omega) \quad ; \quad \underset{\sim}{W}_0^{1,4}(\Omega) = \left(W_0^{1,4}(\Omega) \right)^2 \right\} \tag{4.2}$$

with

$$\|u\|_{1,4,\Omega} = \left\{ \int_\Omega |\nabla u|^4 \, dX \right\}^{1/4} \quad ; \quad |\nabla u|^4 \equiv \sum_{i,j=1}^{2} |u_{i,j}|^4 \tag{4.3}$$

It is possible to prove the following (see [6])

<u>*Theorem 2.*</u> Let $A\colon \underset{\sim}{W}_0^{1,4}(\Omega) \to \underset{\sim}{W}^{-1,4/3}(\Omega)$ be characterized by

$$\langle A(u),v\rangle = \sum_{i,j=1}^{2} \int_\Omega \frac{\partial\sigma(u)}{\partial u_{i,j}} v_{i,j}\, dX$$

where $u_{i,j} \equiv \partial u_i / X^j$ are the generalized displacement gradients and σ is given by (4.1). Then

(i) A is bounded

(ii) A is hemicontinuous

(iii) A is coercive if $E_2 > 0$ and $E_3 \geqslant 0$.

Moreover, if $u,v \in B_\mu(0) \subset \underset{\sim}{W}_0^{1,4}(\Omega)$, then, $\forall\ \varepsilon > 0$, there exists a constant $\gamma = \gamma(\varepsilon,\mu)$ for which

$$\begin{aligned}\langle A(u) - A(v), u - v\rangle \geqslant (E_2 - \varepsilon)\, \|u - v\|^4_{1,4,\Omega} \\ + (E_1 + E_3 - 4E_2)\, \|u - v\|^2_{1,2,\Omega} \\ - \gamma(\varepsilon,\mu)\, \|u - v\|^{4/3}_{0,4,\Omega}\end{aligned} \tag{4.4}$$

□

The existence theorem is now obvious.

Theorem 3. Let A be defined as in Theorem 2, and let any of the conditions (i'), (ii'), or (iii') hold. Then there exists at least one solution to the problem

$$\langle A(u),v\rangle = \langle f,v\rangle \qquad \forall\ v \in \underset{\sim}{W}_0^{1,4}(\Omega) \tag{4.5}$$

where $\langle f,v\rangle$ is given by (2.10) if

(i) $E_2 > 0$, $E_3 \geqslant 0$ (coerciveness) and

(ii) $E_1 + E_3 - 4E_2 > 0$ (satisfiaction of (3.6)) (4.6)

■

Several remarks are in order:

1. This theory is apparently compatible with the linearized theory of elasticity. When the displacements are small, the first two terms on the right side of (4.4) dominate the third. The operator A is then strongly monotone and solutions are unique.

As the data is "increased," $\|u\|^{1}_{1,4,\Omega}$ and $\gamma(\varepsilon,\mu,M)$ increases until the right side of (4.4) changes from positive to negative. At this point a primary bifurcation is possible.

2. We have used an existence theorem to obtain conditions (specifically (4.6), on the form of the energy function σ.

3. We have not assumed strong ellipticity. Both Antman (e.g., [7]) and Ball [8] assume strong ellipticity in their work. However, Knowles and Sternberg [9,10] have recently shown that the strong ellipticity condition can be violated by physically reasonable solutions to the equations of plane elastostatics. In the linear theory, the existence of a Gårding inequality implies strong ellipticity for sufficiently smooth u. This does not seem to be the case for our nonlinear operator. If not, the framework of this theory could provide a means for handling those solutions which violate the strong ellipticity condition.

V. APPROXIMATIONS

In approximating (2.10), we introduce the usual finite-element spaces $S_h(\Omega)$ of piecewise polynomials of degree k. If h is the mesh parameter, then these spaces have the following well-known interpolation property: if $\underset{\sim}{w} \in \underset{\sim}{W}^{\ell,p}(\Omega)$, there exists a $\tilde{\underset{\sim}{w}}_h \in \left(S_h(\Omega)\right)^n$ such that

$$\|\underset{\sim}{w} - \tilde{\underset{\sim}{w}}_h\|_{,p} \leqslant C_3 h^{\mu} \|\underset{\sim}{w}\|_{\ell,p} \tag{5.1}$$
$$\mu = \min(k, \ell-1)$$

Here

$$\|w\|^{p}_{\ell,p} = \int_{\Omega} \sum_{i=1}^{n} \sum_{\alpha \leqslant \ell} \left|D^{\alpha} w_i\right|^{p} dX \tag{5.2}$$

If $\underset{\sim}{w}$ is the solution of (10), its finite element approximation $\underset{\sim}{w}_h \subset \mathring{\underset{\sim}{S}}_h(\Omega) = \left(S_h(\Omega)\right)^3 \cap \underset{\sim 0}{W}^{1,p}(\Omega)$ satisfies

$$\langle A(\underset{\sim}{w}_h),\ \underset{\sim}{v}_h\rangle = \int_\Omega \rho_0 \underset{\sim}{f} \cdot \underset{\sim}{v}_h \, dX \qquad \forall\ \underset{\sim}{v}_h \in \underset{\sim}{S}_h(\Omega) \tag{5.3}$$

and the orthogonality condition

$$\langle A(\underset{\sim}{w}_h) - A(\underset{\sim}{w}_h),\ \underset{\sim}{v}_h\rangle = 0 \qquad \forall\ \underset{\sim}{v}_h \in \overset{\circ}{\underset{\sim}{S}}_h(\Omega) \tag{5.4}$$

The error $\underset{\sim}{e}_h \equiv \underset{\sim}{w} - \underset{\sim}{w}_h$ satisfies the inequality

$$\begin{aligned} \|\underset{\sim}{e}_h\|_{1,p} &\leqslant \|\underset{\sim}{w} - \tilde{\underset{\sim}{w}}_h\|_{1,p} + \|\underset{\sim}{w}_h - \tilde{\underset{\sim}{w}}_h\|_{1,p} \\ &\leqslant C_3 h^\mu \|\underset{\sim}{w}\|_{\ell,p} + \|\underset{\sim}{w}_h - \underset{\sim}{w}_h\|_{1,p} \end{aligned} \tag{5.5}$$

Let $\underset{\sim}{E}_h \equiv \underset{\sim}{w}_h - \tilde{\underset{\sim}{w}}_h$. Then

$$\begin{aligned} \|\underset{\sim}{E}_h\|^p_{1,p} &\leqslant C_0^{-1} \langle A(\underset{\sim}{w}_h) - A(\tilde{\underset{\sim}{w}}_h),\ \underset{\sim}{w}_h - \underset{\sim}{w}_h\rangle + C_0^{-1}\,\gamma(\mu)\,\|E_h\|^{p'}_{0,p} \\ &= C_0^{-1} \langle A(\underset{\sim}{w}) - A(\underset{\sim}{w}_h),\ \underset{\sim}{w}_h - \underset{\sim}{w}_h\rangle + C_0^{-1}\,\gamma(\mu)\,\|\underset{\sim}{E}_h\|^{p'}_{0,p} \\ &\leqslant C_0^{-1} \|\underset{\sim}{E}_h\|_{1,p} \|\underset{\sim}{w} - \tilde{\underset{\sim}{w}}_h\|_{1,p}\, G(\underset{\sim}{w},\ \tilde{\underset{\sim}{w}}_h) + C_0^{-1}\,\gamma(\mu)\,\|\underset{\sim}{E}_h\|^{p'}_{0,p} \\ &\leqslant C_0^{-1} \|\underset{\sim}{E}_h\|_{1,p}\, C_3 h^\mu \|\underset{\sim}{w}\|_{\ell,p}\, G(\underset{\sim}{w},\ \underset{\sim}{w}_h) + C_0^{-1}\,\gamma(\mu)\,\|\underset{\sim}{E}_h\|^{p'}_{0,p} \end{aligned} \tag{5.6}$$

where $G(\underset{\sim}{w},\tilde{\underset{\sim}{w}}_h) = C_1\left(1 + \|\underset{\sim}{w}\|^{p-2}_{1,p} + \|\tilde{\underset{\sim}{w}}_h\|^{p-2}_{1,p}\right)^{\frac{p-2}{p}}$. Next we observe that $h \to 0$

$$G(\underset{\sim}{w},\ \underset{\sim}{w}_h) = G_0(\underset{\sim}{w}) + O(h^\mu)^{\frac{p-2}{p}}$$

$$G_0(\underset{\sim}{w}) = C_1\left(1 + 2\,\|\underset{\sim}{w}\|^{p-2}_{1,p}\right)$$

Also, by the Poincaré inequality,

$$\|\underset{\sim}{E}_h\|_{0,p} \leqslant C_4\,\|\underset{\sim}{E}_h\|_{1,p} \qquad C_4 > 0$$

Hence, as $h \to 0$ the previous inequalities combine to give

$$\|\underset{\sim}{E}_h\|^{p-1}_{1,p} \leqslant C_0^{-1}\, C_3 h^\mu\, \|\underset{\sim}{w}\|_{\ell,p}\, G_0(\underset{\sim}{w}) + C_0^{-1}\,\gamma(\mu)\, C_4\, \|\underset{\sim}{E}_h\|^{p'/p}_{1,p} \tag{5.7}$$

We reach, at this point, a problem not yet resolved. We must determine a number σ such that

$$C_5 x^{\sigma} \leqslant x^{p-1} - C_0^{-1}\,\gamma(\mu)\,C_4 x^{p'/p} \quad , \quad x \geqslant 0 \tag{5.8}$$

where C_5 is a constant greater than 0. If σ is known, we have

$$\|\underset{\sim}{E}_h\|_{1,p} \leqslant C_6 h^{\mu/\sigma}\, H(\underset{\sim}{w})$$

where

$$C_6 = (C_5 C_0)^{-1/\sigma} \quad , \quad H(\underset{\sim}{w}) = \left(\|\underset{\sim}{w}\|_{\ell,p}\, G_0(\underset{\sim}{w})\right)^{1/\sigma}$$

The final error estimate is then

$$\|e_h\|_{1,p} \leqslant C_7\left(h^{\mu}\,\|\underset{\sim}{w}\|_{\ell,p} + h^{\mu/\sigma}\, H(\underset{\sim}{w})\right) \tag{5.9}$$

Thus, if $\sigma > 0$, the method converges. However, this estimated rate of convergence is generally not optimal.

VI. REFERENCES

[1] Wang, C.-C. and Truesdell, C., *Introduction to Rational Elasticity*, Noordhoff international Publishing, Leyden, 1973.

[2] Truesdell, C., "*Das ungelöste Hauptproblem der endlichen Elastizität-scheorie*", Zietschrift für Angewandte Mathematik und Mechanik, 36, (1959), pp. 97-103. English translation, "The Main Unsolved Problem of the Theory of Finite Elasticity", reprinted in Continuum Mechanics III: Foundations of Elasticity Theory, edited by C. Truesdell, Gordon and Breach, New York, (1965), pp. 101-108.

[3] Oden, J. T., "*Existence Theorems for a Class of Problems in Nonlinear Elasticity*", J. Math. Anal. Appl. (to appear).

[4] Oden, J. T. and Reddy, C. T., "*Existence Theorems for a Class of Problems in Nonlinear Elasticity: A Model Problem in Finite Plane Strain*", TICOM Report 77-7 (in press), The University of Texas at Austin, 1977.

[5] Brezis, H., "*Équations et Inéquations Non Linéaires dans les Espaces Vectoriels en Dualité*", Ann. Inst. Fourier, Grenoble, 18 (1968), pp. 115-175.

[6] Lions, J. L., *Quelques Méthodes de Résolution des Problèmes aux Limites Non Linéaires*, Dunod, Paris, 1969.

[7] Antman, S. S., "*Ordinary Differential Equations of Non-Linear Elasticity II: Existence and Regularity Theory for Conservative Boundary Value Problems*", Arch. Rational Mech. Anal., 61 (1976), pp. 353-393.

[8] Ball, J. M., "*Convexity Conditions and Existence Theorems in Nonlinear Elasticity*", Arch. Rational Mech. and Anal., 63 (1977), pp. 337-403.

[9] Knowles, J. K. and Sternberg, E., "*On the Ellipticity of Nonlinear Elastostatics for a Special Material*", Journal of Elasticity, 5 (1975), pp. 341-361.

[10] Knowles, J. K. and Sternberg, E., "*On the Failure of Ellipticity of the Equations for Finite Elastostatic Plane Strain*", Arch. Rational Mech. Anal., 63 (1977), pp. 321-336.

EXISTENCE THEOREMS FOR SEMILINEAR ABSTRACT AND DIFFERENTIAL EQUATIONS WITH NONINVERTIBLE LINEAR PARTS AND NONCOMPACT PERTURBATIONS

W. V. Petryshyn*
Rutgers University

INTRODUCTION†

Let X and Y be real Banach spaces. In this paper we are concerned with the problem of the existence of a solution of

$$Lx + Nx = f \qquad (x \in X) \tag{1}$$

for a given f in Y, where $L \in L(X,Y)$ is Fredholm of index zero and $N\colon X \to Y$ is a nonlinear map of a particular type such that $N(x) = o(||x||)$ as $||x|| \to \infty$. It is known that many BV Problems for ordinary and partial differential equations can be formulated as abstract operator equations of the above type if X and Y are chosen judiciously.

If either N or the generalized inverse of L is compact, then one can treat such problems by using the Leray-Schauder degree theory or the Schauder fixed point theorem (see [12, 14]). When compactness is not available, which is often the case with differential operators when N depends on the highest order terms or the underlying domain in R^n is unbounded, the recent surjectivity theorems for monotone type operators or the degree theory

---

* Supported in part by the NSF Grant MCS-06352 A01.

† See Section 1 for the definitions of various notions and the precise statements of the results mentioned in the Introduction.

ISBN 0-12-434160-8

for condensing vector fields proved to be useful in treating certain classes of semilinear equations (see [3,9,14]). In this paper we show how the theory of mappings of A-proper type can be successfully used in treating the above existence problem under conditions which are weaker than those of other authors.

The purpose of this paper is two-fold. In Sections 1 and 2 we present some abstract existence results for (1) when $T = L + N$ is weakly A-proper and in Section 3 we apply them to the solvability of certain differential equations. Using the Brouwer degree theory and some properties of A-proper mappings from [24], it is first shown in Lemma 1.1 that if L is A-proper and injective, then (1) is solvable for each f in Y. It is then shown in Theorem 1.1 that if $N(L) \neq \{0\}$, then one can always construct a map $C \in L(X,Y)$ such that $L + \lambda C$ is A-proper and injective for each $\lambda \neq 0$, while $L + \lambda C + N$ remains weakly A-proper, and so for each $k \in \mathbb{N}$ the equation

(2) $$Lx_k + \lambda_k Cx_k + Nx_k = f \qquad (\lambda_k \to 0 \text{ as } k \to \infty)$$

has a solution $x_k \in X$ for each f in Y. Thus, if $\{x_k\}$ is bounded for some f in Y, then (1) is solvable provided $T(B)$ is closed whenever B is a closed ball in the space X. In Theorem 1.2 the boundedness of $\{x_k\}$ is established under various asymptotic positivity conditions which have their origin in the existence result of Landesman and Lazer [18] and which has since been extended in various directions by a number of authors (see [31,21,15,6,20,11,13,8,3,9,0] and others cited there).

It turns out that the class of weakly A-proper mappings $T = L + N$ is not only well suited for the problem at hand but it is also large enough so as to allow us to extend and unify the abstract results of [8] (including some in [20,11,13]) with some of [9] (and [3] if the linear part is bounded), which were obtained by these authors by using various types of arguments. Indeed, it is shown in Section 2 that if $L \in L(X,Y)$ is A-proper and $N\colon X \to Y$ is either compact, weakly continuous or of type (KM), then $T = L + N$ is weakly A-proper and $T(B)$ is closed provided that if N is of type (KM) some additional conditions hold. In view of this, Corollary 2.1 extends Theorems 1, 2 and 3 of DeFigueiredo [8], and some results in [20,11,13], while Theorems 1.1 and 1.2 extend some results of Brezis and Nirenberg [3] to weakly A-proper maps. It is also shown that if $L \in L(X,Y)$ is Fredholm of index 0 and $N(L + C)^{-1}$ is ball-condensing, then one can construct a special scheme $\tilde{\Gamma}$ for (X,Y) so that L and $L + N$ are A-proper w.r.t. $\tilde{\Gamma}$ and thus Propositions 2.4 and 2.5 of Fitzpatrick [9] also follow from Theorem 1.2. Now the author of [9] uses the degree theory for condensing vector fields to study (1) and therefore his results are also valid for set-condensing maps $F = N(L + C)^{-1}$. It should be added that Fitzpatrick indicated in [9] that, by using the generalized degree theory from [5], he will continue to study the solvability of (1) under the assumption that $I - F$ is P-compact. There is a hope that under this stronger assumption one might obtain the solvability of (1) under slightly weaker growth conditions. However, it should be emphasized that the degree argument for maps acting in Banach spaces cannot be used to study the solvability of (1) involving the more general class of weakly A-proper maps treated in this paper!

In case (1) is a differential BV Problem our conditions on $T = L + N$ allow N to depend on the highest order terms without the condition (used in [9]) that N is Lipschitzian with respect to them. Consequently, our existence results allow application to broader classes of differential BV Problems including those whose underlying domain in R^n is unbounded (see Problem 2).

In Section 3 we show what type of BV Problems for ODE's and PDE's can be treated by our existence results. Problems 1 and 4 in this section were first studied in [9]. We include them here to indicate the difference in assumptions for the respective abstract existence results to be applicable. Thus, in Problem 1 we establish the solvability in $X \subset W_2^2([0,1])$ of

$$(3) \qquad -u''(t) - g(t,u',u'') + f(x,u,u') = 0$$
$$u(0) = u(1), \ u'(0) = u'(1),$$

assuming that g is bounded and $(g(t,r,p) - g(t,r,q))(p - q) \geq -\alpha|p - q|^2$ for some $\alpha \in (0,1)$ and all $(t,r) \in [0,1] \times \mathbb{R}$ and $p, q \in \mathbb{R}$. This result improves on that of [9, Theorem 3.3] where it is assumed that $|g(t,r,p) - g(t,r,q)| \leq \alpha|p - q|$ and $g(t,r,s)$ is continuous in r uniformly in $(t,s) \in [0,1] \times \mathbb{R}$.

In Problem 2 the abstract result in modelled on situations which arise when one attempts to treat semilinear elliptic problems on unbounded domains in R^n. That is, we consider

$$(4) \qquad Au - \lambda u + Nu = f \qquad (u \in D(A), \ f \in H, \ \lambda \in \mathbb{R}),$$

where H is a Hilbert space, A is a densely defined, positive definite, self-adjoint operator whose essential spectrum $\sigma_e(A)$ is bounded below, and $N:D(A^{1/2}) \to H$ is such that $N = N_1 + N_2$ where $N_2:H_0 \to H$ is compact, $N_1:H_0 \to H$ is monotone and $||Nu||/||u||_0 \to 0$ as $||u||_0 \to \infty$, where H_0 is the completion of $D(A)$ in $||u||_0 = (Au,u)^{1/2}$. It is shown that when λ is an eigenvalue of A such that $|\lambda|\gamma^{-1} < 1$, where γ is determined by $\sigma_e(A)$, then (4) is solvable provided some asymptotic positivity condition holds. It seems that Problem 2 cannot be treated by any other known existence result.

Problem 3 considers the weak solution of the BV Problem

$$(5) \qquad \begin{cases} \sum_{|\alpha|, |\beta| \leq m} (-1)^{|\alpha|} D^\alpha (a_{\alpha\beta}(x) D^\beta u) + h(x,u(x)) = f(x) \\ \qquad\qquad\qquad\qquad (x \in Q \subset R^n, Q \text{ bounded}) \\ D^\alpha u = 0 \qquad (x \in \partial Q, \ |\alpha| \leq m - 1). \end{cases}$$

giving conditions on $\underline{h}(\pm\infty) = \liminf_{s\to\pm\infty} h(x,s)$ and $\overline{h}(\pm\infty) = \limsup_{s\to\pm\infty} h(x,s)$ ensuring the existence of a weak solution in $X = \overset{\circ}{W}{}_2^m$. The positivity condition we impose are the same as those in [0, Theorem 2.1] and they include the corresponding conditions in [18,31,20,9] and others (see [3,12,14] for an exhaustive list of references dealing with problem of type (5)).

In Problem 4 we consider the solvability of the equation

$$(6)\quad \begin{aligned} &-\Delta u(x) + g(x,\nabla u(x),\Delta u(x)) + h(u(x)) = f(x) && (x \in Q) \\ &\frac{\partial u}{\partial \eta}(x) = 0 && (x \in \partial Q) \end{aligned}$$

assuming that g satisfies the same condition as in Problem 1. Under the same assumption on h as in [9] we show that (6) has a solution in $X = \{u \in W_2^2 \mid \frac{\partial u}{\partial \eta} = 0\}$ for each f in L_2. In [9] Fitzpatrick assumes that $g(x,\nabla u,s)$ is Lipschitzian in s, continuous in ∇u uniformly in $(t,s) \in Q \times \mathbb{R}$, $Y = L_p$ with $p > n$ and $X = \{u \in W_2^p \mid \frac{\partial u}{\partial \eta} = 0\}$. These assumptions are essential for his abstract existence results to be applicable.

I wish to thank P. M. Fitzpatrick for helpful discussions

SECTION 1

Let $\{X_n\}$ and $\{E_n\}$ be sequences of oriented finite dimensional spaces with $X_n \subset X$, V_n the inclusion map of X_n into X, and W_n a linear map of Y onto E_n. We use the same symbol $||\cdot||$ to denote the norms in X, Y and E_n, and "$\to$" and "$\rightharpoonup$" to denote the strong and the weak convergence respectively.

Definition 1.1. The scheme $\Gamma = \{X_n, V_n; E_n, W_n\}$ is called <u>admissible</u> for (X,Y) provided that $\dim X_n = \dim E_n$ for each n, $\text{dist}(x,X_n) \to 0$ as $n \to \infty$ for each $x \in X$, and $\{W_n\}$ is uniformly bounded.

Note that E_n's are not assumed to be subspaces of Y or that $\{X_n\}$ be nested. For later references we describe three typical schemes

(a) *Injective scheme for (X,X^*).* If $E_n = X_n^*$ and $W_n = V_n^*$ then $\Gamma_I = \{X_n, V_n; X_n^*, V_n^*\}$ is admissible for (X,X^*).

(b) *Projective scheme for (X,Y).* If $E_n = Y_n \subset Y$ and $W_n = Q_n$, where $Q_n\colon\ Y \to Y_n$ is a projection such that $||Q_n|| \leq \beta$, then $\Gamma_p = \{X_n, V_n; Y_n, Q_n\}$ is admissible for (X,Y).

(c) *Projective scheme for (X,X).* If $E_n = X_n$ and $W_n = P_n$, where $P_n\colon\ X \to X_n$ is a projection such that $||P_n|| \leq \alpha$ for some $\alpha \geq 1$, then $\Gamma_\alpha = \{X_n, P_n\}$ is admissible for (X,X) and, in fact, Γ_α is projectionally complete in the sense that $P_n(x) \to x$ in X for each $x \in X$.

If X and Y are Hilbert spaces the projections are always assumed to be orthogonal. Example (a) shows that when X is separable then (X,X^*) always has an admissible scheme, while Example (c) shows that (X,X) has always a projectionally complete scheme if, for example, X has a Schauder basis.

It will be seen below that the following class of maps is not only well suited for the study of the existence of solutions of semilinear equations of the form (1) but it is also large enough so as to allow us to extend and unify the abstract existence results of [8,,11,13,20] with some of those in [9] obtained by these authors for special classes of maps satisfying different conditions and using various types of arguments. We also indicate the extension of some results in [2,3] when the linear part is bounded.

Definition 1.2. $T\colon X \to Y$ is said to be weakly A-proper (resp. A-proper) w.r.t. Γ if $T_n\colon X_n \to E_n$ is continuous for each n and if $\{x_{n_j} \mid x_{n_j} \in X_{n_j}\}$ is any bounded sequence such that $\|T_{n_j}(x_{n_j}) - W_{n_j} g\| \to 0$ as $j \to \infty$ for some g in Y, with $T_n = W_n T|_{X_n}$, then there exist a subsequence $\{x_{n_{j(k)}}\}$ and $x \in X$ such that $x_{n_{j(k)}} \rightharpoonup x \in X$ (resp. $x_{k_{j(k)}} \to x$ in X) and $Tx = g$.

The generality of the class of weakly A-proper maps is illustrated by the following examples (others and the definitions of some of the notions used here will be given later).

Example 1. All A-proper maps. This class includes compact and ball-condensing vector fields, maps of type (S) and (KS) as well as of strongly K-monotone type (see [4,24]). Extending a recent result in [29] it was shown independently in [26] and in [19] that the class of A-proper maps also includes the maps $T\colon X \to X$ of the form $T = I - F + N + C$, where F is ball-condensing, N accretive and continuous, and C is compact.

The interesting feature of the latter class is that neither the theory of accretive nor condensing type mappings is applicable to equations involving the above class of maps.

Example 2. Semibounded mappings $T\colon X \to X^*$ of type (M) when X is reflexive and, in particular, pseudo-monotone, semimonotone, monotone, and weakly continuous maps (see Lemma 2.1 below).

The following example (see Section 2) indicates the type of semilinear mappings which are weakly A-proper.

Example 3. Suppose X is reflexive, $L \in L(X,Y)$ is A-proper,

$N: X \to Y$ is of type (KM) and $C: X \to Y$ is completely continuous. Then $T = L + N + C$ is weakly A-proper provided $\phi(x) = (Lx, Kx)$ is weakly lower semicontinuous, where $K: X \to Y^*$ satisfies suitable conditions.

To state our first result we recall that a map $A \in L(X,Y)$ is called Fredholm if it has a finite dimensional null space $N(A)$ of dimension d and a closed range $R(A)$ of finte codimension $d^* = \dim N(A^*)$. The index of A is ind $(A) = d - d^*$. We say that a nonlinear map $N: X \to Y$ is <u>asymptotically zero</u> if $||Nx||/||x|| \to 0$ as $||x|| \to \infty$ or, equivalently, $N(x) = o(||x||)$ as $||x|| \to \infty$.

The proof of Theorem 1.1 below is based upon the following lemma which has an independent interest.

Lemma 1.1. <u>Suppose</u> $A \in L(X,Y)$ <u>is A-proper w.r.t.</u> Γ <u>and injective. If</u> $N: X \to Y$ <u>is asymptotically zero and</u> $T = A + N$ <u>is weakly A-proper w.r.t.</u> Γ, <u>then the equation</u>

$$(1.1) \qquad Ax + Nx = f$$

<u>has a solution for each</u> f <u>in</u> Y.

Proof: Since the proof of Lemma 1.1 is very similar to the proof of Theorem 2 in [25], we give here only the outline. Now to each $f \in Y$ there are $r_f > 0$ and $n_f \in N$ such that for all $t \in [0,1]$ and all $n \geq n_f$ we have

$$(1.2) \quad W_n A(x) + (1-t)W_n(Nx - f) \neq 0 \text{ for } x \in X_n \text{ with } ||x|| = r_f.$$

Indeed, if (1.2) were not true for some $f \in Y$, then there would exist sequences $\{x_{n_j} | x_{n_j} \in X_{n_j}\}$ and $t_{n_j} \in [0,1]$ such that

$$||x_{n_j}|| \to \infty \text{ as } j \to \infty \text{ and } W_{n_j} A(x_{n_j}) + (1 - t_{n_j}) W_{n_j} (Nx_{n_j} - f) = 0$$

for each j. Since $\{W_n\}$ is uniformly bounded and $||x_{n_j}|| \to \infty$, it follows from the last equation and our condition on N that $A_{n_j}(z_{n_j}) = (t_{n_j} - 1)W_{n_j}(Nx_{n_j} - f)/||x_{n_j}|| \to 0$ as $j \to \infty$, where

$z_{n_j} = x_{n_j}/||x_{n_j}||$. Since $\{z_{n_j} | z_{n_j} \in X_{n_j}\}$ is bounded and $A_{n_j}(z_{n_j}) \to 0$, the A-properness of A implies the existence of a subsequence $\{z_{n_{j(k)}}\}$ and $z \in X$ such that $z_{n_{j(k)}} \to z$ and $Az = 0$ with $||z|| = 1$, in contradiction to the condition that A is injective. Thus (1.2) holds for each $n \geq n_f$ and therefore, by the finite dimensional degree theory, there exists $x_n \in X_n \cap B(0,r_f)$ such that $T_n(x_n) - W_n f = 0$ for each $n \geq n_f$. In view of this and the weak A-properness of $T = A + N$, there exists a subsequence $\{x_{n_j}\}$ and $x \in X$ such that $x_{n_j} \rightharpoonup x$ and $Tx = f$. Q.E.D.

It is known that, even when $T = A + N$ is A-proper, Eq. (1.1) need not have a solution for each f in Y if $N(A) \neq \{0\}$. Extending a result of [16], it was shown in [25] that if ind $(A) = 0$, T is A-proper and $R(N) \subseteq R(A)$, then (1.1) is solvable if and only if $f \in R(A)$. However, it was shown in [7] that the condition "$R(N) \subseteq R(A)$" is very restrictive at least when applied to partial differential equations with sufficiently smooth coefficients.

Employing a perturbation method used in similar situations by other authors (e.g. [2,3,8,15]) and Lemma 1.1 we first prove a general existence Theorem 1.1 for (1) with $N(L) \neq \{0\}$ and then use it as a basis for obtaining existence results for the general class of weakly A-proper maps $T = L + N$ with N satisfying certain asymptotic positivity conditions. In what follows B will always denote an arbitrary closed ball in X.

Theorem 1.1. Suppose that $L \in L(X,Y)$ is A-proper w.r.t. Γ with ind$(L) = 0$ and $N\colon X \to Y$ is a mapping such that $T = L + N$ is weakly A-proper w.r.t. Γ, $N(x) = o(||x||)$ as $||x|| \to \infty$ and $T(B)$ is closed in Y for each $B \subset X$. Then we can construct a compact map $C \in L(X,Y)$ such that

$$Lx + Nx = f \qquad (x \in X, \quad f \in Y) \tag{1.3}$$

is solvable for a given f in Y provided.

(A) Either the set $S_k^+ \equiv \{x_k \in X \mid Lx_k + \lambda_k Cx_k + Nx_k - f = 0,\ \lambda_k > 0$ and $\lambda_k \to 0$ as $k \to \infty\}$ or the set $S_k^- \equiv \{x_k \in X \mid Lx_k + \lambda_k Cx_k + Nx_k - f = 0,$ $\lambda_k < 0$ and $\lambda_k \to 0$ as $k \to \infty\}$ is bounded by a constant independent of k.

Proof: It was shown by the author (see [24]) that since $L \in L(X,Y)$ is A-proper w.r.t. Γ, it follows that L is Fredholm with ind $(L) \geq 0$ and so by our hypotheses $d = d^*$. Hence there exists a closed subspace X_1 of X and a subspace Y_2 of Y with dim $Y_2 = d$ such that $X = N(L) \oplus X_1$, $L(X_1) = R(L)$, L is injective on X_1, and $Y = Y_2 \oplus R(L)$. Let M be an isomorphism of $N(L)$ onto Y_2, let P be the linear projection of X onto $N(L)$, and let L_λ be a linear mapping of X into Y defined by $L_\lambda = L + \lambda C$, where $C = MP$ and $\lambda \neq 0$. Since L_λ is A-proper and C is compact, L_λ is A-proper w.r.t. Γ for each $\lambda \neq 0$. Moreover, L_λ is injective for each $\lambda \neq 0$. Indeed, if $L_\lambda(x) = Lx + \lambda Cx = 0$, then $Lx = -\lambda Cx$ and since $\lambda \neq 0$ and $R(L) \cap Y_2 = \{0\}$, it follows that $Lx = 0$ and $Cx = 0$. Hence $MPx = 0$ and so $Px = 0$ since M is injective. Consequently, $x \in X_1$ with $Lx = 0$. Thus $x = 0$ since L is injective on X_1, i.e., L_λ is injective for $\lambda \neq 0$ and, in fact, bijective by the results in [24].

Furthermore, the map $T_\lambda = L_\lambda + N$ is weakly A-proper for each fixed $\lambda \neq 0$. Indeed, let $\{x_{n_j} \mid x_{n_j} \in X_{n_j}\}$ be any bounded sequence such that $g_{n_j} \equiv W_{n_j} T_\lambda(x_{n_j}) - W_{n_j}(g) \to 0$ as $j \to \infty$ for some g in Y. Since $\{x_{n_j}\}$ is bounded and C is compact, we may assume, without loss of generality, that $C(x_{n_j}) \to z$ in Y for some z in Y. This and the uniform boundedness of $\{W_n\}$ imply that

$$W_{n_j} L(x_{n_j}) + W_{n_j} N(x_{n_j}) - W_{n_j}(g - \lambda z) = g_{n_j} - \lambda W_{n_j}(Cx_{n_j} - z) \to 0$$

as $j \to \infty$. Hence, by the weak A-properness of $L + N$, there exist a subsequence $\{x_{n_{j(k)}}\}$ and $x \in X$ such that $x_{n_{j(k)}} \longrightarrow x$ as $k \to \infty$ and $Lx + Nx = g - \lambda z$. But, since $C \in L(X,Y)$ is completely continuous, it follows that $C(x_{n_{j(k)}}) \to Cx = z$. Hence $T_\lambda x = g$, i.e., T_λ is weakly A-proper w.r.t. Γ for each $\lambda \neq 0$.

Now let $\{\lambda_k\}$ be a sequence such that either $\lambda_k \to 0^+$ or $\lambda_k \to 0^-$ as $k \to \infty$ (i.e. $\lambda_k > 0$ or $\lambda_k < 0$ and $\lambda_k \to 0$ as $k \to \infty$) and observe that $T_k \equiv L + \lambda_k C + N$ satisfies the conditions of Lemma 1.1 for each k. Hence, in either case, for each f in Y and each k there exists a vector $x_k \in X$ such that

(1.4) $$Lx_k + \lambda_k Cx_k + Nx_k = f \quad \text{for each } k.$$

Now, if f is such that condition (A) holds, then in either case there exists a constant $r_f > 0$ such that $||x_k|| \leq r_f$ for all k. In view of this and the convergence of $\{\lambda_k\}$ to 0, it follows that $\{x_k\} \subset \overline{B}(0,r_f)$ and $Lx_k + Nx_k \to f$ as $k \to \infty$. Since, by assumption $T(\overline{B}(0,r_f))$ is closed, it follows that in either case there exists $x_0 \in \overline{B}(0,r_f)$ such that $Lx_0 + Nx_0 = f$. Q.E.D.

We will now show that the boundedness condition (A) is satisfied if in addition to the growth condition $N(x) = o(||x||)$ as $||x|| \to \infty$ one also assumes that N satisfies certain asymptotic positivity conditions. Such conditions have their origin in the existence result of Landesman and Lazer [18] for PDE's, and have since been considered by a number of authors including Nečaš [20], Fučik [11], Fučik, Kečera and Nečaš [13], DeFigueiredo [8], Fitzpatrick [9], Nirenberg [21], Brezis and Nirenberg [3], Ambrosetti and Mancini [0] and others.

It should be added that we mention explicitly only those authors to whose work a direct reference is made in this paper. However, the problem of the solvability of semilinear abstract and differential equations especially by means of the Lyapunov-Schmidt method has been studied by many authors including Cesari, Mahwin, Cronin, Nirenberg, Kannan, Schechter, Hale, Schur, Berger, Ahmad,

Hess, Gustafson, Sather, Osborne, Lazer, Ambrozetti, Mancini, Prodi, Gupta, Leach, Dancer, Williams and others. For an excellent survey of the results of these and other authors see the monographs by Cesari [6], Fučik [12] and Gaines and Mahwin [14].

As the first consequence of Theorem 1.1 we deduce the following result which, as will be shown below, extends abstract results of [8,11,13,20] and some results of [9]. For bounded linear part we also extend some results in [3].

Theorem 1.2. Suppose Y is a Hilbert space, $L \in L(X,Y)$ is A-proper w.r.t. Γ with $\mathrm{ind}(L) = 0$ and $N: X \to Y$ is such that $T = L + N$ is weakly A-proper w.r.t. Γ and $T(B)$ is closed in Y for each $B \subset X$. Let M and P be as defined in the Proof of Theorem 1.1.

Then, for a given f in Y, Eq. (1.3) is solvable provided that one of the following three conditions holds:

B(1) $N(x) = o(||x||)$ as $||x|| \to \infty$, and

(2) Either B^+: $\overline{\lim}\, (Nx_k, My) > (f,My)$ or B^-: $\underline{\lim}\, (Nx_k,My) < (f,My)$ whenever $\{x_k\} \subset X$ is such that $||x_k|| \to \infty$ and $x_k/||x_k|| \to y \in N(L)$.

C(1) $||Nx|| \leq a||x||^\alpha + b$ for all $x \in X$, some $\alpha \in [0,1)$ and $a > 0$, $b \geq 0$, and

(2) Either C^+: $\overline{\lim}\, (N(t_k\tilde{y}_k + t_k^\alpha z_k),My) > (f,My)$ or C^-: $\underline{\lim}\, (N(t_k\tilde{y}_k + t_k^\alpha z_k),My) < (f,My)$ whenever $y \in N(L) \cap \partial B(0,1)$ and sequences $\{t_k\} \subset R^+$, $\{\tilde{y}_k\} \subset N(L)$ and $\{z_k\} \subset X_1$ are such that $\tilde{y}_k \to y$ in X, $t_k \to \infty$ and $||z_k|| \leq c$ for all k and some $c > 0$.

D(1) Condition C(1) holds, and

(2) Either D^+: $\overline{\lim}\, (Nx_k, My_k) > (f,My)$ or D^-: $\underline{\lim}(Nx_k,My_k) < (f,My)$ whenever $\{x_k\} \subset X$ is such that $||Px_k|| \to \infty$ and $y_k = \dfrac{Px_k}{||Px_k||} \to y \in N(L)$.

Proof: It follows from C(1) that $N(x) = o(||x||)$ as $||x|| \to \infty$. Thus to deduce Theorem 1.2 from Theorem 1.1, it suffices to show that the set S_k^+ is bounded if either B^+, $C^+ - C(1)$ or $D^+ - C(1)$ holds, while S_k^- is bounded if either B^-, $C^- - C(1)$ or $D^- - C(1)$ holds (i.e., S_k^+ or S_k^- is bounded by a constant independent of k).

Let $\{\lambda_k\}$ be such that $\lambda_k \to 0^+$ as $k \to \infty$ and let $\{x_k\} \subset X$ be any sequence such that (1.4) holds for each k. We claim that B^+, $C^+ - C(1)$, or $D^+ - C(1)$ implies the boundedness of $\{x_k\}$. If this were not the case, then since $||N(x_k)||/||x_k|| \to 0$ as $t_k \equiv ||x_k|| \to \infty$ and $\{u_k\} \equiv \{\frac{x_k}{||x_k||}\}$ is bounded, it follows from this and (1.4) that

$$L(u_k) = f/||x_k|| - \lambda_k C(u_k) - N(x_k)/||x_k|| \to 0 \quad \text{as} \quad k \to \infty.$$

Hence, since L is proper, there exists a subsequence which we again denote by $\{u_k\}$, such that $u_k \to y$ in X, $||y|| = 1$ and $L(y) = 0$. Now taking the inner product of (1.4) with $M(y)$ and noting that $Y = N(L^*) \oplus R(L)$, $M\colon N(L) \to M(L^*)$ and $C = MP$ we see that $(Lx_k, My) = 0$, $(Cu_k, My) \to (Cy, My) = (My, My) > 0$ as $k \to \infty$, $\lambda_k (Cx_k, My) + (Nx_k - f, My) \doteq 0$ for each k, and $\lambda_k ||x_k|| (Cu_k, My) > 0$ for all k sufficiently large. This implies that

$$(1.6) \qquad (Nx_k - f, My) < 0 \quad \text{for all large } k.$$

Suppose first that B^+ holds. In view of this, (1.6) implies that $\overline{\lim}\, (Nx_k, My) \leq (f, My)$, in contradiction to B^+. Thus the set S_k^+ is bounded if B^+ holds.

Suppose now that $C^+ - C(1)$ holds. Since $X = N(L) \oplus X_1$ we can write $x_k = v_k + w_k$ with $v_k = Px_k \in N(L)$ and $w_k = (I - P)(x_k) \in X_1$ for each k. Let Q be the orthogonal projection of Y onto $R(L)$. Then, since $Y = N(L^*) \oplus R(L)$ and $C\colon Y \to N(L^*)$, applying Q to (1.4) we get the equality $Lx_k + Q(Nx_k - f) = 0$ or $Lw_k = Q(f - Nx_k)$. Since $L = L|_{X_1}\colon X_1 \to R(L)$ has a bounded inverse, it follows from the last equality and C(1) that

$$(1.7) \qquad ||w_k|| \leq a_1 ||x_k||^{\alpha} + a_2 \quad \text{for some } a_1 > 0 \text{ and } a_2 > 0.$$

Setting $u_k = \frac{x_k}{||x_k||} = \tilde{y}_k + \tilde{w}_k$, we see that $||u_k|| = 1$,

$\tilde{y}_k = \frac{v_k}{||x_k||} \in N(L)$, $\tilde{w}_k = \frac{w_k}{||x_k||} \in X_1$, $\tilde{w}_k \to 0$ in X, $\tilde{y}_k \to y$

$\in N(L)$ with $||y|| = 1$ and $x_k = t_k\tilde{y}_k + w_k$ with $w_k = t_k^\alpha z_k$, where $\{z_k\} \subset X_1$ is bounded in view of (1.7). Thus, (1.6) and condition C^+ imply that

$\overline{\lim}(Nx_k,My)=\overline{\lim}(N(t_k\tilde{y}_k+t_k^\alpha z_k),My)\leq(f,My)$, in contradiction to C^+. Thus the set S_k^+ is also bounded if $C^+ - C(1)$ holds.

Suppose now $D^+ - C(1)$ holds. Since $u_k = \frac{x_k}{||x_k||} = \tilde{y}_k + \tilde{w}_k$ with $\tilde{w}_k \to 0$ and $\tilde{y}_k \to y \in N(L)$ with $||y|| = 1$, it follows that $\frac{||v_k||}{||x_k||} = \frac{||Px_k||}{||x_k||} \to 1$ as $k \to \infty$. Hence $||Px_k|| \to \infty$ and

$y_k = \frac{Px_k}{||Px_k||} = \frac{v_k}{||v_k||} \to y$ as $k \to \infty$. Now taking the inner product of (1.4) with My_k and noting that $(Lx_k,My_k) = 0$ we get the equality.

$$\lambda_k(Cx_k,My_k) + (Nx_k - f,My_k) = 0 \quad \text{for each } k.$$

Since $u_n = x_k/||x_k|| \to y$ and $y_k \to y$ we see that $(Cu_k,My_k) \to (Cy,My) = (My,My) > 0$ as $k \to \infty$. This and the fact that $\lambda_k > 0$ imply that $\lambda_k(Cx_k,My_k) = \lambda_k||x_k||(Cu_k,My_k) > 0$ for all k sufficiently large. In view of this, the last equality implies that $(Nx_k - f,My_k) < 0$ for all large k and therefore $\overline{\lim}\,(Nx_k,My_k) \leq (f,My)$ in contradiction to D^+. Thus the set S_k^+ is also bounded if $D^+ - C(1)$ holds.

In a similar way one shows that the set S_k^- is bounded if either B^-, $C^- - C(1)$ or $D^- - C(1)$ holds. Indeed, if not, then by the same argument we would be led to one of the equations

$\lambda_k(Cx_k,My) + (Nx_k - f,My) = 0$ or $\lambda_k(Cx_k,My_k) + (Nx_k - f,My_k) = 0 \ \forall\ k$ with $\lambda_k(Cx_k,My) = \lambda_k||x_k||(Cu_k,My) < 0$ and $\lambda_k(Cx_k,My_k)$ $= \lambda_k||x_k||(Cu_k,My_k) < 0$ for all sufficiently large k since $\lambda_k < 0$ and $0 < (My,My) = \lim\ (Cu_k,My) = \lim\ (Cu_k,My_k)$. Thus $(Nx_k - f,My) \geq 0$ in the first case and $(Nx_k - f,My_k) \geq 0$ in the second for all large k. Thus, as before, we obtain the contradiction in each case. Hence S_k^- is bounded. Q.E.D.

Note 1. It should be underlined that whenever we say that S_k^+ or S_k^- is bounded we always mean that S_k^+ or S_k^- is bounded by some constant independent of k.

Remark 1.0. It is not hard to show that the asymptotic positivity condition B^+ is equivalent to the hypothesis:

B_1^+: $\underline{\lim}\ (Nx_k,My) > (f,My)$ whenever $\{x_k\} \subset X$ is such that $||x_k|| \to \infty$ and $x_k/||x_k|| \to y \in N(L)$,

while the condition B^- is equivalent to the hypotheses:

(B_1^-) $\overline{\lim}(Nx_k,My) < (f,My)$ whenever $\{x_k\} \subset X$ is such that $||x_k|| \to \infty$ and $x_k/||x_k|| \to y \in N(L)$.

The same can be said about the other conditions.

Remark 1.1. If $N: X \to Y$ is also continuous and $T = L + N$ is A-proper w.r.t. Γ, then the conclusions of Theorems 1.1 and 1.2 hold without the assumption that $T(B)$ is closed since, as has been shown by the author (see [24]), every continuous A-proper map is proper and, in particular $T(B)$ is closed in Y for each $B \subset X$.

Remark 1.2. Assuming that $X \subset Y$ and $Y = N(L) \oplus R(L)$ the asymptotic growth and positivity conditions $B(1)$ - $B(2)$ (in the form stated in Remark 1.0 with $M = I$) were used in [9] for a special case of maps to be discussed in Section 2, while conditions $C(1)$ - $C(2)$ were used earlier in [8] as extensions of the hypotheses used in [13, 20] for $X = Y$. We add that when $N: X \to Y$ satisfies condition $C(1)$, then it is not hard to show that the positivity conditions imposed in [13, Theorem 2.3] and in [9, Proposition 2.5] imply $D(2)$ of Theorem 1.2.

Remark 1.3. In case $Y = X = H$, where H is a separable Hilbert space and thus H always has a projectionally complete scheme $\{X_n, P_n\}$, Theorem 1.2 takes a particulary simple form when one assumes that $N(L) = N(L^*)$ since in that case $M = I$ and the perturbation equation (1.4) takes the form

(1.8) $$Lx_k + \lambda_k Px_k + Nx_k = f,$$

where P is the orthogonal projection of H onto $N(L)$. It will be shown in Section 2 that in this case Theorem 1.2 extends a number of results obtained earlier by other authors for special classes of weakly A-proper mappings.

We conclude this section with the indication how one can use Theorem 1.2 (when $Y = X = H$) to obtain existence results for the class of weakly A-proper mappings under the asymptotic growth and positivity conditions used by Brezis and Nirenberg [3] for a different but related class of maps which we shall indicate at the end of this section (see also Section 2). We may add that our proofs for the existence of solutions are somewhat simpler than those given in [2,3] for characterization of ranges of semilinear operators. To state further results we first recall that if $L \in L(H,H)$ is a Fredholm map with $N(L) = N(L^*)$, then there exists a constant $\alpha_0 > 0$ such that $||Lx|| \geq \alpha_0 ||w||$, where $x = v + w$ with $v \in N(L)$ and $w \in R(L)$. Hence $|(Lx,x)| = |(Lx,w)| \leq ||Lx|| \; ||w|| \leq \frac{1}{\alpha_0}||Lx||^2$, i.e.,

$(Lx,x) \geq -\frac{1}{\alpha_0}||Lx||^2$ for all x in H. In what follows we shall take $\alpha > 0$ to be the largest $\alpha_0 > 0$ such that $(Lx,x) \geq -\frac{1}{\alpha}||Lx||^2$ for all x in H.

Corollary 1.1. Suppose $L \in L(H,H)$ is A-proper with $N(L) = N(L^*)$ and $T = L + N$ is weakly A-proper. Then, for a given $f \in H$, Eq. (1.3), $Lx + Nx = f$, is solvable provided the following additional conditions hold:

(1.9) $\quad N(x) = o(||x||)$ as $||x|| \to \infty$ and $T(B)$ is closed for each $B \subset H$.

(1.10) $\quad (Nx - Ny,x) \geq \frac{1}{\gamma}||Nx||^2 - C(y)\ \forall\, x,y \in H$ and some positive $\gamma < \alpha$, where $C(y)$ is independent of x.

(1.11) $\quad \liminf_{t\to\infty} (N(tx),x) > (f,x)\ \forall\, x \in N(L),\ ||x|| = 1.$

Proof: In view of Theorem 1.1(A), to prove Corollary 1.1 it suffices to show that the set S_k^+ is bounded. If not, then there would exist sequences $\{\lambda_k\} \subset R^+$ and $\{x_k\} \subset X$ with $\lambda_k \to 0^+$ as $k \to \infty$ such that x_k satisfies (1.8) for each k and $||x_k|| \to \infty$ as $k \to \infty$. In view of this and the first part of (1.9) we may assume that $y_k \equiv x_k/||x_k|| \to y \in N(L)$ with $||y|| = 1$.

Taking the inner product of (1.8) with x_k we get

(1.12) $\quad (Lx_k,x_k) + \lambda_k(Px_k,x_k) + (Nx_k,x_k) = (f,x_k).$

Since $\lambda_k(Px_k,x_k) = \lambda_k||Px_k||^2 > 0$ it follows from (1.12) that

(1.13) $\quad (Lx_k,x_k) + (Nx_k,x_k) \leq (f,x_k)$ for each k.

Now it follows from (1.10) that for each fixed $t > 0$ and each k

(1.14) $\quad (Nx_k,x_k) \geq \frac{1}{\gamma}||Nx_k||^2 + (N(ty),x_k) - C(ty).$

On the other hand applying $Q = I - P$ to (1.8) we get

$Lx_k = Q(f - Nx_k)$ and therefore $||Lx_k||^2 \leq ||f - Nx_k||^2 = ||f||^2 - 2(Nx_k,f) + ||Nx_k||^2$. Hence

$$(1.15)\quad (Lx_k,x_k) \geq -\frac{1}{\alpha}||Lx_k||^2 \geq -\frac{1}{\alpha}||Nx_k||^2 + \frac{2}{\alpha}(Nx_k,f) - \frac{1}{\alpha}||f||^2.$$

In view of (1.14) - (1.15) and the assumption that $\gamma < \alpha$, it follows from (1.13) that

$$(1.16)\quad (N(ty),x_k) - C(ty) + \frac{2}{\alpha}(Nx_k,f) - \frac{1}{\alpha}||f||^2 \leq (f,x_k)$$

for each k.

It follows from (1.16) that

$$(N(ty),y_k) \leq (f,y_k) + \frac{C(ty)}{||x_k||} + \frac{\alpha^{-1}||f||^2}{||x_k||} + \frac{2\alpha^{-1}||f||\ ||Nx_k||}{||x_k||}$$

Since $y_k \to y$ and $||Nx_k||/||x_k|| \to 0$ as $k \to \infty$, taking the limit superior in the last inequality as $k \to \infty$ leads to

$(N(ty),y) \leq (f,y)$ for each $t > 0$ and $y \in N(L)$ with $||y|| = 1$.

Hence $\liminf_{t\to\infty} (N(ty),y) \leq (f,y)$ with $y \in N(L)$ and $||y|| = 1$, in contradiction to (1.11). Thus S_k^+ is bounded. Q.E.D.

Since $T + F$ remains weakly A-proper when F is completely continuous (i.e. $Fu_j \to Fu$ whenever $u_j \rightharpoonup u$ in H), the same arguments used to prove Corollary 1.1 yield also the validity of the following practically useful result.

Corollary 1.2. Suppose L and N satisfy conditions of Corollary 1.2 except for (1.9) which is replaced by

(1.9') $Nx = o(||x||)$ as $||x|| \to \infty$ and $T = L + N$ is demiclosed (i.e. if $u_j \rightharpoonup u$ in H and $Tu_j \to h$, then $Tu = h$)

If F is completely continuous, $R(F)$ is bounded (weaker condition suffices) and $\lim \frac{(Fx_k,x_k)}{||x_k||} = \eta \geq 0$ whenever $||x_k|| \to \infty$

$k \to \infty$, then the equation $Lx + Nx + Fx = f$ is solvable.

As our final result in this section we have

Corollary 1.3. Suppose $L \in L(H,H)$ is A-proper with $N(L) = N(L^*)$ and $T = L + N$ is weakly A-proper. Then, for a given f in H, Eq. (1.3) is solvable provided the following additional conditions hold:

(1.17) $||Nx|| / ||x||^{1/2} \to 0$ as $||x|| \to \infty$ and $T(B)$ is closed for each $B \subset H$.

(1.18) $\underline{\lim}\,(Nx_k, u_k) > (f,y)$ whenever $\{x_k\} \subset X$ is such that $||x_k|| \to \infty$ and $u_k \equiv x_k/||x_k|| \to y \in N(L)$ with $||y|| = 1$.

Proof: By Theorem 1.1(A), it suffices to show that the set S_k^+ is bounded if (1.17) and (1.18) hold. If S_k^+ were not bounded, then we could find $\{x_k\}$ satisfying (1.8) such that $||x_k|| \to \infty$ and $u_k \equiv \frac{x_k}{||x_k||} \to y \in N(L)$ with $||y|| = 1$. Taking the inner product of (1.8) with x_k and noting that $\lambda_k(Px_k, x_k) > 0$ we get (1.13). Since $||Lx_k||^2 \le ||f - Nx_k||^2 \le 2\{||f||^2 + ||Nx_k||^2\}$ we see that

$$(Lx_k, x_k) \ge -\frac{1}{\alpha}||Lx_k||^2 \ge -\frac{2}{\alpha}(||f||^2 + ||Nx_k||^2) \quad \text{for all } k.$$

It follows from this and (1.13) that

$$(Nx_k, x_k) \le (f, x_k) + \frac{2}{\alpha}(||f||^2 + ||Nx_k||^2).$$

Dividing the last inequality by $||x_k||$ we get the relation

$$(Nx_k, u_k) \le (f, u_k) + \frac{2}{\alpha}\left\{ \frac{||f||^2}{||x_k||} + \frac{||Nx_k||^2}{||x_k||} \right\}.$$

Since $u_k \to y \in N(L)$, it follows from the last inequality and the first part of (1.17) that $\underline{\lim}\,(Nx_k, u_k) \le (f,y)$, in contradiction to (1.18).

Q.E.D.

Remark 1.4. If $(Lx,x) \geq 0$ for $x \in H$, then instead of the condition that $Nx = o(||x||^{1/2})$ as $||x|| \to \infty$ it sufficies to assume that $Nx = o(||x||)$ as $||x|| \to \infty$.

Remark 1.5. Corollary 1.1 is related to [3, Corollary II.7] where it is assumed that $L\colon\ D(L) \subset H \to H$ is unbounded but $(\lambda P + L)^{-1}\colon\ H \to H$ is compact and $N\colon\ H \to H$ is monotone. The same can be said about Corollary 1.3 and the corollary of [2, Theorem 13] attributed in [2] to [3]. See [3] for the discussion of the condition (1.10) and its usefulness in applications.

Remark 1.6. It was shown in [3] that if $N\colon\ H \to H$ is any mapping, then one can define a "recession function" of N for any $y \in H$ by

$$J_N(y) = \liminf_{\substack{t \to \infty \\ v \to y}} (N(tv),v) \tag{1.19}$$

and the function $J_N(y)$ thus defined is lower semicontinuous from H to $[-\infty,\infty]$ with $J_N(\lambda y) = \lambda J_N(y)$ for each $\lambda > 0$ and $y \in H$; moreover, if N is monotone, hemicontinuous, and $||Nx||/||x|| \to 0$ as $||x|| \to \infty$, then $J_N(y) = \liminf_{t \to \infty} (N(ty),y)$.

It follows from this that if in Corollary 1.3 the map N is assumed to be monotone, then since $(Nx_k,u_k) = (N(t_k u_k,u_k)$ for each k with $t_k \to \infty$ and $u_k \to y$, it follows from the definition of $J_N(y)$ that condition (1.18) is equivalent to

$$J_N(y) > (f,y) \qquad y \in N(L), \qquad ||y|| = 1. \tag{1.20}$$

Moreover, since $|(Nx_k,u_k) - (Nx_k,y)| \to 0$ whenever $R(N)$ is bounded, it follows that in this case condition (1.20) is equivalent to the hypothesis (B_1^+) of Remark 1.0, i.e., $\underline{\lim}\ (Nx_k,y) > (f,y)$ whenever $\{x_k\} \subset X$ is such that $||x_k|| \to \infty$ and $u_k \equiv \frac{x_k}{||x_k||} \to y \in N(L)$. Indeed, since $0 = \underline{\lim}\ \{(Nx_k,u_k) - (Nx_k,y)\} \leq \underline{\lim}\ (Nx_k,u_k) - \underline{\lim}\ (Nx_k,y)$ we see that $\underline{\lim}\ (Nx_k,y) \leq \underline{\lim}\ (Nx_k,u_k) = J_N(y)$, i.e., B_1^+ implies (1.20).

On the other hand, since $0 = \underline{\lim}\ \{(Nx_k,y) - (Nx_k,u_k)\}$ $\leq \underline{\lim}\ (Nx_k,y) - \underline{\lim}\ (Nx_k,u_k)$ we get that $J_N(y) = \underline{\lim}(Nx_k,u_k)$ $\leq \underline{\lim}\ (Nx_k,y)$ i.e., (1.20) implies B_1^+.

SECTION 2

In this section we discuss some special classes of maps $N\colon X \to Y$ for which $T = L + N$ is weakly A-proper and $T(B)$ is closed in Y. As special cases we deduce from Theorems 1.1 and 1.2 the existence results of other authors mentioned in the Introduction.

We begin with the following notion which is due to Brezis [1] when $Y = X^*$. In what follows we say that $T\colon X \to Y$ is of <u>type (KM)</u> provided the following conditions hold:

(i) If $x_j \rightharpoonup x$ in X, $Tx_j \rightharpoonup g$ in Y and $\overline{\lim}\ (Tx_j,Kx_j) \leq (g,Kx)$, then $Tx = g$, where K is a suitable map of X into Y^*.

(ii) T is continuous from finite dimensional subspaces of X into Y equipped with weak* topology.

We shall also say that T is <u>semibounded</u> if $\{Tx_j\}$ is bounded whenever $\{x_j\}$ and $\{(Tx_j,Kx_j)\}$ are bounded.

Lemma 2.1. <u>Suppose</u> X <u>is reflexive,</u> $\Gamma = \{X_n,V_n;E_n,W_n\}$ <u>is admissible for</u> (X,Y), $K\colon X \to Y^*$ <u>continuous with</u> $R(K)$ <u>dense in</u> Y^*, <u>and</u> $K_n\colon X_n \to D(W_n^*)$ <u>is such that</u> $\{K_n(z_n)\}$ <u>is uniformly bounded whenever</u> $\{z_n | z_n \in X_n\}$ <u>is bounded and for each</u> n <u>we have</u>

$$(2.1)\qquad (W_n g, K_n x) = (g,Kx) \quad \text{for all } x \in X_n \text{ and } g \in Y.$$

(a) <u>If</u> $T\colon X \to Y$ <u>is weakly continuous, then</u> T <u>is weakly A-proper w.r.t.</u> Γ <u>and</u> $T(B)$ <u>is closed for each closed ball</u> $B \subset X$.

(b) <u>If</u> $T\colon X \to Y$ <u>is semibounded and of type (KM), then</u> T <u>is weakly A-proper w.r.t.</u> Γ <u>and</u> $T(B)$ <u>is closed provided</u> K <u>is</u>

also weakly continuous.

We omit the proof of Lemma 2.1 since the proof of (a) is similar to the proof of Proposition 2 in [23] while the proof of (b) is essentially the same as the proof of Propositions 11 and 12 in [23].

Remark 2.1. The conditions on K in (b) of Lemma 2.1 are certainly satisfied if $K \in L(X,Y^*)$ and $R(K)$ is dense in Y^*. In applications this is often the case as we shall see in Problem 1 of Section 3.

In view of Lemma 2.1, Theorems 1.1 and 1.2 yield the following

Corollary 2.1. Suppose X, Γ, K and K_n are as in Lemma 2.1, $L \in L(X,Y)$ is A-proper w.r.t. Γ with $\text{ind}(L) = 0$, $N\colon X \to Y$ is such that $N(x) = o(||x||)$ as $||x|| \to \infty$, and one of the following three conditions holds:

(H1) N is compact

(H2) N is weakly continuous

(H3) N is semibounded and of type (KM) and $\phi(x) = (Lx,Kx)$ is weakly lower semicontinuous (i.e., $x_j \rightharpoonup x \Rightarrow \phi(x) \leq \underline{\lim}\, \phi(x_j)$).

Then (1.3), $Lx + Nx = f$ has a solution for a given f in Y provided that condition (A) of Theorem 1.1 holds or Y is a Hilbert space and conditions (B), (C) or (D) of Theorem 1.2 hold.

Proof: (H1) Since $N\colon X \to Y$ is compact, $T = L + N$ is a continuous A-proper mapping. Hence in this case the conclusion of Corollary 2.1 follows from Theorem 1.1 and Remark 1.1 if (A) holds or from Theorem 1.2 and Remark 1.1 if Y is a Hilbert space and (B), (C) or (D) holds.

(H2) Since $T = L + N\colon X \to Y$ is weakly continuous, it follows from Lemma 2.1(a) that T is weakly A-proper w.r.t. Γ

and $T(B)$ is closed in Y for each $B \subset X$. Thus, Corollary 2.1 (H2) follows from Theorem 1.1 if (A) holds and from Theorem 1.2 if Y is a Hilbert space and (B), (C) or (D) holds.

(H3) In view of Lemma 2.1(b), to establish the assertion of Corollary 2.1 when (H3) holds, it suffices to show that $T = L + N$ is semibounded and of type (KM) and then use Theorems 1.1 or 1.2. The fact that T is semibounded follows easily from the boundedness of L and semiboundedness of N. To show that T is of type (KM), we let $x_j \rightharpoonup x$ in X, $Tx_j \rightharpoonup h$ in Y and $\overline{\lim}\, (Tx_j, Kx_j) \leq (h, Kx)$. Since $Lx_j \rightharpoonup Lx$ in Y, it follows that $Nx_j \rightharpoonup h - Lx$ and thus, by our condition on ϕ we have

$$\overline{\lim}\, (Nx_j, Kx_j) = \overline{\lim}\, \{(Tx_j, Kx_j) - (Lx_j, Kx_j)\}$$

$$\leq \overline{\lim}\, (Tx_j, Kx_j) - \underline{\lim}\, (Lx_j, Kx_j) \leq (h - Lx, Kx).$$

Since N is of type (KM), it follows that $Nx = h - Lx$ or $Tx = h$, i.e., T is of type (KM). Q.E.D.

Remark 2.2. If $K \in L(X, Y^*)$ and $A, C \in L(X, Y)$ are such that $(Ax, Kx) \geq 0$ for $x \in X$ and C is compact, then $\phi(x) = (Lx, Kx)$ is weakly lower semicontinuous, where $L = A + C$ or $L = A - C$.

Special Case. If we set $Y = X^*$, where X is a separable reflexive space, then the natural choice for Γ, K and K_n which satisfy the condition of Lemma 2.1 is $\Gamma = \Gamma_I = \{X_n, V_n; X_n^*, V_n^*\}$, $K = I$ and $K_n = I_n$ with I and I_n denoting the identities in X and X_n respectively. In this case, the map N in (H3) is of type (M) in the sense of Brezis [1]. This class includes all the hemicontinuous monotone mappings and the class of pseudomonotone maps introduced in [1]. Hence Corollary 2.1 is valid for these maps $N: X \to X^*$.

In case $Y = X = H$, where H is a separable Hilbert space, Corollary 2.1 extends essentially Theorems 1, 2 and 3 of [8] as well as the corresponding earlier results of [13,20].

Remark 2.3. The following observations will prove to be useful in applications of the above results to semilinear differential equations.

(i) Suppose H_0 and H are Hilbert spaces such that $H_0 \subseteq H$, H is dense in H and the imbedding of H_0 into H is compact. If $L \in L(H_0,H_0)$ and if there exist constants $c_0 > 0$ and c_1 such that

$$(2.2) \qquad (Lu,u)_0 \geq c_0||u||_0^2 - c_1||u||^2 \quad \text{for all} \quad u \in H_0,$$

then L is A-proper w.r.t. any projectionally complete scheme $\Gamma_0 = \{X_n,P_n\}$ for (H_0,H_0), $\mathrm{ind}(L) = 0$, and the functional $\phi(u) = (Lu,u)_0$ is weakly lower semicontinuous (see [24]). It is known (see [10]) that, in view of Gärding's inequality, the generalized Dirichlet forms on a Sobolev space $\overset{\circ}{W}{}_2^m(Q)$ corresponding to strongly elliptic operators of order $2m$ give rise to mappings $L \in L(\overset{\circ}{W}{}_2^m,\overset{\circ}{W}{}_2^m)$ satisfying (2.2).

(ii) Let X be a real Banach space and let Q be any bounded set in X. Then the ball-measure of noncompactness of Q, $\chi(Q)$, is defined to be $\chi(Q) = \inf\{r > 0 | Q$ can be covered by a finite number of balls with radii $\leq r\}$. A continuous mapping $F\colon X \to X$ is said to be k-ball-contraction (resp. ball-condensing) if $\chi(F(Q)) \leq k\chi(Q)$ for all bounded sets $Q \subset X$ and some $k \geq 0$ (resp. $\chi(F(Q)) < \chi(Q)$ whenever $\chi(Q) \neq 0$). For the survey of the theory of these classes of mappings see [27]. It is known that if $F\colon X \to X$ is k-ball-contractive with $k \in [0,1)$, then $L = I \pm F$ is A-proper w.r.t. $\Gamma = \{X_n,P_n\}$. Moreover, if $X = H$ and $F \in L(H,H)$, then F can be represented as $F = A + C$ with C compact and $||A|| \leq \frac{k+1}{2}$ (see [30]). It follows from this and Remark 2.2 that L is A-proper w.r.t. Γ_1, $\mathrm{ind}(L) = 0$, and $\phi(u) = (Lu,u)$ is weakly lower semicontinuous on H.

To deduce some existence results in [9] from Theorem 1.2 we need

Lemma 2.2. Suppose $\Gamma_\alpha = \{X_n, P_n\}$ is projectionally complete for (X,X). If $L \in L(X,Y)$ is Fredholm with $\text{ind}(L) = 0$, $Y_n = (L + C)(X_n) \subset Y$ for each n, and Q_n is a projection of Y onto Y_n, where $C \in L(X,Y)$ is the compact map constructed in the proof of Theorem 1.1, then L is A-proper w.r.t. the projectionally complete scheme $\Gamma_c = \{X_n, P_n; Y_n; Q_n\}$ for (X,Y).

Proof: It is not hard to show that since $L + C$ is a linear homeomorphism of X onto Y, the scheme Γ_c is projectionally complete for (X,Y). To show that L is A-proper w.r.t. Γ_c, let $\{x_{n_j} | x_{n_j} \in X_{n_j}\}$ be any bounded sequence such that $Q_{n_j} L(x_{n_j}) - Q_{n_j} g \to 0$ for some g in Y, i.e., $Q_{n_j} L(x_{n_j}) \to g$ as $j \to \infty$ because $Q_{n_j} g \to g$ in Y as $j \to \infty$. Since $\{x_{n_j}\}$ is bounded, C compact, and $Q_n h \to h$ for each h in Y, we may assume that $C(x_{n_j}) \to g_1$ for some g_1 in Y and so $Q_{n_j} C(x_{n_j}) \to g_1$ in Y. Therefore, $y_{n_j} \equiv Q_{n_j}(L+C)(x_{n_j}) \to g + g_1 \equiv y$ in Y as $j \to \infty$. But $Q_{n_j}(L + C)(x_{n_j}) = (L + C)(x_{n_j})$ for each j and, therefore, $x_{n_j} = (L + C)^{-1}(y_{n_j}) \to (L + C)^{-1} y \equiv x$ in X. Hence $(L + C)x = g + g_1$ with $g_1 = \lim_j C(x_{n_j}) = C(x)$, i.e., $Lx = g$ and thus L is A-proper w.r.t. Γ_c. Q.E.D.

Corollary 2.2. Suppose $L \in L(X,Y)$ is Fredholm with $\text{ind}(L) = 0$ and Γ_c is the scheme constructed above with $||Q_n|| = 1$ for each n. Suppose $N\colon X \to Y$ is such that $N(L + C)^{-1}\colon Y \to Y$ is ball-condensing.

If, for a given $f \in Y$, condition (A) holds or Y is a Hilbert space and either $B(1) - B(2)$ or $D(1) - D(2)$ of Theorem 1.2 holds, then Eq. (1.3), $Lx + Nx = f$, is solvable.

Proof: In view of Lemma 2.2 and Remark 1.1, Corollary 2.2 will follow from Theorem 1.1 when (A) holds and from Theorem 1.2 when either $(B1) - B(2)$ or $D(1) - D(2)$ holds if we show that

$T = L + N$ is A-proper w.r.t. Γ_c. So let $\{x_{n_j} | x_{n_j} \in X_{n_j}\}$ be any bounded sequence such that $Q_{n_j} L(x_{n_j}) + Q_{n_j} N(x_{n_j}) \to g$ for some g in Y. If we set $y_{n_j} = (L + C)(x_{n_j})$ for each j, then $\{y_{n_j} | y_{n_j} \in Y_{n_j}\}$ is bounded in Y and $y_{n_j} + Q_{n_j} A(y_{n_j}) \to g$ in Y as $j \to \infty$, where $A: Y \to Y$ is a ball-condensing mapping given by $A = (N - C)(L + C)^{-1}$. Since $I + A$ is A-proper w.r.t. $\{Y_n, Q_n\}$, there exist a subsequence $\{y_{n_{j(k)}}\}$ and $y \in Y$ such that $y_{n_{j(k)}} \to y$ in Y as $k \to \infty$ and $y + Ay = g$. Hence $x_{n_{j(k)}} = (L + C)^{-1} y_{n_{j(k)}} \to x \equiv (L + C)^{-1} y$ in X as $k \to \infty$, $(L + C)x = y$ and $y + Ay = (L + C)x + (N - C)x = g$, i.e., $Lx + Nx = g$. Q.E.D.

The second part of Corollary 2.2 includes [9, Propositions 2.4 and 2.5] where it is also assumed that $X \subset Y$, $L: X \to Y$ is such that $Y = N(L) \oplus R(L)$. It should be added that the author of [9] uses the degree theory for condensing vector fields to study the solvability of (1.3) and therefore his results are also valid for set-condensing maps and for the space X which need not have a projectionally complete scheme. However, the arguments of [9] cannot be used to study the solvability of Eq. (1.3) for the more general class of weakly A-proper maps treated in this paper.

SECTION 3

In this section we deduce some consequences of the results of Sections 1 and 2 for semilinear ordinary and partial differential equations. Some of the problems we treat here show that they can not be put into the framework to which the theories of compact and condensing operators or the theory of monotone operators are applicable. However, as we shall see, the results for the more general class of A-proper mappings are applicable.

Problem 1 treats a BVProblem for an ODEquation of second order whose nonlinear part N depends also on u'' and satisfies certain "monotonicity" conditions which preclude the applicability of the condensing type mapping results obtained in [9], where the same problem is treated under Lipschitz type conditions on N.

Problem 2 treats the case where unbounded discontinuous operators are involved. The abstract results below are modelled on situations which arise when one attempts to treat semilinear elliptic problems on unbounded domains in which case the resolvent operators are not compact and, in general, the nonlinear part N is also noncompact.

Problem 3 deals with the existence of weak solutions for elliptic semilinear equations of order $2m$ treated by other authors earlier under somewhat different conditions and by different methods. We include it here so as to illustrate the applicability of the results of Section 1 which we believe provides a somewhat simpler proof for the existence of solutions (see [0,9, 18,31] and others).

Problem 4 treats a semilinear elliptic second order equation with Neumann boundary condition studies in [9] under more restrictive conditions on the nonlinear part. The results in [9] cannot be applied to the problem treated here.

Problem 1. To compare our existence Theorem 3.1 below with that of [9, Theorem 3.3] we first study the solvability of the BVProblem:

$$(3.1) \qquad \begin{cases} -u''(t) - g(t,u'(t),u''(t)) + f(t,u(t),u'(t)) = 0 \\ u(0) = u(1),\ u(0) = u'(1) \end{cases}$$

treated in [9] under rather restrictive condition on g. To state the hypotheses under which (3.1) has a solution in $W_2^2([0,1])$ we let $Y = L_2([0,1])$ and $X = \{u | u \in W_2^2([0,1]),\ u(0) = u(1),\ u'(0) = u'(1)\} \subset Y$, where Y and X are separable real Hilbert spaces with the respective norms $||\cdot||$ and $||\cdot||_2 = (||u||^2 + ||u'||^2 + ||u''||^2)^{1/2}$ and the corresponding inner products $(\cdot,\cdot)$ and $(\cdot,\cdot)_2$. Defining $L\colon X \to Y$ by

$L(u)(t) = -u''(t)$, for $u \in X$ and $t \in [0,1]$, it is easy to show that $L \in L(X,Y)$, $\text{ind}(L) = 0$, $N(L)$ consists of constant functions, $Y = N(L) \oplus R(L)$ and $L + I: X \to Y$ is a linear homeomorphism, where I denotes the inclusion map of X into Y which is compact by Sobolev Imbedding Theorem (see [10]).

Let $\{X_n\} \subset X$ be a sequence of finite dimensional spaces, such that $\text{dist}(u,X_n) = \inf\{||u - v||_2 : v \in X_n\} \to 0$ as $n \to \infty$ for each u in X. Let K be the homeomorphism of X onto Y given by $K = L + I$ and, for each n, let $Y_n = K(X_n) \subset Y$. Then, if $P_n: X \to X_n$ and $Q_n: Y \to Y$ denote orthogonal projections, the scheme $\Gamma_0 = \{X_n, P_n; Y_n, Q_n\}$ is projectionally complete for (X,Y) and, since

$$(3.2) \qquad (Lu, Ku) \geq ||u||_2^2 - ||u||^2 \quad \text{for } u \in X,$$

the map $L: X \to Y$ is A-proper w.r.t. Γ_0 with $\text{ind}(L) = 0$ (see [24]). We impose the following conditions on the functions g and f:

(a1) $g: [0,1] \times R^2 \to R$ is continuous and there are constants $p > 0$ and $c_0 \in (0,1)$ such that $|g(t,s,r)| \leq p$ for all $t \in [0,1]$ and $r,s \in \mathbb{R}$ and $(g(t,s,r_1) - g(t,s,r_2))(r_1 - r_2) \geq -c_0|r_1 - r_2|^2$ for all $t \in [0,1]$ and $s,r_1,r_2 \in \mathbb{R}$.

(b1) $f: [0,1] \times \mathbb{R}^2 \to \mathbb{R}$ is continuous and there exist $a(x) \in Y$, $\beta \in (0,1)$ and $\gamma > 0$ such that $|f(t,s,r)| \leq a(t) + \gamma(|s|^\beta + |r|^\beta)$ for $t \in [0,1]$ and $r,s \in \mathbb{R}$.

We define the operators $G,C: X \to Y$ by $G(u)(t) = -g(t,u'(t),u''(t))$ and $C(u)(t) = f(t,u(t),u'(t))$ for $t \in [0,1]$ and $u \in X$. First note that since X is compactly imbedded into $C^1([0,1])$, the conditions on f in (b1) imply that C is compact, as a map from X to Y, and

$$(3.3) \qquad ||C(u)|| \leq ||a|| + \gamma_1 ||u||_2^\beta \quad \text{for } u \in X.$$

Now it is not hard to show that, in virtue of (a1), $G: X \to Y$ is continuous, $||Gu|| \leq p$ for $u \in X$, and $(Gu - Gv, Ku - Kv) \geq - c_0||u'' - v''||^2 - 2p||u - v|| + (g(t,u',v'') - g(t,v',v''), u'' - v'')$ for $u,v \in X$. It follows from this and (3.2) that all $u,v \in X$ we have $((L + G)u - (L + G)v, Ku - Uv) \geq (1 - c_0)||u - v||^2 - ||u - v||^2 - 2p||u - v|| + (g(t,u',v'') - g(t,v',v''), u'' - v'')$. This and the results in [24] imply

Lemma 3.1. If g and f satisfy conditions (a1) and (b1) respectively, then the operator $T = L + G + C: X \to Y$ is continuous and A-proper w.r.t. Γ_0 and $Nu = o(||u||_2)$ as $||u||_2 \to \infty$ with $N = G + C$.

Our first result in this section, Theorem 3.1 below, improves [9, Theorem 3.3] and the corresponding results in [17] in that our nonlinearity g is not assumed to satisfy the Lipschitz condition of [9], while f is not assumed to satisfy the growth conditions of [17] (no u'' is permitted in the nonlinear part in [17]).

Theorem 3.1. Suppose that (a1) and (b1) hold and that there exist $N_0 > 0$ and $\delta > p$ such that either (H1) or (H1') holds, where

(H1) For $t \in [0,1]$ and $r \in \mathbb{R}$ we have $f(t,r,s) \geq \delta$ when $s \geq N_0$ and $f(t,r,s) \leq -\delta$ when $s \leq -N_0$.

(H1') For $t \in [0,1]$ and $r \in \mathbb{R}$ we have $f(t,s,r) \leq -\delta$ when $s \geq N_0$ and $f(t,r,s) > \delta$ when $s \leq -N_0$.

Then the BVProblem (3.1) has a solution in $W_2^2([0,1])$.

Proof: In virtue of Lemma 3.1 and Theorem 1.2 (with Remarks 1.0 and 1.1), to prove Theorem 3.1 it suffices to show that (H1) implies B_1^+, while (H1') implies B_1^-. Thus, suppose first that (H1) holds. We have to show that if $\{u_n\} \subset X$ is a sequence such that $||u_n||_2 \to \infty$ and $\{u_n/||u_n||_2\}$ converges in X to the function whose constant value is either $+1$ (resp. -1), then

$$\underline{\lim} \int_0^1 g(t,u_n'(t),u''_n(t)) + f(t,u_n(t),u'_n(t))\,dt > 0$$

$$(\text{resp. } \overline{\lim} \int_0^1 \{g(t,u'_n,u''_n) + f(t,u_n,u'_n)\}dt < 0).$$

Suppose first that $y_n \equiv u_n/||u_n||_2 \to +1$ in X as $n \to \infty$. Then, since X is compactly imbedded into $C([0,1])$, $y_n(t) \to 1$ uniformly on $[0,1]$. Hence there exists $n_0 \geq 1$ such that $u_n(t) = ||u_n||y_n(t) \geq N_0$ if $n_0 \geq N_0$ and $t \in [0,1]$ and therefore, by (H1),

$$g(t,u'_n,u''_n) + f(t,u_n,u'_n) \geq f(t,u_n(t),u'_n(t)) - |g(t,u'_n,u''_n)|$$
$$\geq \delta - p > 0$$

for $t \in [0,1]$ and $n \geq n_0$. This implies the above inequality. In a similar way one shows that (H1') implies B_1^- of Remark 1.0.

Remark 3.0. In [9] the solvability of (3.1) was studies by means of the degree theory of condensing vector fields under the more restrictive and essential assumption that

(3.4) $|g(t,s,r_1) - g(t,s,r_2)| \leq c_0|r_1 - r_2|$ for some $c_0 \in (0,1)$

and all $t \in [0,1]$ and $s,r_1,r_2 \in \mathbb{R}$. The author of [9] also assumes that $g(t,s,r)$ is continuous in s uniformly for $(t,r) \in [0,1] \times \mathbb{R}$.

Problem 2. The abstract results below are modelled on situations which arise when one attempts to treat semilinear elliptic problems on unbounded domains (see [29] and others cited there, where global bifurcation phenomenon is treated for such operators). The interesting feature of this problem is that the resolvents of the corresponding elliptic operators are no longer compact, the Sobolev Imbedding theorems in general are no longer valid and unless very strict conditions are imposed, the nonlinear part is not compact. As a second example we consider the solvability of the equation

(3.5) $\quad Au - \lambda u + Nu = f \quad (u \in D(A),\ f \in H,\ \lambda \in \mathbb{R}),$

where H is a real separable Hilbert space and A, N and λ are assumed to satisfy the following hypotheses:

(a2) A is a densely defined, positive definite, self-adjoint operator whose essential spectrum $\sigma_e(A)$ is bounded below, i.e., there is a number $\gamma > 0$ such that, for each $\varepsilon > 0$, $\sigma(A) \cap (-\infty, \gamma - \varepsilon)$ consists of a nonempty set of isolated eigenvalues, each of finite multiplicity, with

$$\lambda_0 < \lambda_1 < \lambda_2 < \; , \ldots, < \gamma.$$

(b2) The number λ is an eigenvalue of A such that $|\lambda|\gamma^{-1} < 1$.

(c2) $N\colon\ D(A^{1/2}) \to H$ is a map such that $N = N_1 + N_2$ where $N_2\colon\ H_0 \to H$ is compact, $N_1\colon\ H_0 \to H$ is continuous and monotone (i.e., $(N_1u - N_1v, u - v) \geq 0 \ \forall u, v \in H_0$) and $||Nu||/||u||_0 \to 0$ as $||u||_0 \to \infty$, where H_0 is the completion of $D(A)$ in the metric $[u,v] = (Au,v)$ and $||u||_0 = (Au,u)^{1/2}$ for all $u,v \in D(A)$.

It is known that H_0 is continuously imbedded into $H, H_0 = D(A^{1/2})$, $||u||_0 = ||A^{1/2}u||$ for all u in H_0, A has an inverse A^{-1} defined on H, $A^{-1}\colon\ H \to H$ is bounded, self-adjoint and positive, its square root $A^{-1/2}\colon\ H \to H$ is also self-adjoint positive and bounded, and the square root $A^{1/2}$, considered as a mapping from H_0 to H, is a linear homeomorphism. Moreover, it follows from condition (a2) that $A^{-1}\colon H \to H$ is γ^{-1}-ball-contractive and $S \equiv A^{-1/2}\colon\ H \to H$ is $\gamma^{-1/2}$-ball-contractive (see Stuart [28]). This and (b2) imply that λA^{-1} is k-ball -contractive with $k = |\lambda|\gamma^{-1} < 1$, the operator $L \equiv I - \lambda A^{-1}$ has a nontrivial null space $N(L)$ and $u \in N(L)$ if and only if $Au - \lambda u = 0$. Furthermore, $F = SNS\colon\ H \to H$ is such that $F = F_1 + F_2$ with $F_2 = SN_2S$ compact and $F_1 = SN_1S$ monotone and continuous by (c2).

Remark 3.1. It is easy to see that $u \in D(A)$ is a solution of (3.5) if and only if $v = A^{1/2}u$ and v is a solution of

$$Lv + Fv = S(f) \quad (v \in H,\ S(f) \in H,\ F = SNS). \tag{3.6}$$

We are now in the situation where we can apply to (3.5), or equivalently to (3.6), some of the results of Sections 1 and 2. To accomplish this we set $Y = X = H$, $\Gamma = \Gamma_1 = \{X_n, P_n\}$ and $T = L + F$. Example 1 shows that L is A-proper w.r.t. Γ_1 and therefore Fredholm with $\mathrm{ind}(L) = 0$ (see [22,24]). In view of the above discussion concerning $F = F_1 + F_2\colon H \to H$, it follows from Lemma 2.8 in [29] (see Example 1) that $T = L + F$ is A-proper w.r.t. Γ_1. Moreover, the last part of (c2) implies that $||Fv||/||v|| \to 0$ as $||v|| \to \infty$. Indeed, since $||Fv|| \leq ||S||\ ||N(Sv)||$ for each v in H, to prove the latter assertion it suffices to show that $||N(Sv)||/||v|| \to 0$ as $||v|| \to \infty$. But, since $u = Sv \in H_0$ and $||u||_0 = ||A^{1/2}Sv|| = ||v||$, it follows from the last assumption in (c2) that

$$\frac{||N(Sv)||}{||v||} = \frac{||Nu||}{||u||_0} \to 0 \quad \text{as} \quad ||v|| = ||u||_0 \to \infty.$$

We are now in the situation to which Theorem 1.2 and Remark 1.1 (i.e., the A-proper mapping version of Theorem 1.2) apply.

Theorem 3.2. Suppose that conditions (a2), (b2) and (c2) hold. Then (3.5) has a solution $u \in D(A)$ for a given f in H if the following condition holds:

(H2) Either $\overline{\lim}(Nu_j, u_0) > (f, u_0)$ or $\underline{\lim}(Nu_j, u_0) < (f, u_0)$ whenever $\{u_j\} \subset H_0$ is such that $||u_j||_0 \to \infty$ and $u_j/||u_j||_0 \to u_0$ in H_0 with $u_0 \in N(A - \lambda I)$.

Proof: Since the solvability of (3.5) is equivalent to the solvability of (3.6) in H, in view of the preceding discussion and Theorem 1.2 with Remark 1.1, it suffices to show that either $\overline{\lim}(Fv_j, v_0) > (Sf, v_0)$ or $\underline{\lim}(Fv_j, v_0) < (Sf, v_0)$ whenever $\{v_j\} \subset H$ is such that $||v_j|| \to \infty$ and $v_j/||v_j|| \to v_0 \in N(L)$.

To show this, let $\{v_j\} \subset H$ be such that $||v_j|| \to \infty$ and $\frac{v_j}{||v_j||} \to v_0 \in N(L)$ Now since $A^{1/2}$ is a linear homeomorphism of H_0 onto H, there exist $u_0 \in H_0$ and $u_j \in H_0$ for each $j \in N$ such that $v_0 = A^{1/2}u_0$ and $v_j = A^{1/2}u_j$ with $||u_j||_0 = ||v_j|| \to \infty$ and

$$\left|\left|\frac{u_j}{||u_j||_0} - u_0\right|\right|_0 = \left|\left|A^{1/2}\left(\frac{u_j}{||u_j||_0} - u_0\right)\right|\right| = \left|\left|\frac{v_j}{||v_j||} - v_0\right|\right| \to 0$$

as $j \to \infty$. Moreover, since $v_0 = A^{1/2}u_0 \in N(L)$ and $L = I - \lambda A^{-1}$, it follows that $0 = L(A^{1/2}u_0) = A^{1/2}u_0 - \lambda A^{-1/2}u_0$ and so $Au_0 - \lambda u_0 = 0$. Thus $\{u_j\} \subset H_0$ is such that $||u_j|| \to \infty$ and $u_j/||u_j||_0 \to u_0$ in H_0 with $u_0 \in N(A - \lambda I)$ and therefore, by (H2), either $\overline{\lim}(Nu_j, u_0) > (f, u_0)$ or $\underline{\lim}(Nu_j, u_0) < (f, u_0)$. Suppose that $\overline{\lim}(Nu_j u_0) > (f, u_0)$. In view of this and the fact that $u_j = Sv_j$, $u_0 = Sv_0$ and S: $H \to H$ is bounded and self-adjoint, we obtain the desired asymptotic positivity

$$(v_0, Sf) = (u_0, f) < \overline{\lim}(u_0, Nu_j) = \overline{\lim}(Sv_0, NSv_j) = \overline{\lim}(v_0, Fv_j)$$

when Theorem 1.2 (with Remark 1.1) is applied to (3.6). Similarly one shows that $\underline{\lim}(Fv_j, v_0) < (f, Sv_0)$ follows from $\underline{\lim}(Nu_j, u_0) < (f, u_0)$.

Q.E.D.

Problem 3. Let $Q \subset R^n$ be a bounded domain with sufficiently smooth boundary so that the Sobolev Imbedding Theorem holds. Let $W_2^m(Q) \equiv W_2^m$ be the Sobolev space of all real functions u such that u and its generalized derivatives $D^\alpha u \in L_2(Q)$ for $|\alpha| \leq m$, where $\alpha = (\alpha_1, \ldots, \alpha_n)$ is the multiindex with $|\alpha| = \alpha_1 + \ldots + \alpha_n$. W_2^m is a separable Hilbert space with the inner product

$$(u, v)_m = \sum_{|\alpha| \leq m} \int_Q D^\alpha u D^\alpha v dx.$$

Let $\mathring{W}_2^m$ be the subspace of W_2^m which is the completion in W_2^m-norm of $C_0^\infty(Q)$, the set of infinitely differentiable functions with compact support in Q.

As an application of Corollary 2.1 we consider the following BVProblem:

$$(3.7)\quad \begin{cases} \sum_{|\alpha|,|\beta|\le m} (-1)^{|\alpha|} D^\alpha (a_{\alpha\beta}(x) D^\beta u) + h(x,u(x)) = f(x) \quad (x \in Q) \\ D^\alpha u = 0 \quad (x \in \partial Q, \quad |\alpha| \le m-1, \end{cases}$$

where $a_{\alpha\beta}(x)$ and $h(x,u)$ satisfy the following conditions:

(a3) $a_{\alpha\beta}(x) \in L^\infty(\overline{Q})$ for $|\alpha|,|\beta| \le m$, $a_{\alpha\beta}(x) \in C(\overline{Q})$ for $|\alpha| = |\beta| = m$, and $\sum_{|\alpha|,|\beta|=m} a_{\alpha\beta}(x)\eta^\alpha\eta^\beta \ge d|\eta|^{2m}\ \forall \eta \in R^n$ and some $d > 0$.

(b3) $h\colon\ Q \times \mathbb{R} \to \mathbb{R}$ is continuous and $|h(x,s)| \le a(x)$ $\forall x \in Q$, $s \in \mathbb{R}$ and some $a \in L_2$.

For a given $f \in L_2(Q)$, by a <u>weak solution</u> of (3.7) we mean a function $u \in \mathring{W}_2^m$ such that

$$(3.8)\qquad a(u,v) + b(u,v) = \langle f,v\rangle \quad \forall v \in \mathring{W}_2^m,$$

where $a(u,v)$ and the $b(u,v)$ are generalized Dirichlet forms on $\mathring{W}_2^m$ given by

$$(3.9)\quad a(u,v) = \sum_{|\alpha|,|\beta|\le m} \langle a_{\alpha\beta} D^\beta u, D^\alpha v\rangle,\ b(u,v) = \langle h(x,u),v\rangle$$

where $\langle\ ,\ \rangle = (\ ,\ )_0$ and $||\cdot|| = ||\cdot||_0$ is the inner product and the norm in $L_2(Q)$ respectively. Condition (a3) implies that $a(u,v)$ is a bounded bilinear form on $\mathring{W}_2^m$ and hence there exists a unique mapping $L \in L(\mathring{W}_2^m, \mathring{W}_2^m)$ such that

$$(3.10)\qquad (Lu,v)_m = a(u,v) \quad \forall u,v \in \mathring{W}_2^m$$

and L satisfies the Garding inequality (see [10]):

(3.11) $(Lu,u)_m \geq c_1||u||_m^2 - c_2||u||^2 \ \forall u \in \overset{\circ}{W}{}_2^m$ and some $c_1 > 0$, $c_2 \geq 0$.

Similarly condition (b3) implies that $b(u,v)$ is a continuous linear functional of v in $\overset{\circ}{W}{}_2^m$ and so it determines a unique bounded continuous map $N: \overset{\circ}{W}{}_2^m \to \overset{\circ}{W}{}_2^m$ such that $(Nu,v)_m = b(u,v)$ for all $u,v \in \overset{\circ}{W}{}_2^m$. It was shown in [24] that since the imbedding of $\overset{\circ}{W}{}_2^m$ into L_2 is compact, it follows from (3.11) that $L: \overset{\circ}{W}{}_2^m \to \overset{\circ}{W}{}_2^m$ is A-proper with respect to any given projectionally complete scheme $\Gamma_0 = \{X_n, P_n\}$ for $\overset{\circ}{W}{}_2^m$. The latter exists since $\overset{\circ}{W}{}_2^m$ is separable. It is obvious that, in view of our conditions on $h(x,s)$ and the Sobolev Imbedding Theorem, N is compact and $||Nu||_m/||u||_m \to 0$ as $||u||_m \to \infty$. Let $w_f \in \overset{\circ}{W}{}_2^m$ be such that $\langle f,v \rangle = (w_f,v)_m$ for all v in $\overset{\circ}{W}{}_2^m$. Then the equation

$$(3.12) \qquad Lu + Nu = w_f \quad (u \in \overset{\circ}{W}{}_2^m,\ w_f \in \overset{\circ}{W}{}_2^m)$$

is equivalent to the conditions (3.8) for the weak solution of (3.7). It follows from Lemma 1.1 that if $N(L) = \{0\}$, then (3.7) has a weak solution for every $f \in L_2(Q)$. However, if $N(L) \neq 0$, then (3.12) need not have a solution for every $w_f \in \overset{\circ}{W}{}_2^m$. To apply our results of Section 1 to the solvability of (3.12) when $N(L) \neq \{0\}$, note first that $L^*: \overset{\circ}{W}{}_2^m \to \overset{\circ}{W}{}_2^m$ is also A-proper w.r.t. Γ_0 since L^* satisfies the same inequality as L. Consequently, by Theorem 2.3B in [24], the A-proper map L is Fredholm with $\text{ind}(L) = 0$. Moreover, $\phi(u) = (Lu,u)_m$ is weakly lower semicontinuous. The last fact will be used elsewhere.

To state our existence results for (3.7) or (3.12) let us introduce the following symbols (see [0]):

$$\underline{h}(\pm\infty) = \liminf_{s \to \pm\infty} h(x,s),\ \overline{h}(\pm\infty) = \limsup_{s \to \pm\infty} h(x,s) \quad (x \in Q)$$

$Q^+(w) = \{x \in Q | w(x) > 0\}, \quad Q^-(w) = \{x \in Q | w(x) < 0\} \quad (w \in N(L))$

Theorem 3.3. Suppose that (a3) and (b3) hold and suppose further that $N(L^*) = N(L)$. Then, given f in $L_2(Q)$, Eq. (3.7) has a weak solution provided that either

$$(3.13) \qquad \int_Q fwdx < \int_{Q^+(w)} \underline{h}(\infty)wdx + \int_{Q^-(w)} \overline{h}(-\infty)wdx \qquad \forall w \in N(L),\ ||w||_m = 1$$

or

$$(3.14) \qquad \int fwdx > \int_{Q^+(w)} \overline{h}(\infty)wdx + \int_{Q^-(w)} \underline{h}(-\infty)wdx \qquad \forall w \in N(L),\ ||w||_m = 1$$

Proof: In view of the preceding discussion and Remark 1.1, Theorem 3.3 will follow from Theorem 1.2 (with $Y = X = \underline{H}$) if we can show that (3.13) implies B_1^+ of Remark 1.0 (i.e., $\underline{\lim} \int_Q h(x,u_n)wdx > \int_Q fwdx$ whenever $\{u_n\}$ is a sequence in $\overset{\circ}{W}{}_2^m$ such that $||u_n||_u \to \infty$ and $u_n/||u_n||_m \to w \in N(L))$ or that (3.14) implies B_1^- of Remark 1.0.

We first claim that (3.13) implies B_1^+ of Remark 1.0 with $Y = X = \overset{\circ}{W}{}_2^m$ and $M = I$. If not, then there would exist a sequence $\{u_n\} \subset \overset{\circ}{W}{}_2^m$ with $t_n = ||u_n||_m \to \infty$ and $w_n = t_n^{-1}u_n \to w \in N(L)$ with $||w||_m = 1$ such that

$$(3.15) \qquad \underline{\lim} \int_Q h(x,u_n(x))w(x)dx \le \int_Q f(x)w(x)dx.$$

Suppose that $\{u_n\}$ and w are as above and note that

$$(3.16) \qquad \begin{cases} \underline{\lim} \int_Q h(x,u_n)wdx = \underline{\lim}\{\int_{Q^+(w)} h(x,u_n(x))w(x)dx \\ \qquad + \int_{Q^-(w)} h(x,u_n(x))w(x)dx\} \\ \ge \underline{\lim} \int_{Q^+(w)} h(x,u_n(x))w(x)dx + \underline{\lim} \int_{Q^-(w)} h(x,u_n(x))w(x)dx. \end{cases}$$

Since $\{u_n\}$, or at least a subsequence, converges pointwise a.e.

on Q to w, it follows that if $x \in Q^+(w)$ then $u_n(x) = t_n w_n(x) \to +\infty$, while if $x \in Q^-(w)$ then $u_n(x) = t_n w_n(x) \to -\infty$ as $n \to \infty$. In view of this and the boundedness of h, an application of the Lebesque-Fatou's lemma leads to

$$\underline{\lim}\int_{Q^+(w)} h(x,u_n)w\,dx + \underline{\lim}\int_{Q^-(w)} h(x,u_n)w\,dx \geq \int_{Q^+(w)} \underline{h}(\infty)w\,dx + \int_{Q^-(w)} \overline{h}(-\infty)w\,dx.$$

This together with (3.15) and (3.16) implies that

$$\int_Q fw\,dx \geq \int_{Q^+(w)} \underline{h}(\infty)w\,dx + \int_{Q^-(w)} \overline{h}(-\infty)w\,dx, \quad ||w||_m = 1, \; w \in N(L),$$

in contradiction to (3.13).

In a similar way one shows that (3.14) implies B_1^- of Remark 1.0.

Theorem 3.3 is related to Theorem 2.1 in [0] which is proved there under different conditions on h by means of the Lyapunov-Schmidt method (see [3,9,18,20,39] and others where similar problems are treated).

Problem 4. In our final problem we use an approach similar to that of Problem 1 to study the solvability of the BVProblem

$$(4.1)\quad \begin{cases} -\Delta u(x) - g(x,\nabla u(x),\Delta u(x)) + h(u(x)) = f(x), & (x \in Q) \\ \dfrac{\partial u}{\partial \eta}(x) = 0, & (x \in \partial Q), \end{cases}$$

which was studied in [9] under more restrictive conditions on g and the Sobolev spaces. We present it here in order to contrast the applicability of our Theorem 1.2 for A-proper maps with the results in [9] for condensing maps. To state the conditions on g and h under which (4.1) has a solution in $W_2^2(Q)$ for each $f \in L_2(Q)$ we let $X = \{u \in W_2^2(Q), \frac{\partial u}{\partial \eta} = 0$ for $x \in \partial Q\}$ and $Y = L_2(Q)$. Let $L\colon X \to Y$ be defined by $Lu = -\Delta u$ for $u \in X$. It is known (see [10]) that $N(L)$ consists of constant functions, $R(L)$ consists of those functions whose mean value is zero, ind$(L) = 0$, $Y = N(L) \oplus R(L)$ and $K = L + I: X \to Y$ is a linear homeomorphism, where $I: X \to Y$ is the inclusion map which is compact by Sobolev Imbedding Theorem. In what follows we shall use $||\cdot||_0$ to denote the equivalent norm in X given by $||u||_0 = ||Ku||$.

Let $\{X_n\} \subset X$ be such that dist$(u, X_n) \to 0$ as $n \to$ for each u in X and let $Y_n = K(X_n)$ for each n. Then, if $P_n: X \to X_n$ and $Q_n: Y \to Y_n$ denote the orthogonal projections, then $\Gamma_0 = \{X_n, P_n; Y_n, Q_n\}$ in complete for (X,Y). A direct calculation shows that

$$(4.2) \qquad (Lu, Ku) \geq ||u||_0^2 - ||u||_1^2 \text{ for all } u \in X.$$

Since W_2^1 is compactly imbedded in W_2^2, it follows from (4.2) that $L: X \to Y$ is A-proper w.r.t. Γ_0. Suppose that the functions g and h satisfy the following conditions:

(a4) $g: Q \times \mathbb{R}^{n+1} \to \mathbb{R}$ is continuous and there exist α $(0,1)$ and $\phi(x) \in L_2(Q)$ such that $|g(x,r,s)| \leq \phi(x)$ for $(x,r,s) \in Q \times \mathbb{R}^n \times \mathbb{R}$ and $(g(x,r,s_1) - g(x,r,s_2))(s_1 - s_2) \geq -\alpha|s_1 - s_2|^2$ for $(x,r) \in Q \times \mathbb{R}^n$ and $s_1, s_2 \in \mathbb{R}$.

(b4) $h: \mathbb{R} \to \mathbb{R}$ is continuous and there are $a,b \subset \mathbb{R}^+$ and $\delta \in (0,1)$ such that $|h(s)| \leq a + b|s|^\delta$ for $s \in \mathbb{R}$ and $\lim_{s\to\infty} h(s) = +\infty$, while $\lim_{s\to-\infty} h(s) = -\infty$.

We define the operators $C, G: X \to Y$ by $Cu = h(u(x))$ and $Gu = g(x, \nabla u(x), \Delta u(x))$ for $u \in X$. First, (b4) implies that C is compact and $||Cu||/||u||_0 \to 0$ as $||u||_0 \to \infty$. Second, (a4) implies that G is continuous, $R(G)$ is bounded and for $u,v \in X$

$$(4.3) \qquad (Gu - Gv, K(u - v)) \geq -\alpha||\Delta(u - v)||^2 - 2p||u - v|| + (g(x,\nabla u,\Delta v) - g(x,\nabla v,\Delta v), \Delta u - \Delta v).$$

where $p = ||\phi||$. From (4.2) and (4.3) follows

$$(4.4) \qquad ((L + G)u - (L + G)v, K(u - v)) \geq (1 - \alpha)||u - v||_0^2 - ||u - v||_1^2 - 2p||u - v|| + (g(x,\nabla u,\Delta v) - g(x,\nabla v,\Delta v), \Delta u - \Delta v) \quad \text{for } u,v \in X.$$

In view of (4.4), the results in [24] imply that $L + G$ is A-proper w.r.t. Γ_0 and so is the map $T = L + N$ with $N = G + C$ because C is compact. We are now in the situation to which Theorem 1.2 applies. Indeed, if $\{u_n\} \subset X$ is such that $t_n \equiv ||u_n||_0 \to \infty$ and $w_n \equiv u_n/||u_n||_0 \to w \in N(L)$ as $n \to \infty$, then since w is a nonzero constant and $\{w_n\}$, or at least a subsequence, converges pointwise a.e. on Q to w, it follows that $h(u_n) = h(t_n w_n(x)) \to +\infty$ a.e. when $w > 0$ and $h(u_n) \to -\infty$ a.e. when $w < 0$. Hence, in either case, $\lim_{n\to\infty}(Nu_n, w) = +\infty$. Thus we have the following improvement of [9, Theorem 3.7].

Theorem 3.4. Suppose that g and f satisfy (a4) and (b4) respectively. Then the problem (4.1) has a solution in X for each $f \in L_2$.

In [9] the solvability of (4.1) for each $f \in L_p$ was proved by means of the degree theory of condensing maps under the more restrictive and essential conditions that $p > n$, $|g(x,r,s_1) - g(x,r,s_2)| \leq \alpha|s_1 - s_2|$ for some $\alpha \in (0,1)$ and all $(x,r) \in Q \times \mathbb{R}^n, s_1, s_2 \in \mathbb{R}$, and that $g(x,r,s)$ is continuous in $r \in \mathbb{R}^n$ uniformly with respect to $(x,s) \in Q \times \mathbb{R}$.

REFERENCES

[0] Ambrosetti, A. and Mancini, G., *Theorems of existence and multiplicity for nonlinear elliptic problems with noninvertible linear part*, (to appear).

[1] Brezis, H., *Équations et inéquations non-linéaires dans les espaces vectoriels en dualité*, Ann. Inst. Fourier (Grenoble) 18 (1968), fasc. 1, pp. 115-175.

[2] ________, *Quelques proprietes des operateurs monotones et des semi-groupes non lineaires*, Lecture Notes in Math., #543 (1976), pp. 58-82.

[3] Brezis, H. and Nirenberg, L., *On some linear operators and their ranges*, Ann. Sc. Norm. Sup. Pisa, (to appear).

[4] Browder, F. E., *Nonlinear operators and Nonlinear Equations of Evolution in Banach spaces*, Proc. Symp. in Pure Math, AMS, 18 (1976), Providence, R.I.

[5] Browder, F. E. and Petryshyn, W. V., *Approximation methods and the generalized topological degree for nonlinear mappings in Banach spaces*, J. Functional Anal., 3 (1969), pp. 217-245.

[6] Cesari, L., *Alternative Methods in Nonlinear Analysis*, in "Intern. Conf. Diff. Equations", H. A. Antosiewics, ed., Acad. Press, 1975, pp. 95-148 (Ch. II).

[7] Dancer, E. N., *Some remarks on a theorem of Kachurovskii*, (to appear).

[8] DeFigueiredo, D. G., *On the range of nonlinear operators with linear asymptotes which are not invertible*, Comment. Math. Univ. Carolinae 15 (1974), pp. 415-428.

[9] Fitzpatrick, P. M., *Existence results for equations involving noncompact perturbations of Fredholm mappings with applications to differential equations*, J. Math. Anal. Appl. (to appear).

[10] Friedman, A., *Partial Differential Equations*, Holt. Rinehart and Winston, New York, 1969.

[11] Fučik, S., *Surjectivity of operators involving linear noninvertible part and nonlinear compact perturbation*, Funkciolaj Ekvacioj, 17 (1974), pp. 73-83.

[12] ________, *Ranges of Nonlinear Operators*, Lecture Notes I, II, Charles University Praha (1977).

[13] Fučik, S., Kučera, M., and Nečas, J., *Ranges of nonlinear asymptotically linear operators*, J. Diff. Equations 17 (1975), pp. 375-394.

[14] Gaines, R. E. and Mahwin, J. L., *Coincidence degree and Nonlinear Differential Equations*, Lecture Notes No. 568, Springer Verlag, Berlin, 1977.

[15] Hess, P., *On a theorem of Landesman and Lazer*, Indiana Univ. Math. J. 2 (1974), pp. 827-830.

[16] Kačurovskii, R. I., *On the Fredholm theory for nonlinear operator equations*, Dokl. Akad. Nauk SSSR 192 (1970), pp. 969-972.

[17] Kannan, R. and Schur, J., *Boundary value problems for even order nonlinear differential equations*, Bull. AMS, 82, 1 (1976).

[18] Landesman, E. M. and Lazer, A. C., *Nonlinear perturbations of linear elliptic boundary value problems at resonance*, J. Math. Mech. 19 (1970), pp. 609-623.

[19] Milojević, P. S., *A generalization of Leray-Schauder Theorem and surjectivity results for multivalued A-proper and pseudo-A-proper mappings*, Nonlin. Anal. Theory & Appl. 1 (3) (1977), pp. 263-276.

[20] Nečas, J., *On the range of nonlinear operators with linear asymptotes which are not invertible*, Comment. Math. Univ. Carolinae, 14 (1973), pp. 63-72.

[21] Nirenberg, L., *An application of generalized degree to a class of nonlinear problems*, Trois, Coll. Anal. Fonction-elle, Liege (1970)(Math. Vander, pp. 57-74).

[22] Nussbaum, R. N., *The radius of the essential spectrum*, Duke Math. J. 38 (1970), pp. 473-478.

[23] Petryshyn, W. V., *On nonlinear equations involving pseudo-A-proper mappings and their uniform limits with applications*, J. Math. Anal. Appl. 38 (1972), pp. 672-720.

[24] ________, *On the approximation-solvability of equations involving A-proper and pseudo-A-proper mappings*, Bull. AMS, 81 (1975), pp. 223-312.

[25] ________, *Fredholm alternative for nonlinear A-proper mappings with applications to nonlinear elliptic boundary value problems*, J. Funct. Anal. 18 (1975), pp. 288-317.

[26] ________, *On the solvability of nonlinear equations involving abstract and differential equations*, M. Z. Nashed, ed., Lecture Notes, Springer-Verlag, Berlin (to appear).

[27] Sadovskii, B. N., *Ultimately compact and condensing mappings*, Uspehi Mat. Nauk 27 (1972), pp. 81-146.

[28] Stuart, C. A., *Some bifurcation theory for k-set-contractions*, Proc. London Math. Soc. 27 (1973), pp. 531-550.

[29] Tolland, J. F., *Global bifurcation theory via Galerkin method*, Nonlinear Anal., Theory, Methods and Appl. 1 (1977), pp. 305-317.

[30] Webb, J. R. L., *On a characterization of k-set-contractions*, Accad. Naz. Dei Lincei Ser. 8, 50 (1971), pp. 358-361.

[31] Williams, S. A., *A sharp sufficient condition for solutions of a nonlinear elliptic boundary value problem*, J. Diff. Equations, 8 (1970), pp. 580-586.

ITERATIVE METHODS FOR ACCRETIVE SETS*

Simeon Reich
University of Chicago
and
University of Southern California

Let E be a real Banach space. A subset A of $E \times E$ with domain $D(A)$ and range $R(A)$ is said to be accretive if for all $[x_i, y_i] \in A$, $i = 1, 2$, and $r > 0$, $|x_1 - x_2| \leqslant |x_1 - x_2 + r(y_1 - y_2)|$. Let I denote the identity. An accretive set A is m-accretive (hypermaximal accretive in Browder's terminology [3]) if $R(I + A) = E$. (It then follows that $R(I + rA) = E$ for all positive r.) In 1970 Martin [12] showed that an accretive and continuous $A: E \to E$ is m-accretive. He used an existence theorem for the initial value problem $u' + Au = 0$, $u(0) = x$. Crandall and Pazy [8] have recently proved Martin's theorem without using differential equations by considering the following iterative scheme (and employing some ideas due to Kobayashi [11]):

(1) $$x_{n+1} - x_n + h_{n+1}(y_{n+1} + x_{n+1}) = w_{n+1}, \quad n \geqslant 0$$

where $[x_n, y_n] \in A$, $h_n > 0$, $\sum_{n=1}^{\infty} h_n = \infty$, and $\sum_{n=1}^{\infty} |w_n| < \infty$. Under certain conditions $\{x_n\}$ exists and converges strongly to the zero of $I + A$. Our main purpose in the present paper is to study an extension of this scheme that will enable us to find a zero of A itself:

* Partially supported by the NSF, Grant MCS 74-07495.

ISBN 0-12-434160-8

(2) $\quad x_{n+1} - x_n + h_{n+1}(y_{n+1} + p_{n+1}x_{n+1}) = w_{n+1}, \quad n \geqslant 0,$

where $[x_n, y_n] \in A$, $p_n, h_n > 0$, $\{p_n\}$ decreases to $p \geqslant 0$, $\sum_{n=1}^{\infty} p_n h_n = \infty$, and $\sum_{n=1}^{\infty} |w_n| < \infty$. We will also discuss related iterative procedures and indicate some applications.

For x and y in E, let $(x,y)_{\pm} = \lim_{t\to 0\pm} (|x+ty| - |x|)/t$.

Proposition 1. Let A be accretive and suppose that $\{x_n\}$ satisfies (2). If $x_{\infty} = \lim_{n\to\infty} x_n$ exists, then $(x - x_{\infty}, y + px_{\infty})_+ \geqslant 0$ for all $[x,y] \in A$.

Proof. Let r be positive and $y \in Ax$. Since A is accretive, we have $|x - x_{n+1}| \leqslant |x - x_{n+1} + rh_{n+1}(y - y_{n+1})/$ $/(r + h_{n+1} + rh_{n+1}p_{n+1})|$. Multiplying by $(r + h_{n+1} + rh_{n+1}p_{n+1})/r$ and replacing $y - y_{n+1}$ with $(x_{n+1} - x_n + h_{n+1}p_{n+1}x_{n+1} - w_{n+1} + h_{n+1}y)/h_{n+1}$, we get $(1 + h_{n+1}p_{n+1} + h_{n+1}/r)|x - x_{n+1}| \leqslant |x - x_n| + |w_{n+1}|$ $+ h_{n+1}|x - x_{n+1} + r(y + p_{n+1}x)|/r$, and $(1 + h_{n+1}p_{n+1})|x - x_{n+1}|$ $\leqslant |x - x_n| + |w_{n+1}| + h_{n+1}(|x - x_{n+1} + r(y + p_{n+1}x)|$ $- |x - x_{n+1}|)/r$. It follows that for $i > k$,

$$(3)\quad |x - x_i| \leqslant |x - x_k| \prod_{j=k+1}^{i} (1+p_j h_j)^{-1} + \sum_{j=k+1}^{i} \Big(\prod_{m=j}^{i} (1+p_m h_m)^{-1}\Big)|w_j|$$
$$+ r^{-1} \sum_{j=k+1}^{i} \Big(\prod_{m=j}^{i} (1+p_m h_m)^{-1} h_j (|x - x_j + r(y+p_j x)| - |x - x_j|)\Big).$$

The first two terms on the right hand side of this inequality tend to zero as $i \to \infty$. Also, $\sum_{j=k+1}^{i} \prod_{m=j}^{i} (1 + p_m h_m)^{-1} p_j h_j \leqslant 1$, and $\limsup_{j\to\infty} (x - x_j, y + p_j x)_+ \leqslant (x - x_{\infty}, y + px)_+$. Consequently, $p|x - x_{\infty}| \leqslant (x - x_{\infty}, y + px)_+$ and the result follows.

Let $cl(D)$ denote the closure of a subset D of E. In addition to the assumptions of Proposition 1, suppose that A is maximal accretive in $cl(D(A))$, or that A is closed and $\liminf_{h\to 0+} d(x,R(I + h(A + pI)))/h = 0$ for all x in $cl(D(A))$, where $d(x,D) = \inf\{|x - y|: y \in D\}$. Then $x_\infty = J_{1/p}0$ (where J_r denotes the resolvent of A) if $p > 0$ and $x_\infty \in A^{-1}(0)$ if $p = 0$. If (2) is replaced by

$$(4)\ x_{n+1} - x_n + h_{n+1}(y_{n+1} + p_{n+1}(x_{n+1} - z)) = w_{n+1},\quad n \geq 0,$$

and $p > 0$, then $\lim_{n\to\infty} x_n = J_{1/p}z$. In order to further identify x_∞ in case $p = 0$, recall [4, 14] that if C is a closed convex subset of E and F a closed subset of C, then a retraction $P: C \to F$ is called sunny if $P((1 - t)Px + tx) = Px$ whenever $t \geq 0$ and x and $(1 - t)Px + tx$ belong to C.

Proposition 2. Let A be accretive and E smooth. Suppose that for each z in a convex $cl(D(A))$ there exists a convergent sequence $\{x_n\}$ that satisfies (4) with $p = 0$. Then $\lim_{n\to\infty} x_n = Pz$ where P is the unique sunny nonexpansive retraction of $cl(D(A))$ onto $A^{-1}(0)$.

Proof. Assume for a moment that $z = 0$. Using (3) with $x \in A^{-1}(0)$ and $y = 0$ ($A^{-1}(0)$ is not empty by Proposition 1), we obtain $|x - x_\infty| \leq (x - x_\infty, x)_+$. Hence $(x - x_\infty, x_\infty)_+ \geq 0$ and $(x - x_\infty(z), x_\infty(z) - z)_+ \geq 0$ for all z in $cl(D(A))$ and x in $A^{-1}(0)$. The result now follows from [4, Theorem 1] and [14, Lemma 2.7].

Note that if $R(I + rA) \supset cl(D(A))$ for all positive r, and $cl(D(A))$ is convex, then there always exists a sequence $\{x_n\}$ that satisfies (4) (with $w_n = 0$ for all n). We also observe that if $p_n = p > 0$ for all n, then a sequence $\{x_n\}$ that satisfies (2) (or (4)) is always strongly convergent (even if $\sum_{n=1}^{\infty} h_n < \infty$). This is essentially due to Crandall and Pazy [8]. The next result provides a sufficient condition for the convergence of $\{x_n\}$ when $p = 0$.

Proposition 3. Suppose that $\{x_n\}$ satisfies (4) with $p = 0$ and that the strong $\lim_{r\to\infty} J_r z$ exists. If

$$\lim_{n\to\infty} (p_{n-1}/p_n - 1)/p_n h_n = 0, \tag{5}$$

then the strong $\lim_{n\to\infty} x_n$ exists and equals $\lim_{r\to\infty} J_r z$.

Proof. We may and shall assume that $z = 0$. Denoting $J_{1/p_n} 0$ by u_n, we have $w_n/h_n - p_n x_n \in (x_n - x_{n-1})/h_n + Ax_n$ and $-p_n u_n \in Au_n$. Therefore [7, Lemma 1.7] implies that $|x_n - u_n| \leqslant |x_{n-1} - u_n|/(1 + h_n p_n) + |w_n|/(1 + h_n p_n)$ $\leqslant (|x_{n-1} - u_{n-1}| + |u_n - u_{n-1}|)/(1 + h_n p_n) + |w_n|$. Induction on n yields $|x_n - u_n| \leqslant \prod_{i=k+1}^{n} (1 + h_i p_i)^{-1} |x_k - u_k|$ $+ \sum_{j=k+1}^{n} \prod_{i=j}^{n} (1 + h_i p_i)^{-1} |u_j - u_{j-1}| + \sum_{j=k+1}^{n} |w_j|$. Since $(u_j - u_i, - p_j u_j + p_i u_i)_+ \geqslant 0$ and $\{p_n\}$ is decreasing, $|u_j - u_i| \leqslant M(p_i - p_j)/p_j$ for some constant M and $j \geqslant i$. In particular, $|u_j - u_{j-1}| \leqslant M(p_{j-1}/p_j - 1)$ and the result follows.

Condition (5) is satisfied, for example, if $p_n = n^{-a}$ and $h_n = n^{-b}$ where $0 < a,b$ and $a + b < 1$ (cf. [1]). In Proposition 3 it can be replaced by (cf. [5]) the assumption that there is a strictly increasing sequence $\{n(k)\}$ of positive integers such that

$$\liminf_{k\to\infty} \{p_{n(k)} \sum_{j=n(k)+1}^{n(k+1)} h_j\} > 0 \tag{6}$$

and

$$\lim_{k\to\infty} (p_{n(k)} - p_{n(k+1)}) \sum_{j=n(k)+1}^{n(k+1)} h_j = 0. \tag{7}$$

There are known conditions [16, 17] that guarantee the existence of $\lim_{r\to\infty} J_r z$:

Proposition 4. If $A \subset E \times E$ is an accretive set such that $cl(D(A))$ is convex, $R(I + rA) \supset cl(D(A))$ for all $r > 0$, and $0 \in R(A)$, then each of the following assumptions implies that the strong $\lim_{r\to\infty} J_r z = Pz$ for all z in $cl(D(A))$, where P is the (unique) sunny nonexpansive retraction of $cl(D(A))$ onto $A^{-1}(0)$:

a) E is smooth and uniformly convex with a duality mapping that is weakly sequentially continuous at zero,

b) E is smooth and the resolvents J_r are condensing,

c) $E = L^p$, $1 < p < \infty$, and the resolvents are compact in measure,

d) E is reflexive and A is linear,

e) $A - pI$ is accretive for some $p > 0$.

The example $E = C[0,1]$, $(Af)(x) = (1 - x)f(x)$, $0 \leqslant x \leqslant 1$, shows that $\lim_{r\to\infty} J_r z$ (and $\lim_{n\to\infty} x_n(z)$) do not always exist, even if A is linear.

Combining the previous propositions, we obtain, for example, the following theorem (cf. [17, Theorems 2.3 and 3.2]).

Theorem 1. Let E be a smooth uniformly convex Banach space with a duality mapping that is weakly sequentially continuous at zero, $A \subset E \times E$ an accretive set such that $cl(D(A))$ is convex, $R(I + rA) \supset cl(D(A))$ for all $r > 0$, and $0 \in R(A)$. Let $\{p_n\}$, $\{h_n\}$ be two positive sequences such that $\{p_n\}$ decreases to zero, $\sum_{n=1}^{\infty} p_n h_n = \infty$, and (5) holds. If z is in $cl(D(A))$, then there exists a sequence $\{x_n\} \subset D(A)$ that satisfies (4) and converges strongly to Pz, where P is the unique sunny nonexpansive retraction of $cl(D(A))$ onto $A^{-1}(0)$.

Although (4) is an implicit scheme, the previous propositions can be applied to explicit schemes. For example, let z belong to E and suppose that a sequence $\{x_n\} \subset D(A)$ can be defined by [5]

(8) $$x_{n+1} = x_n - h_n(y_n + p_n(x_n - z)), \quad n \geq 0$$

where $y_n \in Ax_n$. In addition, suppose that A is single-valued and Lipschitzian and that $\{x_n\}$ and $\{y_n\}$ are bounded. Then

$$|x_{n+1} - x_n + h_n(y_{n+1} + p_n(x_{n+1} - z))| \leq h_n|y_{n+1} - y_n| + h_np_n|x_{n+1} - x_n| \leq Mh_n^2$$ for some constant M, so that $\{x_n\}$

satisfies (4) if

(9) $$\sum_{n=0}^{\infty} h_n^2 < \infty.$$

Thus Theorem 1 implies the following result (cf. [16, Proposition 5.2] and [17, Theorem 2.4]).

Theorem 2. Let C be a bounded closed convex subset of a uniformly convex Banach space E with a duality mapping that is weakly sequentially continuous at zero. Suppose a single-valued accretive and Lipschitzian $A: C \to E$ is of the form $I - T$ where T is a self-mapping of C. Let $\{p_n\}$, $\{h_n\}$ be two positive sequences such that $\{p_n\}$ decreases to zero, $\sum_{n=1}^{\infty} p_nh_n = \infty$, $h_n(1 + p_n) \leq 1$ for all n, and (5) and (9) hold. If z is in C, then the sequence $\{x_n\}$ defined by (8) converges strongly to a zero of A.

If we set $p_n = 0$ for all n in (2) we obtain an iterative procedure with a different kind of behavior (even if the errors $w_n = 0$):

(10) $$x_{n+1} = J_{h_{n+1}} x_n, \quad n \geq 0.$$

In particular, a sequence $\{x_n\}$ that satisfies (10) with $0 \in R(A)$ need not converge strongly, even in Hilbert space. The scheme (10) has already been studied by Rockafellar [18] and Brézis and Lions [2] (in Hilbert space). The following two results have been recently established in [6] where other results concerning (10) can be found.

Theorem 3. Let $A \subset E \times E$ be m-accretive with $0 \in R(A)$. Suppose that the modulus of convexity of E satisfies $\delta(\varepsilon) \geqslant K\varepsilon^p$ for some $p \geqslant 2$ and $K > 0$, and that $\{h_n\}$ is a positive sequence such that $\sum_{n=0}^{\infty} h_n^p = \infty$. Let $\{x_n\}$ be defined by (10).

a) If E has a duality mapping that is weakly sequentially continuous at zero, then $\{x_n\}$ converges weakly to a zero of A.

b) If A is linear, then $\{x_n\}$ converges strongly to a zero of A.

Theorem 4. Let $A \subset E \times E$ be m-accretive and h positive.

a) If E is uniformly convex and A is odd, then for each x in E, $\{J_h^n x\}$ converges strongly to a zero of A.

b) If E is uniformly convex with a duality mapping that is weakly sequentially continuous at zero and $0 \in R(A)$, then for each x in E, $\{J_h^n x\}$ converges weakly to a zero of A.

c) If E is reflexive and A is linear, then for each $x \in E$, $\{J_h^n x\}$ converges strongly to a zero of A.

The following theorem is an extension of [13, Propositions 3 and 4]. It can now be proved without using existence theorems for differential equations (cf. [15]).

Theorem 5. Let D be a closed subset of a Banach space E, and let a continuous $T\colon D \to E$ satisfy

(11) $$\liminf_{h \to 0+} d((1-h)x + hTx, D)/h = 0 \quad \text{for all } x \in D,$$

and

(12) $$(x - y, Tx - Ty)_- \leqslant k|x - y| \quad \text{for all } x,y \in D.$$

a) If $k < 1$, then T has a unique fixed point.

b) If $k = 1$, D is convex, every bounded closed convex

subset of D has the fixed point property for nonexpansive mappings, and $I - T$ is unbounded on unbounded subsets of D, then T has a fixed point.

Proof. Let $A\colon D \to E$ be continuous and consider the following two conditions:

(13) $$\liminf_{h\to 0+} d(x - hAx, D)/h = 0 \quad \text{for all } x \in D,$$

(14) $$\liminf_{h\to 0+} d(x, R(I + hA))/h = 0 \quad \text{for all } x \in D.$$

We observe that (13) implies (14) (and if A is also accretive, then the converse is also true). Letting $A = I - T$ we see that (a) essentially follows from the methods of Crandall and Pazy [8]. To prove (b), let J_r be the resolvent of A. The proof of [15, Theorem 3.1] shows that J_r is defined on D. Since $|AJ_r^n x| = |J_r^{n-1}x - J_r^n x|/r \leq |x - J_r x|/r$, $\{AJ_r^n x\}$ is bounded and so is $\{J_r^n x\}$. Hence the result.

We close with a simple proof of a result that is due to Deimling [9]. It shows that the recent results of Kirk and Schöneberg [10] can be proved by "elementary" means.

Theorem 6. Let D be an open subset of a Banach space E, and let a continuous $T\colon D \to E$ satisfy (12) with $k < 1$. Then $(I - T)(D)$ is open.

Proof. Let $B(x,r)$ be the closed ball of center x and radius r. Suppose $x_0 - Tx_0 = y_0$, $B(x_0,r) \subset D$, $z \in B(y_0,(1 - k)r)$, and define $f\colon B(x_0,r) \to E$ by $f(x) = z + Tx$. f certainly satisfies (12). Since D is open, it actually satisfies $(x - y, f(x) - f(y))_+ \leq k|x - y|$. Now let $|x - x_0| = r$ and $j \in J(x - x_0)$, where J is the (normalized) duality mapping. We have $(f(x) - x, j) = (z - y_0 + x_0 - x + Tx - Tx_0, j)$
$\leq (1 - k)r^2 - r^2 + kr^2 = 0$. Thus f satisfies (11) and has a fixed point by Theorem 5.

REFERENCES

[1] Bakušinskii, A. B. and Poljak, B. T., *On the solution of variational inequalities*, Soviet Math. Dokl. 15(1974), pp. 1705-1710.

[2] Brézis, H. and Lions, P. L., *Infinite products of resolvents*, to appear.

[3] Browder, F. E., *Nonlinear monotone and accretive operators in Banach space*, Proc. Nat. Acad. Sci. U.S.A. 61(1968), pp. 388-393.

[4] Bruck, R. E., *Nonexpansive projections on subsets of Banach spaces*, Pacific J. Math. 47(1973), pp. 341-355.

[5] Bruck, R. E., *A strongly convergent iterative solution of* $0 \in U(x)$ *for a maximal monotone operator* U *in Hilbert space*, J. Math. Anal. Appl. 48(1974), pp. 114-126.

[6] Bruck, R. E. and Reich, S., *Nonexpansive projections and resolvents of accretive operators in Banach spaces*, preprint.

[7] Crandall, M. G. and Evans, L. C., *On the relation of the operator* $\delta/\delta s + \delta/\delta\tau$ *to evolution governed by accretive operators*, Israel J. Math. 21(1975), pp. 261-278.

[8] Crandall, M. G. and Pazy, A., *On the range of accretive operators*, to appear.

[9] Deimling, K., *Zeros of accretive operators*, Manuscripta Math. 13(1974), pp. 365-374.

[10] Kirk, W. A. and Schöneberg, R., *Some results on pseudo-contractive mappings*, to appear.

[11] Kobayashi, Y., *Difference approximation of Cauchy problems for quasi-dissipative operators and generation of nonlinear semigroups*, J. Math. Soc. Japan 27(1975), pp. 640-665.

[12] Martin, R. H., *A global existence theorem for autonomous differential equations in Banach space*, Proc. Amer. Math. Soc. 26(1970), pp. 307-314.

[13] Martin, R. H., *Differential equations on closed subsets of*

a Banach space, Trans. Amer. Math. Soc. 179(1973), pp. 399-414.

[14] Reich, S., *Asymptotic behavior of contractions in Banach spaces*, J. Math. Anal. Appl. 44(1973), pp. 57-70.

[15] Reich, S., *On fixed point theorems obtained from existence theorems for differential equations*, J. Math. Anal. Appl. 54(1976), pp. 26-36.

[16] Reich, S., *Extension problems for accretive sets in Banach spaces*, J. Functional Analysis, to appear.

[17] Reich, S., *An iterative procedure for constructing zeros of accretive sets in Banach spaces*, J. Nonlinear Analysis, to appear.

[18] Rockafellar, R. T., *Monotone operators and the proximal point algorithm*, SIAM J. Control and Optimization 14(1976), pp. 877-898.

MODEL EQUATIONS FOR NONLINEAR DISPERSIVE SYSTEMS

R. E. Showalter
The University of Texas at Austin

Numerous model equations and systems have been proposed to approximate long low water waves on a channel of constant depth. The general procedure is to construct a family of models depending on two (real) parameters which measure the nonlinearity due to wave amplitude and the dispersion arising from wavelength. All computations are retained for terms of *0th* and *1st* order only, so the model is then correct to *2nd* order. The great variety for equivalent models arises in part from the possibility of using the *0th* order approximation to alter the *1st* order terms. Also, there is the choice of dependent variables to be made, so many formally equivalent models are possible. One should of course make choices which lead to a model with (at least) the following desirable properties:

- Appropriate initial-boundary value problems are well-posed
- Response of the system is not overly sensitive to short-wave components
- Solutions of the system are consistent with assumptions leading to the model and also to experimental data

We follow a standard formal expansion procedure [7, p. 464] to obtain a model for two-way propagation of waves; the variables are scaled so that the relative size of each term is explicit. In the scaled variables $u, v, \S, \tau$ corresponding respectively to horizontal velocity at the height y, surface disturbance,

ISBN 0-12-434160-8

position, and time, the model is given by

$$\text{I.}\quad \begin{cases} u_\tau + v_\xi - \frac{\mu^2}{2}(1 - y^2)u_{\xi\xi\tau} + \varepsilon uu_\xi = 0, \\ v_\tau + u_\xi - \frac{\mu^2}{2}(y^2 - \frac{1}{3})v_{\xi\xi\tau} + \varepsilon(uv)_\xi = 0. \end{cases}$$

Here ε corresponds to amplitude, $1/\mu$ to wavelength, and ε,μ are positive and small. If we further restrict consideration to predominantly rightward-moving waves, the corresponding model is

$$\text{II.}\quad \begin{cases} u_\tau + v_\xi - \frac{\mu^2}{2}(1 - y^2)u_{\xi\xi\tau} + \varepsilon\, uu_\xi = 0, \\ v_\tau + u_\xi - \frac{\mu^2}{2}(y^2 - \frac{1}{3})v_{\xi\xi\tau} + \varepsilon(v^2)_\xi = 0. \end{cases}$$

Finally, we observe that if (and only if!) we additionally require that this wave be of predominantly constant shape, the system reduces to either of the formally equivalent equations

$$\text{P.}\qquad u_\tau + u_\xi - \frac{\mu^2}{6}u_{\xi\xi\tau} + \frac{3}{2}\varepsilon\, uu_\xi = 0,$$

$$\text{KdV.}\qquad u_\tau + u_\xi + \frac{\mu^2}{6}u_{\xi\xi\xi} + \frac{3}{2}\varepsilon\, uu_\xi = 0.$$

For details in the above we refer to [6]. Similar results for systems of the type of I and II are given in [2], and discussion of P and KdV are given in [1,3,4,7].

Our preference for systems I and II over competing model systems arises from the observation that they can be posed in Sobolev spaces as nonlinear evolution equations which are locally Lipschitz. The Cauchy problem is trivially well-posed locally in time; the abstract form of II is dissipative, so we obtain global-in-time solutions. The details appear in [6].

Numerical studies of I and II are currently underway [5] and early experience shows that straightforward finite-difference schemes are effective in resolving systems I and II. This arises from the relative insensitivity of the models to short-wave components.

REFERENCES

[1] Benjamin, T. B., Bona, J. L., and Mahony, J. J., *Model*

equations for long waves in nonlinear dispersive systems, Phil. Trans. Roy. Soc., London, 272(A), 1972, pp. 47-78.

[2] Bona, J. L., Smith, R., *A model for two-way propogation of water waves in a channel*,

[3] Newell, A. C., (editor), *Nonlinear Wave Motion, Lectures in Applied Mathematics*, 15, Amer. Math. Society, Providence, 1974.

[4] Peregrine, D. H., *Calculations of the development of an undular bore*, J. Fluid Mech., 25, 1966, pp. 321-334.

[5] Rooney, M., *Thesis*, University of Texas, 1977, in preparation.

[6] Showalter, R. E., *Well-posed problems for some nonlinear dispersive systems*, J. Math. Pures et App., 56, 1977.

[7] Whitham, G. B., *Linear and Nonlinear Waves*, Wiley, New York, 1974.

SECOND ORDER DIFFERENTIAL EQUATIONS IN BANACH SPACE

C. C. Travis
Health and Safety Research Division
Oak Ridge National Laboratory[1]

and

G. F. Webb[2]
Vandebilt University

I. INTRODUCTION

In recent years there has been an extensive effort to develop a general theory of differential equations in Banach space. Interest in this subject dates from the work of Hille and Yosida (1948) on the Cauchy problem for first order equations with unbounded operator coefficients. The associated theory of linear and nonlinear semigroups of operators has also received widespread attention and is the subject of several excellent monographs. Concurrent with the development of a theory of abstract first order equations, numerous contributions have been made to the theory of second order differential equations in Banach space. Among the earliest and most significant, we find the work of S. Kurepa, M. Sova, G. DaPrato, E. Giusti, H. Fattorini, and I. Segal.

---

[1] Operated by the Union Carbide Corporation for the Energy Research and Development Adminstration.

[2] Supported in part by the National Science Foundation Grant NSF MCS 75-06332 A01.

ISBN 0-12-434160-8

It is our intention in this paper to survey the theory of second order differential equation in Banach space. We will not attempt to give a complete discussion of this theory, but rather provide an introduction to certain selected aspects of the subject. Proofs of results presented will not be given, but references where proofs may be found will be indicated. The main part of our exposition will concern linear second order differential equations, a subject intimately linked with the theory of strongly continuous cosine families of bounded linear operators in Banach space. As a second part of our discussion, we will present results concerning nonlinear second order equations having a special semilinear form.

The organization of the paper is as follows: Section II concerns the basic theory of strongly continuous cosine families in Banach space, and the fundamental generation theorem of Sova-Da Prato-Giusti-Fattorini; Section III concerns the problem of converting an abstract second order linear differential equation to an abstract first order differential system; Section IV concerns perturbations of the infinitesimal generator of a strongly continuous cosine family, and the approximation theorem of Konishi-Goldstein; Section V concerns the special properties of compactness, uniform continuity, and inhomogeneous equations for strongly continuous cosine families; and lastly, Section VI concerns abstract semilinear second order initial value problems in which the linear term is a cosine family generator.

We now state some notation that will be used throughout the paper. We let X be a Banach space with norm $\|\cdot\|$. Let $B(X;X)$ denote the Banach algebra of bounded everywhere defined linear operators from X to X with norm $|\cdot|$. If A is a linear operator from X to X, then $\rho(A)$ denotes the resolvent set of A and $R(\lambda;A)$ denotes the resolvent of A; that is, $\lambda \in \rho(A)$ if and only if $\lambda I - A$ is one to one and onto, and $R(\lambda;A) \overset{\text{def}}{=} (\lambda I - A)^{-1} \in B(X;X)$. If A is a closed linear operator in X with domain $D(A)$, then $[D(A)]$ denotes the Banach space consisting of the set $D(A)$ endowed with the graph norm $\|x\| = \|Ax\| + \|x\|$, $x \in D(A)$. We say that the linear operator A from X to X commutes with $L \in B(X;X)$ if and only if for each $x \in D(A)$, $Lx \in D(A)$ and $ALx = LAx$. Finally, we will presume upon the reader a basic acquaintance with the theory of strongly continuous semigroups (and groups) of bounded linear operators in Banach space.

II. BASIC THEORY OF STRONGLY CONTINUOUS COSINE FAMILIES

The theory of strongly continuous cosine families of bounded linear operators in Banach space is kindred in spirit to the theory of strongly continuous semigroups of bounded linear operators in Banach space, and it is equally appealing because of its great generality and simplicity. Strongly continuous cosine families of bounded linear operators are related to abstract linear second order differential equations in the same manner that strongly continuous semigroups of bounded linear operators are related to abstract linear first order differential equations. Roughly speaking, every second order differential equation of the form $u'' = Au$ which is well-posed in a certain sense gives rise to a strongly continuous cosine family of bounded linear operators with infinitisimal generator A, and conversely, every strongly continuous cosine family with infinitesimal generator A may be associated with the well-posed second order differential equation $u'' = Au$. In the discussion below, we will make these ideas more precise.

Let A be a linear, but possible unbounded, operator from the Banach space X to itself, and let $x,y \in X$. By a <u>linear second order initial value problem</u> in X, we mean

(2.1) $\quad u''(t) = Au(t), \quad t \in R,$

(2.2) $\quad u(0) = x,$

(2.3) $u'(0) = y$.

A solution $u(t) = u(t;x,y)$ of the initial value problem (2.1) - (2.3) is a function $u\colon R \to X$ which is twice continuously differentiable on R in the norm of X, is contained in $D(A)$ for $t \in R$, and satisfies (2.1) - (2.3). Notice that if (2.1) and (2.2) are to be satisfied, then x is necessarily in $D(A)$. Since A may be unbounded, and thus $D(A)$ may not be all of X, we cannot expect to obtain a solution to the initial value problem for every $x \in X$. We say that the initial value problem (2.1) - (2.3) is well-posed provided that (2.1) - (2.3) has a unique solution for each $x \in D(A)$ and $y = 0$, and the mapping $x \to u(t;x,0)$ has an extension which belongs to $B(X;X)$.

A one parameter family $C(t)$, $t \in R$, of bounded linear operators mapping the Banach space X into itself is called a strongly continuous cosine family if and only if

(2.4) $C(t + s) + C(t - s) = 2C(t)C(s)$ for all $s,t \in R$;

(2.5) $C(0) = I$;

(2.6) $C(t)x$ is continuous in t on R for each fixed $x \in X$.

Proposition 2.1. Let the initial value problem (2.1) - (2.3) be well-posed, and for each $x \in D(A)$ and $t \in R$, let $C(t)x = u(t;x,0)$, where $u(t;x,0)$ is the unique solution of (2.1) - (2.3) with $y = 0$. For each $t \in R$, let $C(t)$ be the extension in $B(X;X)$ of $C(t)$ on $D(A)$. Then $C(t)$, $t \in R$, is a strongly continuous cosine family of bounded linear operators in X.

Proposition 2.1 is easily established using the definition of a strongly continuous cosine family. Formula (2.4), which is known as the cosine identity, is a consequence of the uniqueness of solutions to the initial value problem, while (2.6) is a consequence of the definition of well-posedness. In the propositions below, we establish a converse to Proposition 2.1. We will first, however, present a further development of the elementary properties of strongly continuous cosine families.

We associate with the strongly continuous cosine family $C(t)$, $t \in R$ in X the strongly continuous <u>sine family</u> $S(t)$, $t \in R$, defined by

(2.7) $$S(t)\,x = \int_0^t C(s)\,x\,ds, \quad x \in X, \quad t \in R,$$

and the two sets

(2.8) $E_1 = \{ x \in X:\ C(t)x$ is once continuously differentiable in t on $R\}$

(2.9) $E_2 = \{ x \in X:\ C(t)x$ is twice continuously differentiable in t on $R\}$.

The infinitesimal generator of a strongly continuous cosine family $C(t)$, $t \in R$, is the operator $A\colon X \to X$ defined by

(2.10) $$Ax = d^2/dt^2\, C(0)\, x$$

with $D(A) = E_2$.

The following basic results concerning strongly continuous cosine families were established in [27], [5], and [30].

Proposition 2.2. Let $C(t)$, $t \in R$, be a strongly continuous cosine family with associated sine family $S(t)$, $t \in R$. The following are true:

(2.11) $C(t) = C(-t)$ and $S(t) = -S(-t)$ for all $t \in R$;

(2.12) $C(s)$, $S(s)$, $C(t)$, and $S(t)$ commute for all $s, t \in R$;

(2.13) $S(t)x$ is continuous in t on R for each fixed $x \in X$;

(2.14) $S(s + t) + S(s - t) = 2S(s)C(t)$ for all $s, t \in R$;

(2.15) $S(s + t) = S(s)C(t) + S(t)C(s)$ for all $s, t \in R$;

(2.16) there exist constants $M \geqslant 1$ and $\omega \geqslant 0$ such that

$$|C(t)| \leqslant Me^{\omega|t|} \quad \text{for all} \quad t \in R;$$

(2.17) $$|S(t) - S(\hat{t})| \leqslant M \left| \int_{\hat{t}}^{t} e^{\omega|s|}\, ds \right| \quad \text{for all} \quad t, \hat{t} \in R.$$

We shall say that a strongly continuous cosine family $C(t)$, $t \in R$, is of type (M,ω) if it satisfies $|C(t)| \leqslant Me^{\omega|t|}$ for all $t \in R$.

<u>*Proposition 2.3.*</u> Let $C(t)$, $t \in R$, be a strongly continuous cosine family with infinitesimal generator A and associated sine family $S(t)$, $t \in R$. The following are true:

(2.18) $D(A)$ is dense in X and A is a closed operator in X;

(2.19) if $x \in X$ and $r, s \in R$, then $z \overset{\text{def}}{=} \int_r^s S(u)\, x\, du \in D(A)$ and $Az = C(s)x - C(r)x$;

(2.20) if $x \in X$ and $r, s \in R$, then $z \overset{\text{def}}{=} \int_0^r \int_0^s C(u)C(v)\, x\, dudv \in D(A)$ and $Az = (C(s+r)\, x - C(s-r)\, x)/2$;

(2.21) if $x \in X$, then $S(t)\, x \in E_1$;

(2.22) if $x \in E_1$, then $C(t)\, x \in E_1$;

(2.23) if $x \in E_1$, then $S(t)\, x \in D(A)$ and $d/dt C(t)\, x = AS(t)\, x$;

(2.24) if $x \in D(A)$, then $C(t)\, x \in D(A)$ and $d^2/dt^2 C(t)\, x = AC(t)\, x = C(t)Ax$;

(2.25) if $x \in E_1$, then $\lim_{t \to 0} AS(t)x = 0$;

(2.26) if $x \in E_1$, then $S(t)\, x \in D(A)$ and $d^2/dt^2\, S(t)x = AS(t)x$;

(2.27) if $x \in D(A)$, then $S(t)x \in D(A)$ and $AS(t)x = S(t)Ax$;

(2.28) $C(t + s) - C(t - s) = 2AS(t)\, S(s)$ for all $t, s \in R$.

<u>*Proposition 2.4.*</u> Let $C(t)$, $t \in R$, be a strongly continuous cosine family with infinitesimal generator A and associated

sine family $S(t)$, $t \in R$. Then the second order initial value problem (2.1) - (2.3) is well-posed. Moreover, if $x \in E_2$ and $y \in E_1$, then the unique solution $u(t;\ x,y)$ of (2.1) - (2.3) is given by $u(t;\ x,y) = C(t)x + S(t)y$.

Except for the claim of uniqueness of solutions to the initial value problem, Proposition 2.4 is a consequence of Proposition 2.3. The uniqueness of solutions is established by Proposition 2.4 in [29]. If $x \notin E_2$ or $y \notin E_1$, we may think of $C(t)x + S(t)y$ as a generalized solution to (2.1) - (2.3).

The following proposition is due to Sova [27], Theorem 2.24. It states that there is a unique correspondence between strongly continuous cosine families and their infinitesimal generators; that is, two different cosine families cannot have the same infinitesimal generator.

Proposition 2.4. Let $C_1(t)$, $t \in R$, and $C_2(t)$, $t \in R$, be strongly continuous cosine families with infinitesimal generators A_1 and A_2, respectively. If $D(A_1) \subset D(A_2)$ and $A_1x = A_2x$ for all $x \in D(A_1)$, then $C_1(t) = C_2(t)$ for $t \in R$.

As an analogue to the definition of the infinitesimal generator of a strongly continuous semigroup of operators, it is natural to try to replace definition (2.10) of the infinitesimal generator of a strongly continuous cosine family by the central difference approximation

$$\lim_{t \to 0} (C(-2t)x - 2C(0)x + C(2t)x)/4t^2,$$

of the second derivative of $C(t)$, $t \in R$, at $t = 0$. In the case of an arbitrary function, it is well-known that the central difference definition of the second derivative is not equivalent to the classical definition. However, as the following proposition demonstrates, these two definitions are equivalent for functions satisfying the cosine functional equation (2.4).

Proposition 2.5. Let $C(t)$, $t \in R$, be a strongly continuous cosine family with infinitesimal generator A. Then $x \in D(A)$

if and only if $\lim_{t\to 0}(C(2t)x - x)/2t^2$ exists, and

$$Ax = \lim_{t\to 0}(C(2t)x - x)/2t^2$$

for $x \in D(A)$.

The proof of Proposition 2.5 was first given in [27], Proposition 2.18. Another proof is presented in [30], Proposition 2.3.

The following proposition, whose proof may be found in [5], Lemma 5.6, gives a characterization of resolvent of the infinitesimal generator of a strongly continuous cosine family.

Proposition 2.6. Let $C(t)$, $t \in R$, be a strongly continuous cosine family of type (M,ω) with infinitesimal generator A and associated sine family $S(t)$, $t \in R$. Then for $\mathrm{Re}\,\lambda > \omega$, λ^2 is in the resolvent set $\rho(A)$ of A and

$$\lambda R(\lambda^2;A)x = \int_0^\infty e^{-\lambda t}\, C(t)x\, dt, \quad x \in X, \tag{2.29}$$

$$R(\lambda^2;A)x = \int_0^\infty e^{-\lambda t}\, S(t)x\, dt, \quad x \in X. \tag{2.30}$$

We now state a necessary and sufficient condition that an operator A be the infinitesimal generator of a strongly continuous cosine family. This condition, which is the analogue of the Hille-Yosida-Phillips-Miyadera generation theorem ([4], Theorem 13, p. 624) of operator semigroup theory, was established independently by M. Sova ([27], Theorem 3.1 and 3.2), G. DaPrato and E. Guisti [3], and H. Fattorini ([5], Theorem 3.1).

Proposition 2.7. A closed densely defined linear operator A in X is the infinitesimal generator of a strongly continuous cosine family $C(t)$, $t \in R$, of type (M,ω) if and only if $R(\lambda^2;A)$ exists for $\lambda > \omega$, is strongly infinitely differentiable, and satisfies

$$|(d/d\lambda)^n(\lambda R(\lambda^2;A))| \leqslant Mn!/(\lambda - \omega)^{n+1} \tag{2.31}$$

for $\lambda > \omega$ and $n = 1, 2, \ldots$.

We conclude this section with some simple examples of strongly continuous cosine families and their infinitesimal generators.

Example 2.1. ([6], p. 51). Let $A \in B(X;X)$ and define $C(t) = \sum_{k=0}^{\infty} A^k t^{2k}/(2k)!$, $t \in R$. Then $C(t)$, $t \in R$, is a strongly continuous cosine family in X with infinitesimal generator A and associated sine family $S(t) = \sum_{k=0}^{\infty} A^k t^{2k+1}/(2k+1)!$, $t \in R$. If $X = R$ and $a > 0$, then $C(t) = \cosh t\sqrt{a}$ if $Ax = ax$ and $C(t) = \cos t\sqrt{a}$ if $Ax = -ax$.

Example 2.2. ([27], Example 2.27). Let $X = C_{2\pi}(R;C)$ be the Banach space of continuous, complex valued, 2π-periodic functions on R with supremum norm, and let

$$(2.32) \qquad (C(t)f)(x) = (f(x+t) + f(x-t))/2, \; f \in X, \; t \in R, \; x \in R.$$

Then $C(t)$, $t \in R$, is a strongly continuous cosine family in X with infinitesimal generator

$$(2.33) \qquad Af = f'',$$

$$D(A) = E_2 = \{f \in X: \; f'' \in X\},$$

and associated sine family

$$(2.34) \qquad (S(t)g)(x) = \int_{x-t}^{x+t} g(s)ds/2, \quad g \in X, \quad t \in R, \quad x \in R,$$

where

$$E_1 = \{g \in X: \; g' \in X\}.$$

For $f \in E_2$ and $g \in E_1$, the formula $(C(t)f)(x) + (S(t)g)(x)$ gives the classical D'Alembert solution of the 1-dimensional wave equation

$$(2.35) \qquad w_{tt}(x,t) = w_{xx}(x,t), \quad t \in R, \quad x \in R,$$

$$w(x,0) = f(x), \quad w_t(x,0) = g(x).$$

III. THE PROBLEM OF CONVERSION TO A FIRST ORDER SYSTEM

It is natural to try to convert the linear second order initial value problem (2.1) - (2.3) to a first order differential system. Formally, this conversion can be easily accomplished by defining $\mathsf{A}: X \times X \to X \times X$ as

$$\mathsf{A}(x,y) = (y,Ax),$$

$$D(\mathsf{A}) = D(A) \times X,$$

and considering the first order system

$$U'(t) = \mathsf{A}U(t), \quad t \in R, \tag{3.1}$$

$$U(0) = (x,y) \in D(A) \times X, \tag{3.2}$$

where $U: R \to X \times X$. The difficulty with this appraoch is that the initial value problem (3.1) - (3.2) may not be well-posed in the sense that A does not necessarily generate a strongly continuous group in $X \times X$. To see that this is the case, let $x \in D(A)$ and $y \in X$, and suppose that A is the infinitesimal generator of a strongly continuous cosine family. Then the solution of the initial value problem (3.1) - (3.2) must be given by $U(t) = (C(t)x + S(t)y, AS(t)x + C(t)y)$. But, since $AS(t)$ is in general unbounded, the mapping $(x,y) \to U(t)$ does not necessarily possess an extension in $B(X \times X; X \times X)$.

It is, however, always possible to convert the second order initial value problem (2.1) - (2.3) to a well-posed first order system in the Banach space $E_1 \times X$. The result is due to J. Kisynski [16] and is stated below.

Proposition 3.1. Let $C(t)$, $t \in R$, be a strongly continuous cosine family with infinitesimal generator A and associated sine family $S(t)$, $t \in R$. Then E_1 under the norm

$$\|x\|_{E_1} = \|x\| + \sup_{0 \leq t \leq 1} \| d/dt\, C(t)\, x \| \tag{3.3}$$

becomes a Banach space and $U(t)$, $t \in R$, defined by

$$U(t)(x,y) = (C(t)x + S(t)y, AS(t)x + C(t)y), \tag{3.4}$$

for $(x,y) \in E_1 \times X$, is a strongly continuous group in $E_1 \times X$ with infinitesimal generator

$$\mathsf{A}(x,y) = (y, Ax)$$

$$D(\mathsf{A}) = D(A) \times E_1.$$

Although the conversion of the second order equation (2.1) - (2.3) to a first order system presented in Proposition 3.1 can always be achieved, it is inconvenient for some problems. In certain problems, we begin with considerable information concerning the infinitesimal generator A, but very little information about the strongly continuous cosine family $C(t)$, $t \in R$, or the subspace E_1. It is desirable, therefore, to have a conversion of (2.1) - (2.3) to a first order system which only involves information about the infinitesimal generator A. The possibility of such a conversion was first investigated by H. Fattorini in [5] and [6], and later by the authors in [30]. The idea is to obtain a square root B of A and to formulate the first order system in the space $[D(B)] \times X$. We will describe this conversion in the discussion below. We first require the following results concerning the existence of a square root of the infinitesimal generator of a strongly continuous cosine family due to H. Fattorini [5], p. 95.

Proposition 3.2. Let $C(t)$, $t \in R$, be a strongly continuous cosine family of type (M, ω) with infinitesimal generator A. If $b \geqslant \omega$ and $A_b \overset{\text{def}}{=\!=} A - b^2 I$, then A_b is the infinitesimal generator of a strongly continuous cosine family $C_b(t)$, $t \in R$, of type $(M_1, \omega + b)$, and there exists a closed linear operator $B_b\colon X \to X$ such that $B_b^{\,2} = A_b$ and B_b commutes with every operator in $B(X,X)$ which commutes with A_b.

We remark that the square root B_b of A_b is defined by Fattorini in the following way: define

$$(3.5) \qquad J_b\, x \overset{\text{def}}{=\!=} \pi^{-1} \int_0^\infty \lambda^{-1/2}\, R(\lambda; A_b)(-A_b\, x)\, d\lambda$$

for $x \in D(A)$. Fattorini [5], p. 95, shows that the integral (3.5) exists, has a closed extension, and that $B_b \overset{\text{def}}{=} \overline{iJ_b}$. Moreover, if $\lambda^2 \in \rho(A_b)$, then $\lambda \in \rho(B_b)$ and $R(\lambda;B_b) = (\lambda I + B_b)R(\lambda^2;A)$. Also, if $b > \omega$, then $0 \in \rho(B_b)$ and we see that it is always possible to find or translate A_b of A such that A_b has a square root with the property that $0 \in \rho(B_b)$.

We now introduce a condition similar to one used by Fattorini in [5], p. 96.

A strongly continuous cosine family $C(t)$, $t \in R$ with infinitesimal generator A and associated sine family $S(t)$, $t \in R$, is said to satisfy condition (F) if and only if the following are true:

(3.6) there exists a closed linear operator B in X such that $B^2 = A$ and B commutes with every operator in $B(X;X)$ which commutes with A;

(3.7) $S(t)$ maps X into $D(B)$ for each $t \in R$ (which implies that $BS(t) \in B(X;X)$ for each $t \in R$ since B is closed);

(3.8) $BS(t)x$ is continuous in t on R for each fixed $x \in X$.

The usefulness of condition (F) in the problem of converting the second order initial value problem (2.1) - (2.3) to a first order system can be seen in the following proposition:

Proposition 3.3. Let A and B be linear operators in X, let $B^2 = A$, let B commute with every operator in $B(X;X)$ which commutes with A, and let $0 \in \rho(B)$. Then the following are equivalent:

(3.9) A is the infinitesimal generator of a strongly continuous cosine family $C(t)$, $t \in R$, satisfying condition (F);

(3.10) B is the infinitesimal generator of a strongly continuous group $T(t)$, $t \in R$;

(3.11) $\mathcal{B}(x,y) \stackrel{\text{def}}{=} (By,Bx)$, with $D(\mathcal{B}) = D(B) \times D(B)$, is the infinitesimal generator of a strongly continuous group $U(t)$, $t \in R$, in $X \times X$;

(3.12) $\mathcal{A}(x,y) \stackrel{\text{def}}{=} (y,Ax)$, with $D(\mathcal{A}) = D(A) \times D(B)$, is the infinitesimal generator of a strongly continuous group $V(t)$, $t \in R$, in $[D(B)] \times X$.

The proof of Proposition 3.3 is given in [3], Proposition 2.6. In the proof that (3.9) ⇒ (3.10), which was first given in [5], Theorem 6.6, one defines $T(t) = C(t) + BS(t)$. In the proof that (3.10) ⇒ (3.11), which was first given in [5], Theorem 6.9, one defines $U(t)(x,y) = (C(t)x + BS(x)y,\ BS(t)x + C(t)y)$, where $C(t) \stackrel{\text{def}}{=} (T(t) + T(-t))/2$ and $S(t) = B^{-1}(T(t) - T(-t))/2$. In the proof that (3.11) ⇒ (3.12), one defines $V(t)$ by

$$V(t) = \begin{bmatrix} B^{-1} & 0 \\ 0 & I \end{bmatrix} U(t) \begin{bmatrix} B & 0 \\ 0 & I \end{bmatrix}.$$

In the proof that (3.12) ⇒ (3.9), one defines $C(t)x = \pi_2 V(t)(0,x)$ and $S(t)x = \pi_1 V(t)(0,x)$, where π_1, π_2 are the projections of $[D(B)] \times X$ onto its first and second components, respectively. For $(x,y) \in D(B) \times X$, $V(t)$ is given by $V(t)(x,y) = (C(t)x + S(t)y,\ BS(t)Bx + C(t)y)$.

Since, for $x \in D(A)$, the set E_1 is precisely the set of $y \in X$ for which the initial value problem (2.1) - (2.3) has a twice continuously differentiable solution, the exact determination of E_1 is of considerable importance. The connection between E_1 and the square root of the infinitesimal generator of the strongly continuous cosine family is given in the proposition below.

<u>*Proposition 3.4.*</u> Let $C(t)$, $t \in R$, be a strongly continuous cosine family with infinitesimal generator A, and let B be an

operator in X such that B commutes with every operator in $B(X;X)$ which commutes with A, zero is in the resolvent set of B, and $B^2 = A$. Then the following are equivalent:

(3.13) $D(B) \subset E_1$;

(3.14) $C(t)$, $t \in R$, satisfies condition (F);

(3.15) $D(B) = E_1$.

The proof that (3.13) ⇒ (3.14) is due to the authors and also I. Miyadera (private communication), and will be published elsewhere. The proof that (3.14) ⇒ (3.13) is found in [5], Remark 6.11, and the proof that (3.14) ⇒ (3.15) is found in [30], Proposition 2.8.

It has been shown in [6], Theorem 2.2, that after some suitable translation of its infinitesimal generator, every strongly continuous cosine family in the Banach space L^p, $1 < p < \infty$, satisfies condition (F). A natural question to ask is whether or not every strongly continuous cosine family satisfies condition (F) (after a suitable translation of its infinitesimal generator). The answer is no and the counterexample is due to J. Kisynski [16], Example 2 and B. Nagy [24], Theorem 3. Consider the strongly continuous cosine family $C(t)$, $t \in R$, presented in Example 2.2 of Section 2, but restricted to the Banach subspace of odd 2π-periodic complex valued functions on R. In [16] it is shown that there does not exist a strongly continuous group $G(t)$, $t \in R$, such that the representation $C(t) = (G(t) + G(-t))/2$ is valid. But if $C(t)$, $t \in R$, defined on this Banach space satisfies condition (F), then by Proposition 3.3 $C(t)$, $t \in R$, has such a representation, where $G(t)$, $t \in R$, is the strongly continuous group defined by $G(t) \overset{\text{def}}{=} C(t) + BS(t)$.

There are many advantages to be gained from converting the linear initial value problem (2.1) - (2.3) to an equivalent well-posed first order system. The primary advantage being that a well-posed first order system corresponds to a strongly continuous group or semigroup of operators, thereby making available for application the extensive theory of operator semigroups. There are some cases, however, in which it is more advantageous to treat the second order equation (2.1) - (2.3) directly, as we shall see in Section 6.

We conclude this section with a discussion of some examples.

Example 3.1. Let $C(t)$, $t \in R$, be the cosine family of Example 2.2. If B is defined by $Bf = f'$, $D(B) = \{f \in C_{2\pi}(R;C):$ $f' \in C_{2\pi}(R;C)\}$, then B satisfies the conditions (3.6), (3.7), and (3.8), and we have that the strongly continuous cosine family $C(t)$, $t \in R$, satisfies condition (F). Notice that $(BS(t)f)(x) = (f(x + t) - f(x - t))$ for $f \in C_{2\pi}(R;C)$, $x \in R$, and consequently that $|BS(t)| = 1$ if $t \neq 0$ and $|BS(t)| = 0$ if $t = 0$. Thus the operator $BS(t)$ is not continuous in the operator topology as a function of t on R. Notice also that $B \neq \overline{iJ_0}$, where J_0 is defined as in (3.5), since $J_0 f$ must be real-valued when f is real-valued. Thus, it is advantageous to state condition (F) in terms of an arbitrary square root B of A rather than the specific square root obtained using (3.5).

The following examples are treated by J. Goldstein in [12], Section 8.

Example 3.2. Let L be a self-adjoint operator in a complex Hilbert space H such that $0 \in \rho(L)$. Define $B = iL$ and notice that $B^* = -iL^* = -B$. Thus, B is skew-adjoint and by Stone's Theorem ([34], Theorem 1, p. 345), B generates a strongly continuous unitary group in H. It follows from the equivalence of (3.9) and (3.10) that $B^2 = -L^2$ is the infinitesimal generator of a strongly continuous cosine family in H satisfying condition (F).

Example 3.3. Let L be a positive self-adjoint operator in the complex Hilbert space H such that $0 \in \rho(L)$. Then L has a positive self-adjoint square root $L^{1/2}$ with the property that $0 \in \rho(L^{1/2})$. It follows from Example 3.2 that $-L$ is the infinitesimal generator of a strongly continuous cosine family in H satisfying condition (F).

IV. PERTURBATION AND APPROXIMATION RESULTS FOR STRONGLY CONTINUOUS COSINE FAMILIES

Our objective in this section is to discuss some perturbation and approximation results for strongly continuous cosine families. A perturbation result is a result of the following type: given an infinitesimal generator A of a strongly continuous cosine family $C(t)$, $t \in R$, in X, and a sequence of infinitesimal generators A_n, $n = 1, 2, \ldots$, of corresponding strongly continuous cosine families $C_n(t)$, $t \in R$, $n = 1, 2, \ldots$, find sufficient conditions on the operators A_n, $n = 1, 2, \ldots$, such that $C_n(t)$ converges in some sense to $C(t)$. One approach to establishing perturbation and approximation results for strongly continuous cosine families is to convert (2.1) - (2.3) to a well-posed first order system, and apply the perturbation and approximation theory of strongly continuous groups. This approach might be especially useful if condition (F) is satisfied. However, not every result known to be true for strongly continuous cosine families can be obtained in this manner. Our purpose here is to obtain results for strongly continuous cosine families directly.

The perturbation results we state below are established using methods analogous to those used by R. S. Phillips in the perturbation theory of strongly continuous semigroups (see [4], Theorem 19, p. 631). The first proposition can be found in [32].

Proposition 4.1. Let A be the infinitesimal generator of the strongly continuous cosine family $C(t)$, $t \in R$, with associated sine family $S(t)$, $t \in R$, and let P be a closed linear operator in X such that

(4.1) $S(t)(X) \subset D(P)$ for all $t \in R$;

(4.2) $PS(t)x$ is continuous in t on R for each fixed $x \in X$.

Then, $A + P$ is the infinitesimal generator of a strongly continuous cosine family in X.

We remark that conditions (4.1) and (4.2) are obviously satisfied provided $P \in B(X;X)$. The stability of strongly continuous cosine family generators under bounded perturbations was also established by B. Nagy in [24], Theorem 1.

As a consequence of Proposition 4.1, the following proposition is established in [32], Corollary 1.

Proposition 4.2. Let $C(t)$, $t \in R$, be a strongly continuous cosine family in X with infinitesimal generator A, let P be a closed operator in X, and let P satisfy

(4.3) $D(A) \subset D(P)$;

(4.4) for each $t \in R$, there exists a constant k_t such that $\| PS(t) x \| \leqslant k_t \| x \|$ for all $x \in D(A)$.

Then, $A + P$ is the infinitesimal generator of a strongly continuous cosine family in X.

The next two results are established in [32], Corollary 2 and Corollary 3. Proposition 4.3 is similar to a result of J. Goldstein [12], Theorem 8.9, p. 91.

Proposition 4.3. Let $C(t)$, $t \in R$, be a strongly continuous cosine family in X with infinitesimal generator A, and let $C(t)$, $t \in R$, satisfy condition (F). Let P be a closed linear operator in X such that

(4.5) $D(B) \subset D(P)$, where B is as in (3.6), (3.7) and (3.8).

Then, $A + P$ is the infinitesimal generator of a strongly continuous cosine family in X.

Proposition 4.4. Let $C(t)$, $t \in R$, be a strongly continuous cosine family in X with infinitesimal generator A, and let $C(t)$, $t \in R$, satisfy condition (F). Let P be a closed linear operator in X such that

(4.6) $D(A) \subset D(P)$

(4.7) there exist constants a and b such that $\|Px\| \leqslant a\|x\| + b\|Bx\|$ for all $x \in D(A)$, where B is as in (3.6), (3.7) and (3.8).

Then, $A + P$ is the infinitesimal generator of a strongly continuous cosine family in X.

The following result, which uses methods similar to those used to extablish the propositions above, has been obtained by T. Takenaka and N. Okazawa in [29].

Proposition 4.5. Let $C(t)$, $t \in R$, be a strongly continuous cosine family of type (M,ω) in X with infinitesimal generator A. Let P be a linear operator in X such that

(4.8) $D(A) \subset D(P)$ and $PR(\lambda^2;A) \in B(X;X)$ for some $\lambda < \omega$;

(4.9) there exists a constant K_0 such that

$$\int_0^1 \|PC(s)\,x\|\,ds \leqslant K_0\,\|x\| \quad \text{for each } x \in D(A).$$

Let $K_\infty = \lim_{\lambda \to \infty} K_\lambda$, where

$$K_\lambda = \sup\left\{ \int_0^1 e^{-\lambda s}\,\|PC(s)\,x\|\,ds:\ \|x\| \leqslant 1,\ \ x \in D(A)\right\},$$

and let ε satisfy $|\varepsilon| < K_\infty^{-1}$. Then $A + \varepsilon P$ is the infinitesimal generator of a strongly continuous cosine family $C_\varepsilon(t)$, $t \in R$, in X and, moreover,

(4.10) $\lim_{\varepsilon \to 0} |C_\varepsilon(t) - C(t)| = 0$ uniformly on finite intervals of R.

The approximation result stated below is an analog to the Trotter-Kato semigroup approximation theorem (see [34], p. 269). This result was obtained by Y. Koniski [17] and J. Goldstein [11].

Proposition 4.6. For each $n = 0, 1, 2, \ldots$, let $C_n(t)$, $t \in R$, be a strongly continuous cosine family in X with infinitesimal generator A_n. Suppose there exist constants $M \geqslant 1$ and $\omega \geqslant 0$ such that $|C_n(t)| \leqslant Me^{\omega t}$ for $t \in R$ and $n = 0, 1, 2, \ldots$, Then the following are equivalent:

(4.11) $\lim_{n\to\infty} C_n(t)x = C_0(t)x$ for each $t \in R$ and $x \in D(A)$;

(4.12) $\lim_{n\to\infty} R(\lambda^2;A_n)x = R(\lambda^2;A_0)x$ for each $\lambda > \omega$ and $x \in X$.

In [11], Theorem 2, it is shown that a sufficient condition for (4.11) to hold is that there exists a linear subspace D in X such that $D \subset D(A_n)$ for $n = 0, 1, 2, \ldots,$ A_n is the closure of its restriction to D, and $\lim_{n\to\infty} A_n x = A_0 x$ for all $x \in D$. An example is given in [11] to show that this sufficient condition is not necessary.

V. SPECIAL PROPERTIES OF STRONGLY CONTINUOUS COSINE FAMILIES: COMPACTNESS, UNIFORM CONTINUITY, INHOMOGENEOUS EQUATIONS

In this section we disciss some special properties of strongly continuous cosine families. The first of these is compactness of the sine family, a property which is very useful in the study of semilinear second order equations. It is well-known (see, e.g., [25]) that if a strongly continuous semigroup $T(t)$, $t \geqslant 0$, in X is compact for some $t_0 > 0$, then it is compact for every $t \geqslant t_0$. To see that an analogous property does not hold for strongly continuous cosine families, consider the strongly continuous cosine family in X whose infinitesimal generator is the operator $-I$. This cosine family is given by $C(t) = (\cos t)I$ and its associated sine family is given by $S(t) = (\sin t)I$. $C(t)$ is compact at odd integer multiples of $\pi/2$, while $S(t)$ is compact at integer multiples of π. If X is an infinite dimensional space, then these are the only values of t for which $C(t)$ and $S(t)$ are compact. The following properties are established in [31], Lemmas 2.1 and 2.2 and Proposition 2.3.

Proposition 5.1. Let $C(t)$, $t \in R$, be a strongly continuous cosine family in X with associated sine family $S(t)$, $t \in R$.

The following are true:

(5.1) if $C(t)$ is compact for t in an interval of positive length, then $C(t)$ is compact for all $t \in R$, and in particular, the identity is compact and X is necessarily finite dimensional;

(5.2) if $S(t)$ is compact for t in an interval of positive length, then $S(t)$ is compact for all $t \in R$.

Proposition 5.2. Let $C(t)$, $t \in R$, be a strongly continuous cosine family in X with infinitesimal generator A and associated sine family $S(t)$, $t \in R$. The following are equivalent:

(5.3) $S(t)$ is compact for every $t \in R$;

(5.4) $R(\lambda;A)$ is compact for some λ, and hence every λ, in $\rho(A)$.

In the example below, we demonstrate that the Cauchy problem for the wave equation

$$(5.5) \qquad \begin{aligned} &w_{tt}(x,t) = w_{xx}(x,t), \quad 0 \leqslant x \leqslant \pi, \quad t \in R, \\ &w(o,t) = w(\pi,t) = 0, \quad t \in R, \\ &w(x,0) = f(x), \quad w_t(x,0) = g(x), \quad 0 \leqslant x \leqslant \pi, \end{aligned}$$

gives rise to a compact sine family in $L^2[0,\pi]$.

Example 5.1. If we write equation (5.5) abstractly in $X = L^2[0,\pi]$, it becomes

$$(5.6) \qquad \begin{aligned} &u''(t) = Au(t), \quad t \in R, \\ &u(o) = f, \; u'(o) = g, \end{aligned}$$

where $u: R \to X$, and A is the operator from X to X defined by

$$(5.7) \qquad Af = f''$$

$$D(A) = \{f \in X: \; f, \; f' \text{ are absolutely continuous}, \; f'' \in X, \; f(o) = f(\pi) = 0\}.$$

A can also be written as

$$Af = -\sum_{n=1}^{\infty} n^2 (f,f_n)\, f_n, \quad f \in D(A),$$

where $f_n(s) = (\sqrt{2}/\pi) \sin n\,x$, $n = 1, 2, \ldots,$ is the orthonormal set of eigenvectors of A. Since $-A$ is positive and self-adjoint in X, we have from Example 3.3 in Section 3 that A is the infinitesimal generator of a strongly continuous cosine family $C(t)$, $t \in R$, in X satisfying condition (F). It is easily seen that $C(t)$, $t \in R$, is given by

$$C(t)f = \sum_{n=1}^{\infty} \cos nt\,(f,f_n)\, f_n, \quad f \in X,$$

and the associated sine family is given by

$$S(t)f = \sum_{n=1}^{\infty} (\sin nt/n)\,(f,f_n)f_n, \quad f \in X.$$

The resolvent of A is given by

$$R(\lambda;A)f = \sum_{n=1}^{\infty} (f,f_n)f_n/(\lambda + n^2),$$

for $f \in X$ and $-\lambda \neq 1^2, 2^2, \ldots$. The compactness of $R(\lambda;A)$ follows from the fact that the eigenvalues of $R(\lambda;A)$ are $\lambda_n = 1/(\lambda + n^2)$, $n = 1, 2, \ldots$. and $\lim_{n\to\infty} \lambda_n = 0$.

The next property of strongly continuous cosine families we discuss is the uniform continuity of $C(t)$, $t \in R$; that is, the continuity of $C(t)$ as a function in t from R to $B(X;X)$. As in semigroup theory, this property is equivalent to the boundedness of the infinitesimal generator (see [4], Corollary 9, p. 621).

Proposition 5.3. Let $C(t)$, $t \in R$, be a strongly continuous cosine family with infinitesimal generator A and associated sine family $S(t)$, $t \in R$. The following are equivalent:

(5.8) $\lim_{t\to\infty} C(t) = I$ in $B(X,X)$;

(5.9) $\lim_{t\to 0} S(t)/t = I$ in $B(X,X)$

(5.10) $A \in B(X;X)$;

(5.11) $C(t)(X) \subset E_1$ for all t in an interval of positive length;

(5.12) there is an interval $[a,b]$ of positive length such that for all $t \in [a,b]$, $S(t)(X) \subset D(A)$ and $AS(t)$ is strongly continuous.

The proof of Proposition 5.3 is given in [31], Proposition 4.1. The equivalence of (5.8) and (5.10) was first established by S. Kurepa in [20].

The last property of strongly continuous cosine families we treat in this section is the regularity of solutions to the inhomogeneous initial value problem

(5.13) $$u''(t) = Au(t) + f(t), \quad t \in R,$$
$$u(o) = x, \quad u'(o) = y,$$

where A is the infinitesimal generator of a strongly continuous cosine family and f is a given continuous function from R to X. The inhomogeneous linear equation (5.13) has implications in the investigation of nonlinear second order abstract equations, as we will see in the next section. We define a function u from R to X to be a <u>strong solution</u> of (5.13) provided that u is twice continuously differentiable on R, $u(t) \in D(A)$ for $t \in R$, and (5.13) is satisfied. The following proposition is established in [31], Proposition 2.4:

<u>*Proposition 5.4.*</u> Let $C(t)$, $t \in R$, be a strongly continuous cosine family with infinitesimal generator A and let $f\colon R \to X$ be continuous. If u is a strong solution of (5.13), then

(5.14) $$u(t) = C(t)x + S(t)y + \int_0^t S(t-s)\, f(s)\, ds.$$

Equation (5.14), however, is more general than equation (5.13), and every function of the form (5.14) need not be a strong solution of (5.13). For this reason, we will define functions of the form (5.14) as <u>mild solutions</u> of equation (5.13). In the discussion below, we will state conditions on f that guarantee that every mild solution is a strong solution.

Proposition 5.5. Let $C(t)$, $t \in R$, be a strongly continuous cosine family with infinitesimal generator A and associated sine family $S(t)$, $t \in R$, and let f be continuous from R to X. The initial value problem (5.13) has a strong solution for every $x \in D(A)$ and $y \in E_1$ if and only if the X-valued function

$$g(t) = \int_0^t S(t - s)\, f(s)\, ds, \quad t \in R, \tag{5.15}$$

is twice continuously differentiable in t on R. A sufficient condition that the function g defined by (5.15) be twice continuously differentiable in t on R is that f be once continuously differentiable in t on R.

Proposition 5.6. Let $C(t)$, $t \in R$, be a strongly continuous cosine family with infinitesimal generator A and associated since family $S(t)$, $t \in R$, and let f be continuous from R to X. The initial value problem (5.13) has a strong solution for every $x \in D(A)$ and $y \in E_1$ if and only if the X-valued function

$$g(t) = \int_0^t S(t - s)\, f(s)\, ds, \quad t \in R, \tag{5.16}$$

has the property that $g(t) \in D(A)$ for all $t \in R$ and $Ag(t)$ is continuous as a function in t from R to X. A sufficient condition that the function g defined by (5.16) have the property that $g(t) \in D(A)$ for all $t \in R$ and $Ag(t)$ is continuous as a function in t from R to X is that $f(t) \in D(A)$ for all $t \in R$ and $Af(t)$ is continuous as a function in t from R to X.

Propositions 5.6 and 5.7 are analogous to results concerning the inhomogeneous first order equation from semigroup theory (see

[25]). The proof of these propositions is given in [31], Proposition 3.4, Corollary 3.5, Proposition 3.6, Corollary 3.7.

VI. ABSTRACT SECOND ORDER SEMILINEAR EQUATIONS

In this section we will discuss second order semilinear initial value problems. For the sake of simplicity, we will restrict our attention to autonomous equations. We will consider equations of the form

$$u''(t) = Au(t) + F(u(t), u'(t)), \tag{6.1}$$

$$u(o) = x, \quad u'(o) = y, \tag{6.2}$$

where A is the infinitesimal generator of a strongly continuous cosine family $C(t)$, $t \in R$, in X, F is a mapping from $X \times X$ to X, and the unknown solution u maps some interval about 0 to X.

We seek mild solutions of (6.1) - (6.2), that is, solutions of the integral equation

$$u(t) = C(t)x + S(t)y + \int_0^t S(t - s)\, F(u(s), u'(s))\, ds. \tag{6.3}$$

We lose no generality by considering solutions to (6.3) since if $F(u(t), u'(t))$ is continuous in t, then by virtue of Proposition 5.4, every solution of the initial value problem (6.1) - (6.2) is also a solution of (6.3). Moreover, equation (6.2) is easier to work with than (6.1) - (6.2) because of the nice properties of the bounded operators $C(t)$, $t \in R$, and $S(t)$, $t \in R$, as opposed to the unbounded operator A in equation (6.1). This approach to problem (6.1) - (6.2) traces back to the work of I. Segal in [26]. In [26] cosine family theory was not employed, but rather (6.1) - (6.2) was converted to a first order system and the theory of strongly continuous groups was applied.

We will allow for a variety of hypotheses in the semilinear initial value problem (6.1) - (6.2): (i) hypotheses on the cosine family $C(t)$, $t \in R$, such as condition (F) or compactness of the sine family; (ii) hypotheses on F such as continuity conditions or the asumption that F depends only on its first variable; (iii) hypotheses on the initial values such as membership in the set $D(A)$ or E_1. Many subtleties arise from these various hypotheses with regard to the existence, uniqueness, continuous dependence upon initial values, smoothness, and continuability of solutions. We will give a sampling of results that can be obtained. The proofs can be found in [30], Section 3. In what follows, we suppose that $C(t)$, $t \in R$, is a strongly continuous cosine family in X with infinitesimal generator A and associated sine family $S(t)$, $t \in R$.

Proposition 6.1. Let D be an open subset of $X \times X$ and let $F\colon D \to X$ satisfy

(6.4) $$\|F(x,y) - F(\hat{x},\hat{y})\| \leqslant L(r)\ (\|x - \hat{x}\| + \|y - \hat{y}\|)$$

for (x,y), $(\hat{x},\hat{y}) \in D$ satisfying $\|x\|$, $\|\hat{x}\|$, $\|y\|$, $\|\hat{y}\| \leqslant r$, where $L(r)$ is a nondecreasing function from R^+ to R^+. For each $(x,y) \in D$ such that $x \in E_1$, there exist $t_0 > 0$ and a unique continuously differentiable function $u\colon (-t_1, t_1) \to X$ satisfying (6.3). Further, if $D = X \times X$ and $L(t)$ is constant, then t_1 can be taken as $+\infty$.

The method of proof employed to obtain the local existence and uniqueness of solutions claimed in Proposition 6.1 is the classical method of successive integrations. Regarding the continuous dependence of solutions on initial values, we have the following result.

Proposition 6.2. Suppose the hypothesis of proposition 6.1 and let $t_1 > 0$. There exist constants $K = K(t_0)$ and $\gamma = \gamma(t_1)$ such that if u and $\hat{u}$ satisfy (6.3) for $|t| < t_1$ with $u(o) = x \in E_1$, $u'(o) = y$, $\hat{u}(o) = \hat{x} \in E_1$, $\hat{u}'(o) = \hat{y}$, (x,y), $(\hat{x},\hat{y}) \in D$, then for $|t| < t_1$,

(6.5) $$\|u(t) - \hat{u}(t)\| + \|u'(t) - \hat{u}'(t)\| \leqslant K(\|x - \hat{x}\| + \|A(x - \hat{x})\| + \|y - \hat{y}\|)\ e^{\gamma|t|}.$$

The proof of Proposition 6.2 uses Gronwall's Lemma and the estimates in (2.16) and (2.17). With additional hypothesis on F, we can claim that mild solutions are strong solutions.

Proposition 6.3. Suppose the hypotheses of Proposition 6.1 and, in addition, that F is continuously differentiable on D and satisfies

$$(6.6)\qquad \|F_1(x,y) - F_1(\hat{x},\hat{y})\| \leq L(r)(\|x - \hat{x}\| + \|y - \hat{y}\|),$$

$$\|F_2(x,y) - F_2(\hat{x},\hat{y})\| \leq L(r)(\|x - \hat{x}\| + \|y - \hat{y}\|),$$

for (x,y), $(\hat{x},\hat{y}) \in D$ satisfying $\|x\|$, $\|y\|$, $\|\hat{x}\|$, $\|\hat{y}\| \leq r$, where $L(r)$ is a nondecreasing function from R^+ to R^+ and F_1 and F_2 denote the derivative of F with respect to its first and second variable, respectively. Then the mild solution u of equation (6.3), whose existence is established by Proposition 6.1, is a strong solution of (6.1) - (6.2), provided $x \in D(A)$ and $y \in E_1$.

The next result concerns the continuability of local solutions of equation (6.3).

Proposition 6.4. Suppose the hypotheses of Proposition 6.1 and suppose $D = X \times X$. If $x \in E_1$, $y \in X$, and u is a solution of (6.3) noncontinuable to the right on $[o,b]$, then either $b = +\infty$ or $\lim_{t \to b^-} (\|u(t)\| + \|u'(t)\|) = +\infty$. An analogous result holds for solutions noncontinuable to the left. If F does not depend on its second variable, then we obtain existence of mild solutions under weaker assumptions on the initial values.

Proposition 6.5. Let D be an open subset of X and let $F\colon\ D \to X$ be such that

$$(6.7)\qquad \|F(x) - F(\hat{x})\| \leq L(r)\|x - \hat{x}\|$$

for x, $\hat{x} \in D$ satisfying $\|x\|$, $\|\hat{x}\| \leq r$, where $L(r)$ is a nondecreasing function from R^+ to R^+. For each $x \in D$ and $y \in X$, there exist $t_1 > 0$ and an unique continuous function

$u\colon (-t_1, t_1) \to X$ satisfying

(6.8) $$u(t) = C(t)x + S(t)y + \int_0^t S(t - s)\, F(u(s))\, ds.$$

Further, if $D = X$ and L is a constant, then t_1 can be taken as $+\infty$. If $D = X$, $x \in E_1$, $y \in X$, and u is a solution of (6.8) noncontinuable to the right on $[o,b]$, then either $b = +\infty$ or $\lim_{t \to b^-} \|u(t)\| = +\infty$. An analogous result holds for solutions noncontinuable to the left.

If we know that F does not depend on its second variable and that $S(t)$, $t \in R$, is a compact sine family, then the Lipschitz continuity condition (6.7) on F can be weakened to only continuity. As is to be expected, we lose the uniqueness of solutions.

Proposition 6.6. Let the sine family $S(t)$ be compact for each $t \in R$ and let $F\colon D \to X$ be continuous where D is an open subset of X. For each $x \in D$ and $y \in X$ there exist $t_1 > 0$ and a continuous function $u\colon (-t_1,t_1) \to Y$ satisfying (6.8). Further, if $D = X$, F maps closed bounded sets into bounded sets, $x \in E_1$, $y \in X$, and u is a solution of (6.8) noncontinuable to the right on $[o,b]$, then either $b = +\infty$ or $\lim_{t \to b} = \|u(t)\| = +\infty$. An analogous result holds for a solution noncontinuable to the left.

Proposition 6.6 is established by an application of the Schrauder Fixed Point Theorem. Another weakening of the hypothesis on F can be achieved if we suppose that $C(t)$, $t \in R$, satisfies condition (F).

Proposition 6.7. Let $C(t)$, $t \in R$, satisfy condition (F) and let B be a square root of A satisfying (3.6) - (3.8). Let D be an open subset of $[D(B)] \times X$ and let $F\colon D \to X$ be such that

(6.9) $$\|F(x,y) - F(\hat{x},\hat{y})\| \leqslant L(r)\ (\|B(x - \hat{x})\| + \|y - \hat{y}\|)$$

for (x,y), $(\hat{x},\hat{y}) \in D$ satisfying $\|Bx\|$, $\|B\hat{x}\|$, $\|y\|$, $\|\hat{y}\| \leqslant r$, where $L(r)$ is a nondecreasing function from R^+ to R^+. For each $(x,y) \in D$, there exist $t_1 > 0$ and a unique continuously differentiable function $u: (-t_1,t_1) \to X$ satisfying (6.3). Further, if $D = D(B) \times X$ and $L(r)$ is constant, then t_1 can be taken as $+\infty$.

The proof of Proposition 6.7 uses the fact that when $C(t)$, $t \in R$, satisfies condition (F), the second order equation (6.1) - (6.2) can be converted to a well-posed first order system in the product space $[D(B)] \times X$. The condition (6.9) on F gives rise to a Lipschitz condition on the nonlinear part of the new first order system, and the proof proceeds by the method of successive integrations. We observe that there is no loss of generality in considering equation (6.1) - (6.2) even in the case where it is necessary to translate the infinitesimal generator to obtain a cosine family satisfying condition (F). We replace A by $A - bI$ and $F(u(t), u'(t))$ by $F(u(t), u'(t)) + bu(t)$, and the hypothesis of Proposition (6.7) will still be satisfied. It is possible to obtain results under the hypothesis of Proposition (6.7) which are analogous to the results of Propositions 6.2, 6.3, and 6.4, but we omit their statement here. We remark that by virtue of Proposition 3.4, we have that $E_1 = D(B)$ whenever condition (F) is satisfied.

VII. REFERENCES

[1] Barbu, V., *A class of boundary problems for second order abstract differential equations*, J. Fac. Sci. Univ. Tokyo I A Math. 19 (1972), pp. 295-319.

[2] Barbu, V., *Nonlinear Semigroups and Differential Equations in Banach Spaces*, Noordhoff International, Leyden, The Netherlands, 1975.

[3] Da Prato, G. and Giusti, E., *Una caratterizzazione dei generatori di funzioni coseno astratte*, Boll. Unione Mat. Italiana 22 (1967), pp. 357-362.

[4] Dunford, N. and Schwartz, J. T., *Linear Operators, Part I: General Theory*, Interscience, New York, 1958.

[5] Fattorini, H. O., *Ordinary differential equations in linear topological spaces, I*, J. Differential Equations 5 (1968), pp. 72-105.

[6] Fattorini, H. O., *Ordinary differential equations in linear topological spaces, II*, J. Differential Equations 6 (1969), pp. 50-70.

[7] Fattorini, H. O., *Uniformly bounded cosine functions in Hilbert space*, Indiana Univ. Math. J. 20 (1970), pp. 411-425.

[8] Giusti, E., *Funzioni coseno periodiche*, Boll. Unione Mat. Italiana 22 (1967), pp. 478-485.

[9] Goldstein, J. *Semigroups and hyperbolic equations*, J. Functional Analysis 4 (1968), pp. 50-70.

[10] Goldstein, J., *On a connection between first and second order differential equations in Banach spaces*, J. Math., Anal., Appl. 30 (1970), pp. 246-251.

[11] Goldstein, J., *On the convergence and approximation of cosine functions*, Aequationes Math. 11 (1974), pp. 201-205.

[12] Goldstein, J., *Semigroups of Operators and Abstract Cauchy Problems*, Monograph, Tulane University, New Orleans, 1970.

[13] Kato, T., *Perturbation Theory for Linear Operators*, Springer-Verlag, New York, 1966.

[14] Kisyński, J., *On operator-valued solutions of d'Alembert's functional equation, I.*, Colloquium Math. 23 (1971), pp. 107-114.

[15] Kisyński, J., *On operator-valued solutions of d'Alembert's functional equation, II.*, Studia Math. 42 (1971), pp. 43-66.

[16] Kisyński, J., *On cosine operator functions and one-parameter groups of operators*, Studia Math. 44 (1972), pp. 93-105.

[17] Konishi, Y., *Cosine functions of operators in locally convex spaces*, J. Fac. Sci. Univ. Tokyo I A Math. 18 (1971/72),

pp. 443-463.

[18] Krein, S. G., *Linear Differential Equations in Banach Spaces*, Amer. Math. Soc. Translations of Math. Monographs, Vol. 29, Providence, 1971.

[19] Kurepa, S., *On some functional equations in Banach spaces*, Studia Math. 19 (1960, pp. 149-158.

[20] Kurepa, S., *A cosine functional equation in Banach algebras*, Acta. Sci. Math. Szeged 23 (1962), pp. 255-267.

[21] Lakshmikantham, V., *Differential Equations in Abstract Spaces*, Academic Press, New York, 1972.

[22] Nagy, B., *On the generators of cosine operator functions*, Publicationes Mathematicae 21 (1974), pp. 151-154.

[23] Nagy, B., *On cosine operator functions in Banach spaces*, Acta Scientarum Mathematicarum Szeged 35 (1974), pp. 281-289.

[24] Nagy, B., *Cosine operator functions and the abstract Cauchy problem*, Periodica Mathematica Hungarica Vol. 7 (3) (1976), pp. 15-18.

[25] Pazy, A., *Semi-Groups of Linear Operators and Applications to Partial Differential Equations*, Lecture Notes, University of Maryland, 1974.

[26] Segal, I., *Non-linear semi-groups*, Ann. Math. 78 (1963), pp. 339-364.

[27] Sova, M., *Cosine operator functions*, Rozprawy Matematiyczne 49 (1966), pp. 1-47.

[28] Sova, M., *Semigroups and cosine functions of normal operators in Hilbert Spaces*, Casopis Pest. Mat. 93 (1968), pp. 437-458.

[29] Takenaka, T. and Okazawa, N., *A Phillips-Miyadera type perturbation theorem for cosine functions of operators*, to appear Tohoku Math. J.

[30] Travis, C. and Webb, G., *Cosine families and abstract non-linear second order differential equations*, to appear.

[31] Travis, C. and Webb, G., *Compactness, regularity, and uniform continuity properties of strongly continuous cosine families*, to appear.

[32] Travis, C. and Webb, G., *Perturbation of strongly continuous cosine family generators*, to appear.

[33] Travis, C. and Webb, G., *An abstract second order semilinear volterra integrodifferential equation*, to appear.

[34] Yosida, K., *Functional Analysis*, Fourth Ed., Springer-Verlag, New York, 1974.

A CHARACTERIZATION OF THE RANGE OF A NONLINEAR VOLTERRA INTEGRAL OPERATOR

Thomas Kiffe
and
Michael Stecher
Texas A&M University

I. INTRODUCTION

In this paper we will be concerned with the existence and uniqueness of solutions to the abstract Volterra integral equation

$$(1.1)\quad \mu(t) + \int_0^t a(t-s)A\mu(s)ds + \int_0^t g(t-s,\mu(s))ds = f(t),$$
$$0 \leqslant t \leqslant T.$$

Our setting will be a real Hilbert space H, $a(t)$ will denote a real-valued function, A will be a positive, linear, self-adjoint operator on H, and $g(s,x)$ will be a function from $R^+ \times H$ to H, where R^+ is the set of all non-negative real numbers. By a strong solution of (1.1) on the interval $[0,T]$ we mean a function $\mu(t)$ satisfying (1.1) subject to the conditions

$$(1.2)\qquad \mu \in L^2[0,T;H], \quad \mu(t) \in D(A) \quad \text{a.e.}$$

$$(1.3)\qquad A\mu \in L^2[0,T;H].$$

Before considering the existence and uniqueness problem for (1.1) we will restrict ourselves to the reduced equation

$$(1.4)\qquad \mu(t) + \int_0^t a(t-s)A\mu(s)ds = f(t)$$

obtained from (1.1) by setting g equal to zero. Our first

ISBN 0-12-434160-8

objective is to give necessary and sufficient conditions on $f(t)$ so that (1.4) has a unique strong solution. Next we will discuss weak solutions of (1.4) and prove a regularity property of such solutions. A function $\mu(t)\colon [0,T] \to H$ will be called a weak solution of (1.1) or (1.4) respectively if there are sequences $\{f_n(t)\}$ and $\{\mu_n(t)\}$ such that μ_n is a strong solution of (1.1) or (1.4) respectively with f replaced by f_n, $f_n \to f$ in $L^2[0,T;H]$ and $\mu_n \to \mu$ in $L^2[0,T;H]$. It will be seen that weak solutions of (1.4) are precisely the solutions of the equation

$$\mu(t) + A \int_0^t a(t - s)\mu(s)ds = f(t). \tag{1.5}$$

The last part of the paper will be concerned with strong and weak solutions of (1.1). Several examples will be presented to illustrate the results contained in this paper.

Abstract linear Volterra equations have been considered by several authors [2, 3, 5]. The goals and techniques of those papers are quite different from ours.

II. STATEMENT AND DISCUSSION OF RESULTS

Throughout this paper we will assume that

(2.1) $a(t)$ is a real valued, locally absolutely continuous function on $[0,\infty)$, $a'(t)$ is of locally bounded variation on $[0,\infty)$ and $a(0) > 0$,

(2.2) The even extension of $a(t)$ to $(-\infty,\infty)$ is positive definite in the sense of Bochner, and

(2.3) A is a linear, self-adjoint operator from $D(A) \subset H$ into H, where H is a real Hilbert space, and $D(A)$ is dense in H.

A will denote the usual extension of A to $L^2[0,T;H]$, i.e., $\mu \in D(\mathsf{A})$ if and only if $\mu(t) \in D(A)$ a.e. and $A\mu \in L^2[0,T;H]$. We define the operator $V\colon L^2[0,T;H] \to L^2[0,T;H]$ by

$V\mu(t) = \int_0^t a(t - s)\mu(s)ds$. By $W_0^{1,2}[0,T;H]$ we will denote the set of all absolutely continuous functions $f: [0,T] \to H$ satisfying $f' \in L^2[0,T;H]$ and $f(0) = 0$.

Proposition 1. Let (2.2) and (2.3) be satisfied. Then $VA: D(A) \to L^2[0,T;H]$ is a linear, positive operator.

Using the notation given above we can write (1.4) as $(I + VA)\mu = f$ where I is the identity operator on $L^2[0,T;H]$. By $R(I + VA)$ we will denote the range of $I + VA$, i.e., the set of all $f \in L^2[0,T;H]$ for which (1.4) has a unique strong solution.

Theorem 1. Let (2.1), (2.2) and (2.3) be satisfied. Then $R(I + VA) = D(A) + W_0^{1,2}[0,T;H]$.

In Section 7 we will give an example of a function $f \in L^2[0,T;H]$ which is not an element of $R(I + VA)$. Hence (1.4) need not have a strong solution for arbitrary $f \in L^2[0,T;H]$. Weak solutions of (1.4) are closely related to solutions of (1.5), which can be written as $(I + AV)\mu = f$. The next theorem characterizes the operator $I + AV$ and presents the relationship between (1.4) and (1.5).

Theorem 2. Let (2.1), (2.2) and (2.3) be satisfied.

(i) For each $f \in L^2[0,T;H]$, (1.4) has a unique weak solution $\mu(t)$ satisfying $\|\mu\|_{L^2[0,T;H]} \leq \|f\|_{L^2[0,T;H]}$;

(ii) $\mu(t)$ is a weak solution of (1.4) if and only if $\mu(t)$ is a solution of (1.5);

(iii) $(I + AV)^{-1}$ is a contraction defined on all of $L^2[0,T;H]$ and maps $D(A) + W_0^{1,2}[0,T;H]$ onto $D(A)$.

Part (ii) of Theorem 2 is a regularity result for weak solutions of (1.4). It says that if $\mu(t)$ is a weak solution of (1.4), then $\int_0^t a(t - s)\mu(s)ds \in D(A)$. Our last theorem concerns (1.1) and basically states that Lipschitz perturbations of (1.4) do not affect the solvability of this equation. In Theorem 3 we use the notation $G\mu(t) = \int_0^t g(t - s, \mu(s))ds$.

Theorem 3. Let (2.1), (2.2) and (2.3) be satisfied. Suppose that

(2.4) $$g(s,x): R^1 \times H \to H$$

and there exist functions $b(s)$ and $c(s)$ in $L^1_{loc}[0,\infty)$ such that

(2.5) $$|g(s,x) - g(s,y)| \leq b(s)|x - y| \quad \text{and}$$
$$|g_1(s,x) - g_1(s,y)| \leq c(s)|x - y|$$

for all $x,y \in H$, where the subscript denotes differentiation with respect to the first variable.

Then

(i) (1.1) has a unique strong solution for every $f \in D(A) + W_0^{1,2}[0,T;H]$, i.e., $R(I + VA + G) = D(A) + W_0^{1,2}[0,T;H]$;

(ii) (1.1) has a unique weak solution for each $f \in L^2[0,T;H]$ and $\mu(t)$ is a weak solution of (1.1) if and only if $(I + AV + G)\mu = f$.

III. PROOF OF PROPOSITION 1

Since A is positive and self-adjoint there is a self-adjoint operator B such that $B^2x = Ax$ for every $x \in D(A)$. Proposition 1 follows from the following identities. Let $\mu \in D(A)$. Then

$$\int_0^T <\mu(t), \int_0^t a(t-s)A\mu(s)ds> dt = \int_0^T <\mu(t), A\int_0^t a(t-s)\mu(s)ds> dt$$
$$= \int_0^T <\mu(t), B^2\int_0^t a(t-s)\mu(s)ds> dt$$
$$= \int_0^T <B\mu(t), B\int_0^t a(t-s)\mu(s)ds> dt$$
$$= \int_0^T <B\mu(t), \int_0^t a(t-s)B\mu(s)ds> dt.$$

Since $a(t)$ is positive definite, the last integral is non-negative, so VA is a positive operator on $D(A)$ into $L^2[0,T;H]$.

IV. PROOF OF THEOREM 1

By the definition of a strong solution of (1.4) and since $a(t)$ satisfies (2.1) it is clear that $R(I + VA) \subset D(A) + W_0^{1,2}[0,T;H]$. To show the opposite inclusion we first note that [4, Theorem 1] implies that (1.4) has a unique strong solution if $f \in W_0^{1,2}[0,T;H]$ since A is a sub-differential [1, Proposition 2.15]. By linearity it only remains to show that (1.4) has a strong solution if $f \in D(A)$. Let $J_\lambda = (I + \lambda A)^{-1}$ and $A_\lambda = AJ_\lambda = \frac{I-J_\lambda}{\lambda}$ for $\lambda > 0$. Since A_λ is Lipschitz, the approximating equation

$$\mu_\lambda(t) + \int_0^t a(t - s)A_\lambda\mu_\lambda(s)ds = f(t) \tag{4.1}$$

has a solution $\mu_\lambda \in L^2[0,T;H]$ for each $\lambda > 0$. Multiply (4.1) by μ_λ and integrate from 0 to T. Since A_λ is self-adjoint Proposition 1 gives us

$$\int_0^T |\mu_\lambda(t)|^2 dt \leqslant \int_0^T < f(t),\mu_\lambda(t) > dt. \tag{4.2}$$

Combining (4.2) and Young's inequality we have

$$\|\mu_\lambda\|_{L^2[0,T;H]} \leqslant \|f\|_{L^2[0,T;H]}. \tag{4.3}$$

Applying A_λ to (4.1) and repeating the above calculations we get, since $f \in D(A)$

$$\|A_\lambda\mu_\lambda\|_{L^2[0,T;H]} \leqslant \|Af\|_{L^2[0,T;H]}. \tag{4.4}$$

Thus there is a sequence $\{\lambda_n\}$ tending to zero and functions $u,v \in L^2[0,T;H]$ such that

(4.5)
$$\mu_{\lambda_n} \rightharpoonup \mu \text{ weakly in } L^2[0,T;H]$$
$$A_{\lambda_n}\mu_{\lambda_n} \rightharpoonup v \text{ weakly in } L^2[0,T;H].$$

It is easy to see that μ and v satisfy $\mu(t) + \int_0^t a(t-s)v(s)ds = f(t)$, for all $0 < t < T$. To complete the proof of Theorem 1 it suffices to show that

(4.6) $$\mu \in D(\mathrm{A}) \text{ and } v = \mathrm{A}\mu.$$

To this end, let $J_\lambda = (I + \lambda \mathrm{A})^{-1}$ and $\mathrm{A}_\lambda = \mathrm{A}J_\lambda = \frac{(I - J_\lambda)}{\lambda}$. Then J_λ and A_λ are also the extensions to $L^2[0,T;H]$ of J_λ and A_λ respedtively. By (4.4) and (4.5) we have $\{\mathrm{A}_{\lambda_n}\mu_{\lambda_n}\}$ is uniformly bounded in $L^2[0,T;H]$ and $\mu_{\lambda_n} \rightharpoonup \mu$, $_{\lambda_n}\mu_{\lambda_n} \rightharpoonup v$. Since $\lambda \mathrm{A}_{\lambda_n}\mu_{\lambda_n} = \mu_{\lambda_n} - J_{\lambda_n}\mu_{\lambda_n}$, we have $\mu_{\lambda_n} - J_{\lambda_n}\mu_{\lambda_n} \to 0$ so $J_{\lambda_n}\mu_{\lambda_n} \rightharpoonup \mu$. Since $\mathrm{A}_{\lambda_n}\mu_{\lambda_n} = \mathrm{A}J_{\lambda_n}\mu_{\lambda_n}$ and A is weakly closed in $L^2[0,T;H] \times L^2[0,T;H]$ we have $\mu \in D(\mathrm{A})$ and $v = \mathrm{A}\mu$. The uniqueness of $\mu(t)$ follows directly from (2.2) and A being positive.

V. PROOF OF THEOREM 2

Let $f \in L^2[0,T;H]$ be arbitrary. Since $W_0^{1,2}[0,T;H]$ is dense in $L^2[0,T;H]$ there is a sequence $\{f_n\} \subseteq W_0^{1,2}[0,T;H]$ such that $f_n \to f$ in $L^2[0,T;H]$. Let $\mu_n(t)$ denote the unique strong solution of (1.4) with f replaced by f_n. Write (1.4) with μ_n, f_n and with μ_m, f_m, subtract, multiply by $\mu_n - \mu_m$ and integrate. By (2.2) we have

(5.1) $$\int_0^T |\mu(t) - \mu_m(t)|^2 dt \leq \int_0^T \langle \mu_n(t) - \mu_m(t), f_n(t) - f_m(t) \rangle \, dt.$$

By Young's inequality we have

(5.2) $$\|\mu_n - \mu_m\|_{L^2[0,T;H]} \leq \|f_n - f_m\|_{L^2[0,T;H]}.$$

Thus $\{\mu_n\}$ is a Cauchy sequence in $L^2[0,T;H]$ and hence converges in $L^2[0,T;H]$ to a weak solution $\mu(t)$ of (1.4). By (4.3) we have $\|\mu_n\|_{L^2[0,T;H]} \leq \|f_n\|_{L^2[0,T;H]}$ so $\|\mu\|_{L^2[0,T;H]} \leq \|f\|_{L^2[0,T;H]}$. Uniqueness follows directly from (5.2).

To prove (ii), let $\mu(t)$ be a weak solution of (1.4). Then there exist sequences $\{f_n\}$ and $\{\mu_n\}$ such that $f_n \to f$, $\mu_n \to \mu$ in $L^2[0,T;H]$ and $\mu_n(t) + \int_0^t a(t-s)A\mu_n(s)ds = f_n(t)$. Since $\mu_n \in D(A)$ we also have

$$\mu_n(t) + A\int_0^t a(t-s)\mu_n(s)ds = f_n(t). \tag{5.3}$$

Since $\mu_n \to \mu$ we have $\int_0^t a(t-s)\mu_n(s)ds \to \int_0^t a(t-s)\mu(s)ds$ in $L^2[0,T;H]$. Since A is a closed operator, letting $n \to \infty$ in (5.3) gives us

$$\mu(t) + A\int_0^t a(t-s)\mu(s)ds = f(t), \quad \int_0^t a(t-s)\mu(s)ds \in D(A) \tag{5.4}$$

i.e., $\mu(t)$ is a solution of (1.5). To prove the opposite implication of (ii) we note that (1.5) has at least one solution for each $f \in L^2[0,T;H]$, namely the corresponding weak solution of (1.4). But since A and V are positive operators and A is self-adjoint (1.5) has at most one solution for $f \in L^2[0,T;H]$.

To prove (iii) we note that $(I + AV)\mu = f$ has a solution $\mu(t)$ for each $f \in L^2[0,T;H]$ and by (i) and (ii) $\|\mu\|_{L^2[0,T;H]} \leq \|f\|_{L^2[0,T;H]}$. Thus $(I + AV)^{-1}$ is a contraction defined on all of $L^2[0,T;H]$. The second part of (iii) follows from the fact that if $f \in D(A) + W_0^{1,2}[0,T;H]$ then (1.4) has a strong solution $\mu(t) \in D(A)$. This function is a solution of (1.5) since A and V commute on $D(A)$.

VI. PROOF OF THEOREM 3

Fix $T_1 < T$ so small that

(6.1) $$\int_0^{T_1} b(s)ds < \tfrac{1}{2}.$$

Let $f \in L^2[0,T;H]$ and define $K_f\colon\ L^2[0,T_1;H] \to L^2[0,T_1;H]$ by

(6.2) $$K_f\mu = (I + AV)^{-1}f - (I + AV)^{-1}G\mu.$$

By (2.5) and (6.1) we have

(6.3) $$\|K_f\mu - K_f v\|_{L^2[0,T_1 H]} < \tfrac{1}{2}\|\mu - v\|_{L^2[0,T_1;H]}$$

so K_f is a strict contraction on $L^2[0,T_1;H]$, and hence has a unique fixed point $\mu(t) \in L^2[0,T_1;H]$ satisfying

(6.4) $$\mu(t) + A\int_0^t a(t-s)\mu(s)ds + \int_0^t g(t-s,\mu(s))ds = f(t),$$
$$\int_0^t a(t-s)\mu(s)ds \in D(A), \quad 0 \leqslant t \leqslant T_1$$

To prove (i), suppose $f \in D(A) + W_0^{1,2}[0,T;H]$. By (2.5) $G\mu \in W_0^{1,2}[0,T;H]$ and hence by Theorem 2, $(I + AV)^{-1}f$ and $(I + AV)^{-1}G\mu$ are elements of $D(A)$, where μ is the fixed point of K_f. Thus $\mu \in D(A)$ and since A and V commute on $D(A)$, μ is a strong solution of (1.1) on $[0,T_1]$. Since any strong solution of (1.1) is a fixed point of K_f, (1.1) has a unique strong solution. Thus $R(I + VA + G) \supseteq D(A) + W_0^{1,2}[0,T;H]$. The opposite inclusion follows readily from (2.5).

To prove (ii) we observe that if μ is a fixed point of K_f and v is a fixed point of K_g, then

(6.5) $$\|\mu - v\|_{L^2[0,T_1;H]} \leqslant 2\|f - g\|_{L^2[0,T_1;H]}.$$

Since $D(A) + W_0^{1,2}[0,T\ ;H]$ is dense in $L^2[0,T\ ;H]$ it follows from (6.5) that (1.1) has a unique weak solution for every $f \in L^2[0,T_1;H]$. The fact that μ is a weak solution of (1.1) if and only if $(I + AV + G)\mu = f$ is proven exactly as in Theorem 2. Thus Theorem 3 has been proven for the interval $[0,T_1]$. The usual continuation argument for convolution Volterra equations can now be applied to extend these results to the whole interval $[0,T]$ since T_1 is only restricted by (6.1). This completes the proof of Theorem 3.

VII. EXAMPLES

The abstract results contained in this paper can be used to prove the existence and uniqueness of a solution to the problem

$$\mu(x,t) - \int_0^t \mu_{xx}(x,s)ds + \int_0^t |\mu(x,s)|ds = f(x,t), \quad 0 < x < 1, \tag{7.1}$$

$$t > 0, \quad \mu(0,t) = \mu(1,t) = 0.$$

For this problem, we set $H = L^2[0,1]$, $A\mu = -u''$, $D(A) = \{\mu \in H\colon \mu \text{ is absolutely continuous, } u'' \in H,\ \mu(0) = \mu(1) = 0\}$ and $a(t) \equiv 1$. It is well known that A satisfies (2.3) and $a(t)$ is positive definite. Thus if $f \in D(A) + W_0^{1,2}[0,T;H]$ then (7.1) has a strong solution. If $f(x,t) \in L^2((0,1) \times (0,1))$ then (7.1) has a unique weak solution $\mu(x,t)$ satisfying $\int_0^t \mu(x,s)ds \in D(A)$.

Our second example shows that (1.4) need not have a strong solution for arbitrary $f \in L^2[0,T;H]$. Let H, A and $a(t)$ be as in the first example and let

$$f(x,t) = \sum_{n=1}^{\infty} \frac{\cos nt \sin n\pi x}{n^{3/4}} + \pi^2 \sum_{n=1}^{\infty} \frac{\sin nt \sin n\pi x}{n^{3/4}}.$$

It is easy to check that $f(x,t) \in L^2((0,1) \times (0,1))$ and if (1.4) is to have a solution $\mu(x,t)$, then

$$\mu(x,t) = \sum_{n=1}^{\infty} \frac{\cos nt \sin n\pi x}{n^{7/4}}.$$

Simple calculations show that $\mu \in L^2((0,1) \times (0,1))$ and that $\mu(x,t)$ satisfies (1.5) so $\mu(x,t)$ is a weak solution of (1.4).

Since

$$A\mu = \pi^2 \sum_{n=1}^{\infty} n^{1/4} \cos nt \sin n\pi x$$

it is clear that $A\mu \notin L^1((0,1) \times (0,1))$, so (1.4) does not have a strong solution for this $f(x,t)$. This example also illustrates that if we weaken the condition $A\mu \in L^2[0,T;H]$ to $A\mu \in L^1[0,T;H]$ for a strong solution, (1.4) still need not have a solution for arbitrary $f \in L^2[0,T;H]$.

VIII. REFERENCES

[1] Brezis, H., *Operateurs maximaux monotones et semi-groupes de contractions dans les espaces de Hilbert,* North Holland, Amsterdam, 1973.

[2] Friedman, A., *Monotonicity of solutions of Volterra integral equations in Banach space,* Trans. Amer. Math. Soc., 138(1969), pp. 129-148.

[3] Friedman, A. and Shinbrot, M., *Volterra integral equations in Banach space,* Trans. Amer. Math. Soc., 126(1967), pp. 131-179.

[4] Londen, S. O., *On an integral equation in a Hilbert space,* SIAM J. Math. Anal. (to appear).

[5] Clement, Ph. and Nohel, J. A., *Abstract linear and nonlinear Volterra equations preserving positivity,* MRC Technical Summary Report, #1716, University of Wisconsin, Madison, 1977.

DISCONTINUOUS PERTURBATIONS OF ELLIPTIC BOUNDARY VALUE PROBLEMS AT RESONANCE

P. J. McKenna
University of Wyoming

I. DISCONTINUOUS PERTURBATIONS AT RESONANCE

In this paper, we consider an abstract operator equation of the type

(1) $$Ex = Nx$$

where E is a linear bounded operator with kernel from a Banach space X into a Banach space Y, and N is a multivalued map from X to Y which can be approximated in a weak sense by continuous compact maps from X to Y.

We show that in the presence of an additional geometric condition on N and its approximations, there exists a solution to the equation (1) in the sense that there exists $x \in X$ and $\gamma \in N(x)$ so that $Ex = \gamma$.

Then, in section 3, we apply these results to the study of equations of the type $Ex = g(u) + f(t)$ where E is a strongly elliptic linear partial differential operator with coercive boundary conditions and $g \in L^{\infty}(\Omega)$.

This type of equation has been extensively studied since the original paper of Landesman and Lazer [3] and an extensive bibliography now exists. For a comprehensive discussion in the framework of the alternative method see [1].

It is necessary to re-interpret g in a manner suggested by J. Rauch in order to extend these results to the case where g is allowed to be discontinuous.

ISBN 0-12-434160-8

II. THE ABSTRACT FRAMEWORK

Let X and Y, $X \subset Y$ be reflexive Banach spaces with norms $\| \|_X$, $\| \|_Y$, contained in a Hilbert space H with inner product $<,>$. We assume E is a bounded linear operator, $E: X \to Y$ and that

E_1) $\ker E = X_0$ is finite dimensional

E_2) $\operatorname{Ran} E \subset (\ker E)^{\perp}$ where the orthogonal complement is taken in H.

Let Q be the restriction to Y of orthogonal projection in H, and we assume there exists a projection $P: X \to X$ with $\operatorname{Ran} P = \ker E$, and a partial inverse H of E such that the usual axioms of the alternative method hold

E_3)
a) $EH(I - Q) = I - Q$
b) $H(I - Q)Ex = (I - P)x$,
c) $EP = QE$

We assume that $N: X \to Y$ is a multivalued map with the properties

N_1) $\exists M > 0$ such that $\|\gamma\|_Y \leq M$ for all $\gamma \in N(x)$, $x \in X$

N_2) There exist continuous compact maps $N_n: X \to Y$ satisfying

i) $\|N_n(x)\| \leq 2M$, all $x \in X$

ii) If $x_n \to x$ weakly in X and $N_n(x_n) \to \gamma$ weakly in Y, then $\gamma \in N(x)$

We make the following geometric assumption on the approximating maps N_n

N_3) $\forall R_1 > 0$, $\exists R_0 > 0$, $\alpha > 0$ so that if $x \in X_0$, $\|x_0\| = R_0$, $x_1 \in X_1$, $\|x_1\| \leq R_1$ then there exists an integer so that $(N_n(x_0 + x_1), x_0) \geq \alpha \|x_0\|$ for $n \geq N$.

Theorem 1. Let E satisfy E_1) and E_2) and let the multivalued map N satisfy N_1) N_2) N_3). Then there exists $x \in X$ and $\gamma \in Y$ so that $Ex = \gamma$ where $\gamma \in N(x)$.

Proof: The proof consists of two stages. First we consider the approximating equations $Ex = N_n(x)$ and show that these equations have solutions x_n which are uniformly bounded in X. Then we show that these solutions x_n converge to a solution x of the original equation.

In the normal framework, of the alternative method [1], the equation $Ex = N_n x$ is equivalent to the following system of equations.

(2) a) $0 = (I - P)x + H(I - Q)N_n x$

b) $0 = QN_n x$

which is equivalent to searching for zeros of the product map.

(3) $(I - T_n)x = x - Px + H(I - Q)N_n x - QN_n x$

where $I - T_n: X \to X$ and

$T_n x = Px - H(I - Q)N_n x + QN_n x$ is compact.

We shall define a region Ω in X and shall show that $d_{LS}(I - T_n, 0, \Omega) = 1$. We shall, if necessary, renorm X and Y so that

$$\|x\|_X = \|Px\|^2 + \|(I - P)x\|^2$$

and

$$\|y\|_Y = \|Qy\|^2 + \|(I - Q)y\|^2$$

This allows us to assume $\|I - P\| = \|I - Q\| = 1$ and clearly does not affect any of the assumptions on E or N.

Let $\Omega = \{x \in X, \quad x = x_0 + x_1, \; x_0 \in X_0, \; x_1 \in X_1$

$$\|x_0\| \leq R_0, \quad \|x_1\| \leq R_1\}$$

where $R_1 = 4\|H\|M$ and R_0 is the corresponding constant in N_3).
We shall consider $I - \lambda T_n$ where $\lambda \in [0, 1]$.

a) First let $x = x_0 + x_1$ $\|x_1\| = R_1$, $\|x_0\| \leqslant R_0$

In this case

$$\|(I - \lambda T_n)x\| = \|x_1 - \lambda H(I - Q)N_n X + (1 - \lambda)x_0 + \lambda QNx\|$$

$$= (\|x_1 + \lambda H(I - Q)N_n x\| + \|(I - \lambda)x_0 + \lambda QNx\|^2)^{1/2}$$

$$\geqslant \|x_1 + H(I - Q)N_n(x)\|$$

$$\geqslant R_1 - \|H\| 2M > 2\|H\| M > \delta > 0.$$

b) Now let $x = x_0 + x_1$, $\|x_1\| \leqslant R_1$ $\|x_0\| = R_0$

In this case

$$((I - \lambda T)x_0, x_0) = (1 - \lambda)\|x_0\|^2 + \lambda(N(x_0 + x_1), x_0)$$

$$\geqslant (1 - \lambda)R_0^2 + \varepsilon\lambda R_0 > \delta > 0$$

for some $\delta > 0$ and all $\lambda \in [0, 1]$.

Consequently, $d_{LS}(I - T_n, 0, \Omega) = 1$ and the maps $I - T_n$ have zeros X_n in Ω.

We now show that x_n converge to a solution of the original equation. Since $\|x_n\| \leqslant (R_0^2 + R_1^2)^{1/2}$, and since X is reflexive, we have $x_n \to x$ weakly in X for some $x \in \overline{\Omega}$. Furthermore $\|N_n(x_n)\|_Y \leqslant 2M$ and thus $N_n(x_n)$ converge weakly to γ for some $\gamma \in Y$ since Y is reflexive. By N_2, $\gamma \in Nx$.

Since E is bounded and continuous, $Ex_n \to Ex$ and thus $Ex = \gamma$ where $\gamma \in Nx$. This concludes the proof of Theorem 1.

In fact a slightly more general theorem can be proved by these methods.

Theorem 2. In the event of the assumptions on E and N of Theorem 1 being satisfied and assuming that N^* is a continuous compact map from X to Y mapping bounded sets to bounded sets, there exists $\varepsilon_0 > 0$ so that the equation $Ex = \gamma + \varepsilon N^*(x)$ is satisfied for some $x \in X$, $\in N(x)$ and all $0 \leqslant \varepsilon < \varepsilon_0$.

Proof: The proof is essentially the same as that of Theorem 1. The only modification necessary is that we replace N_n by the new non-linearity $N_n + \varepsilon N^*$. $I - T$ is then replaced by a corresponding $I - (T_n + \varepsilon T^*)$.

The reader will have noted that in the proof of Theorem 1, we showed that either $\|(I - \lambda T_n)x\|$ or $((I - \lambda T_n)x, x_0)$ were greater than or equal to $\delta > 0$. Since T^* is bounded on Ω, there exists $\varepsilon_0 > 0$ such that the corresponding quantities $\|(I - \lambda(T_n + \varepsilon T^*))x\|$ and $((I - \lambda(T_n + \varepsilon T^*))x, x_0)$ are greater than or equal to $\alpha/2$. Thus the proof of the theorem goes through as before.

Remark 1. Theorem 2 is an example of a theorem of existence across resonance. Cesari first proved that under similar hypotheses on E and N, an equation of the form $Ex + \alpha x = Nx$ has solutions for sufficiently small α.

In [5] it was shown that solutions existed where αx was replaced by a non-linear term $\alpha N^*(x)$ as in Theorem 2. Subsequently, Cesari, in [2], extended his method to prove a similar theorem by the use of Schauder's fixed point theorem.

Remark 2. The hypotheses for Theorems 1 and 2 can be changed in several ways without affecting the substance of the proof. In particular, instead of assuming that the N_n were compact, one could assume that E was unbounded but with a compact inverse H and that the N_n were merely bounded and continuous. The same result would follow by the same methods.

Remark 3. A multivalued map N which can be approximated by continuous maps N_n in the sense of condition (N_2) is called <u>locally maximal</u>. In particular if $N: L^2 \to L^2$, defined by $N(x) = f(x)$ for $f \in L^\infty$, f having maximal monotone graph, then N is locally maximal.

III. APPLICATIONS TO SEMILINEAR ELLIPTIC BOUNDARY VALUE PROBLEMS WITH KERNEL

In this section, we assume E is a self adjoint strongly elliptic partial differential operator on a region G with Dirichlet boundary condition. We assume for simplicity that $\dim \ker E = 1$ and that $\ker E = \{\theta\}$.

We shall consider equations of the form

(4) $$Ex = g(x) - h(t)$$

where $g \in L^{\infty}(G)$ and $f \in L$.

In order to place this problem in the framework of Theorems 1 and 2 we replace the map g by a multivalued map $\hat{g}$ with "the jumps filled in". In other words, in accordance with an idea of Rauch [7] we consider the equation

(5) $$Ex = \hat{g}(x) - f(t)$$

where $\hat{g}$ is defined as follows:

For any $\varepsilon > 0$, and $s \in R$, let

$$\overline{g}_{\varepsilon}(o) = \operatorname*{ess\,sup}_{|t - s| < \varepsilon} g(s)$$

$$\underline{g}_{\varepsilon}(s) \quad \operatorname*{ess\,inf}_{|t - s| < \varepsilon} g(s)$$

For s fixed, $\overline{g}_{\varepsilon}(s)$ is decreasing as $\varepsilon \to 0$ and $\underline{g}_{\varepsilon}(s)$ is increasing as $\varepsilon \to 0$. Let

(6) $$\overline{g}(s) = \lim_{\varepsilon \to 0} \overline{g}\ (s)$$

$$\underline{g}(s) = \lim_{\varepsilon \to 0} g\ (s)$$

We defined the multiple-valued function by

(7) $$\hat{g}(s) = [\underline{g}(s), \overline{g}(s)]$$

For further remarks on $\hat{g}$, see [7].

The usual Landesman-Lazer condition for existence of solutions is that the limits $g(+\infty)$, $g(-\infty)$ exist and the

(8)
$$g(+\infty)\int_{\theta>0} dt + g(-\infty)\int_{\theta<0} \theta dt - \int f\theta dt > \delta > 0$$
$$g(-\infty)\int_{\theta>0} dt + g(+\infty)\int_{\theta<0} \theta dt - \int f\theta dt < -\delta < 0$$

are satisfied. We shall show that these conditions are sufficient to guarantee that (N_3) is satisfied.

Then, we shall show that there exists $x \in \overset{o}{H}{}^{2m}(G)$ and $\gamma(x) \in L^2(\Omega)$ with the property that $\gamma(s) \in \hat{g}(s)$ for almost all $s \in R$, and

$$Ex = \gamma - f(t)$$

In order to apply Theorems 1 and 2, we must define the approximating continuous non-linearities N_n and we must show that in the presence of Ω (8), conditions N_2, N_3 are satisfied.

Following Rauch [7] we construct the non-linearities N_n as follows:

Choose $j \in c_o^\infty(-1, +1)$ such that $j \geqslant 0$ and $\int_{-\infty}^{+\infty} j(s)ds = 1$. Let $j_n(s) = nj(ns)$ and let $g_n = j_n * g$, $N_n(x) = g_n(x) - f(t)$.

Let the space X be $\overset{o}{H}{}^{2m}(G)$ where $2m$ is the order of E and let $Y = H = L^2(G)$.

To satisfy N_2, we need only show that if $u_n \to u$ weakly in $H^{2m}(G)$ and $g_n(u_n) \to \gamma$ weakly in $L^2(G)$, then $\underline{g}(s) \leqslant \gamma(s) \leqslant \overline{g}(s)$. This we now establish.

Since $\overset{o}{H}{}^{2m}(G)$ is compactly embedded in $L^2(G)$ $u_n \to u$ strongly in $L^2(G)$ and thus a.e. in Ω.

Thus $u_n \to u$ uniformly except on a set ω of measure less than η, where η is as small as we please. Furthermore $g_n(u_n)$ converges weakly in $L^2(G)$ to some function γ since $g_n(u_n)$ are bounded.

Thus for any $\varepsilon > 0$, there is an $n_0 > 2/\varepsilon$ such that for $n > n_0$, $|u_n(x) - u(x)| < \varepsilon/2$ for all $x \in G\,\omega$. If $n > n_0$ and $x \in \Omega\backslash\omega$ then $g_n(u_n(x))$ is the average of g over the interval $|s - u_n(x)| < \varepsilon/2$, which is a subset of the interval $(u(x) - \varepsilon, u(x) + \varepsilon)$. Therefore $\underline{g}_\varepsilon(u(x)) \leqslant g_n(u_n(x)) \leqslant \overline{g}_\varepsilon(u(x))$.

Thus $\int_{G\backslash\omega} \underline{g}\ (u(x)h \leqslant \int_{G\backslash\omega} g_n(u_n(x)h \leqslant \int_{G\backslash\omega} \overline{g}_\varepsilon(u(x))h.$

Letting $n \to +\infty$ and using the weak convergence of $g_n(u_n)$, we obtain

$$\int_{G\backslash\omega} \underline{g}_\varepsilon\ (u(x))h \leqslant \int_{G\backslash\omega} \gamma h \leqslant \int_{G\backslash\omega} \overline{g}_\varepsilon\ (u(x))h.$$

Since g_ε are bounded and since ω was of arbitrarily small measure, we obtain by the Lebesque convergence theorem

$$\underline{g}(u) \leqslant \gamma \leqslant \overline{g}(u) \quad \text{a.e.} \quad \text{in} \quad G$$

and thus N_2) is satisfied.

We remark that the construction is identical to that of [7].

Next we show that the condition N_3) is satisfied. This is a variation of a fact first used by Williams [7] and subsequently by McKenna [4, 5], Cesari [1, 2] and others.

We show that if (8) is satisfied then (N_3) is satisfied by considering the inner product $(N_n(x_0 + x_1), x_0)$. Thus we consider

$$\int_G g_n(c\theta + x_1)\theta dt - \int f\theta dt.$$

Without loss of generality, we may assume that $\theta = 0$ only on a set of measure zero. Choose the sets $G_{+,\delta}$, $G_{-,\delta}$ such that $\theta > 0$ on $G_{+,\delta}$ and $\theta < -\delta$ on $G_{-,\delta}$ and define r_0 to be such that if $t > r_0$, then $|g(t) - g(+\infty)| < \eta_1$ and if $t < -r_0$, then $|g(t) - g(-\infty)| < \eta_1$.

We consider

$\int_{G_+} g_n(c\theta + x_1)\ \theta dt$, and show that if $\|x_1\| \leqslant R_1$ then

there exists an R_0 so that if $C > R_0$, then

$$\int_{G_+} g_n(c\theta + x_1)\theta dt > (g(+\infty) - \eta_2 \int_{G_+} \theta dt$$

for small η_2. Let $R_0 > 4r_0/\delta$.

$$\int_{G_+} g_n(c\theta + x_1)\theta = \int_{G_{+,\ \delta}} g_n(c\theta + x_1)\theta + \int_{G_+\setminus G_{+,\ \delta}} g_n(c\theta + x_1)\theta.$$

Since g_n is bounded, the second integral on the right hand side can be made arbitrarily small by choosing δ sufficiently small.

Now consider $\int_{G_{+,\ \delta}} g_n(c\theta + x_1)\theta dt.$

Since $c > 4r_0/\delta$, it follows that $c\theta > 4r_0$. For r_0 sufficiently large, x_1 will be bounded by r_0 except on a set ω of measure r_0. Thus,

$$\int_{G_{+,\ \delta}} g_n(c\theta + x_1)\theta dt = \int_{G_{+,\ \delta\setminus\omega}} g_n(c\theta + x_1)\theta dt - \int_{\omega} g_n(c\theta + x_1)\theta dt$$

for sufficiently large r_0.

However, $g_n(x)$ is the weighted average over the interval $(x - 1/n,\ x + 1/n)$ of g and since $c\theta + x_1 > 3r_0$ on $\Omega_{+,\ \delta}\setminus\omega$, it follows that $|g_n(c\theta + x_1) - g(+\infty)| < \eta_3$, by choice of r_0 and $\Omega_{+,\ \delta}\setminus\omega$.

This implies that for any small η_4 and any R_1, there exists an R_0 such that if $c > R_0$, then

$$\left| \int_{G_+} g_n(c\theta + x_1)\theta dt - g(+\infty) \int_{G_+} \theta dt \right| < \eta_4 .$$

Similarly, we can show that given $R_1 > 0$ and $\eta_4 > 0$ there exists R_0 so that

$$\left| \int_{G_-} g_n(c\theta + x_1)\theta dt - g(-\infty) \int_{G_-} dt \right| < \eta_4 \quad \text{if} \quad c > R_0$$

$$\left| \int_{G_+} g_n(c\theta + x_1)\theta dt - g(-\infty) \int_{G_+} \theta dt \right| < \eta_4 \quad \text{if} \quad c < -R_0$$

$$\left| \int_{G_-} g_n(c\theta + x_1)\theta dt - g(+\infty) \int_{G_-} \theta dt \right| < \eta_4 \quad \text{if} \quad c < -R_0$$

Thus if $\|x_1\| \leqslant R_1$, $c > R_0$

$$(N_n(x_0 + x_1)x_0) = c\{ \int g_n(c\theta + x_1)\theta - \int f\theta \}$$

$$\geqslant c\{g(+\infty) \int_{G_+} \theta + g(-\infty) \int_{G_-} \theta - \int f\theta - 2\eta_4\}$$

$$\geqslant c\{\varepsilon - 2\eta_4) = \frac{\varepsilon c}{2}$$

Also, if $\|x_1\| \leqslant R_1$, $c < -R_0$,

$$(N_n(x_0 + x_1), x_0) = c\{ \int g_n(c\theta + x_1)\theta - \int f\theta \}$$

$$\geqslant c\{g(-\infty) \int_{G_+} \theta + g(+\infty) \int_{G_-} \theta - \int f\theta - 2\eta_4\}$$

$$\geqslant c\{-\delta + 2\eta_4\} = -\frac{\varepsilon c}{2}$$

Thus $(N(x_0 + x_1), x_0) \geqslant \varepsilon\|x_0\|$ for $\|x_1\| \leqslant R_1$ and $\|x_0\| = R_0$.

Therefore in the event of condition (8) being satisfied

condition N_3) is satisfied and Theorem 1 applies. Thus in the presence of (8) the equation (4) has a solution. Furthermore, since Theorem 2 applies, there exists $\varepsilon_0 > 0$ such that

$$Ex = \hat{g}(x) - f(t) + \varepsilon h(u, D^{\alpha}u) \quad (\alpha < 2m - 1)$$

for all $\varepsilon < \varepsilon_0$

proveded that $\| h(u, D^{\alpha}u)\|_{L^2} \leqslant \phi(\| u\|_{H^{2m}})$

for some ϕ: $[0, +\infty) \to [0, +\infty)$

In particular, this theorem implies that in the presence of (8), the equation (4) has a solution if f is maximal monotone, in spite of the fact that E is not monotone.

Remark 1. The existence of limits at $\pm\infty$ was assumed for simplicity. It would have been sufficient to assume that $\overline{g}(+\infty)$, $\underline{g}(+\infty)$, $\overline{g}(-\infty)$, $\underline{g}(-\infty)$ exist. In that case, a sufficient condition would be

$$\underline{g}(+\infty)\int_{G_+}\theta + \overline{g}(-\infty)\int_{G_-}\theta - \int f\theta > 0$$

$$\overline{g}(-\infty)\int_{G_+}\theta + \underline{g}(+\infty)\int_{G_-}\theta - \int f\theta < 0$$

Furthermore, the direction of the inequality can be revised. The only change in the proof of Theorem 1 would be to define

$$Tx = Px - H(I - Q)Nx - QNx$$

instead of $Tx = Px - H(I - Q)Nx + QNx$

Remark 2. Again, Dirichlet boundary conditions were assumed for simplicity. Similar results may be obtained in the case where E is not necessarily self-adjoint, where $\ker E$ is allowed to have dimension greater than one, and where E satisfies coercive boundary conditions. The necessary techniques may be found in

[4] and [7].

Remark 3. In the theorem of existence of solutions across resonance only derivatives of order $\leq 2m - 1$ are allowed to appear in N^*. In [4] we show that in the presence of a mild Lipschitz requirement on N_1, derivatives of order $2m$ may be taken into account.

IV. REFERENCES

[1] Cesari, L., "*Functional Analysis and Nonlinear Differential Equations*", in a volume by the same title, Dekker, (Cesari, Kannan, Schuur, eds.) 1976.

[2] Cesari, L., "*Nonlinear Oscillations Across a Point of Resonance for Non-self-adjoint Systems*", in Nonlinear Analysis, a volume in honor of E. H. Rothe, Academic Press.

[3] Landesman, E. M. and Lazer, A. C., "*Nonlinear Perturbations of Linear Elliptic Boundary Value Problems at Resonance*", J. Math. Mech., 19 (1970) pp. 49-68.

[4] McKenna, P. J., "*Nonselfadjoint Semilinear Equations at Resonance in the Alternative Method*", (to appear).

[5] McKenna, P. J., "*Nonselfadjoint Semilinear Problems in the Alternative Method*", Ph.D. Thesis 1976 (University of Michigan).

[6] McKenna, P. J. and Rauch, J., "*Strongly Nonlinear Perturbations of Elliptic Boundary Value Problems with Kernel*", J. Diff. Equations (to appear).

[7] Rauch, J., "*Discontinuous Nonlinearities and Multiple Valued Maps*", (to appear).

[8] Williams, S., "*A Sharp Sufficient Condition for Solutions of a Nonlinear Elliptic Boundary Value Problem*", J. Diff. Equations, 8(1970) pp. 580-586.

AN EXISTENCE THEOREM FOR WEAK SOLUTIONS OF DIFFERENTIAL EQUATIONS IN BANACH SPACES

A. R. Mitchell* and Chris Smith
University of Texas at Arlington

In this paper we define a measure of weak noncompactness of bounded subsets of any Banach space, E, and then prove Ambrosetti type results [1] relating the measure of weak noncompactness of subsets of $F = C[[a,b],E]$ with the measure of weak noncompactness of the range in E. Also, results relating the weak topologies of E and F are obtained. Using these results we prove a theorem, which extends a result by Szep [7], for the existence of weak solutions to the Abstract Cauchy Problem, $x' = f(t,x)$, $x(t_0) = x_0$.

I. PROPERTIES OF THE WEAK TOPOLOGY AND A MEASURE OF WEAK NONCOMPACTNESS

E is a Banach space and $B = \{x \in E: ||x|| \leq 1\}$ is the closed unit ball in E. Let H be a bounded subset of E. The β measure of the weak noncompactness of H is defined by

$$\beta(H) = inf\{t \geq 0: H \subset K_t + tB \text{ for some weakly compact } K_t \subset E\}$$

The following two lemmas provide the basic properties of β and are found in DeBlasi [3].

Lemma 1.1. Let X and Y be bounded subsets of E. Then the β measure of weak noncompactness has the following properties:

* This work supported by the U.S. Army Research Office, Grant DAAG29-77-G-0062.

ISBN 0-12-434160-8

(1) $X \subset Y$ implies $\beta(X) \leq \beta(Y)$;

(2) $\beta(X) = \beta(\bar{X}^w)$;

(3) $\beta(X) = 0$ iff $\bar{X}^w$ is weakly compact;

(4) $\beta(X \cup Y) = max\{\beta(X), \beta(Y)\}$;

(5) $\beta(X) = \beta(co\ X) = \beta(\overline{co}\ X)$, where $co\ X$ is the convex hull of X;

(6) $\beta(X + Y) \leq \beta(X) + \beta(Y)$;

(7) If X is a singleton, then $\beta(X + Y) = \beta(Y)$;

(8) $\beta(tX) = t\beta(X)$ for $t \geq 0$.

Lemma 1.2. Let $\{X_n\}_{n=1}^{\infty}$ be a sequence of nonvoid weakly closed subsets of E. Suppose that X is bounded, $X_1 \supset X_2 \supset \ldots \supset X_n \supset \ldots$, and $\lim_{n\to\infty} \beta(X_n) = 0$. Then $\bigcap_{n=1}^{\infty} X_n \neq \phi$ and $\bigcap_{n=1}^{\infty} X_n$ is weakly compact.

Let $C \subset E$ be nonvoid, closed, convex, and bounded. A function $f : C \to C$ is said to be β-condensing if there is a k with $0 < k < 1$ such that for every $H \subset C$, $\beta(f(H)) \leq k\beta(H)$. Whenever we say f is weakly continuous we mean that f is weakly-weakly continuous (i.e., the topology on both the domain and the range is the weak topology).

Theorem 1. Let $C \subset E$ be nonvoid, closed, convex, and bounded. If $f : C \to C$ is weakly continuous and β-condensing, then f has a fixed point.

Proof. Let $0 < k < 1$ be such that $\beta(f(H)) \leq k\beta(H)$ for every $H \subset C$. Define

$$X_0 = C$$
$$X_1 = \overline{co}\ f(C) = \overline{co}\ f(X_0)$$
$$X_2 = \overline{co}\ f(X_1)$$
$$\vdots$$
$$X_n = \overline{co}\ f(X_{n-1})$$
$$\vdots$$

Clearly $X_0 \supset X_1 \supset X_2 \supset \ldots \supset X_n \supset \ldots$. Each X_n is closed and since $\beta(X_n) = \beta(f(X_{n-1})) \leq k\beta(X_{n-1})$, $\lim_{n\to\infty} \beta(X_n) = 0$. By Lemma

1.2, $X = \bigcap_{n=1}^{\infty} X_n \neq \phi$ and X is weakly compact. Since each X_n is convex then X is convex. If $x \in X$ then $x \in X_n$ for every $n \geq 0$ and so $f(x) \in f(X_n)$ for every $n \geq 0$. Since $f(X_n) \subset X_{n+1}$ then $f(x) \in X_n$ for every $n \geq 1$. It follows that $f(x) \in X$. Thus, we have $f : X \to X$. Since f is weakly continuous and X is weakly compact then from the Tychonoff Fixed Point Theorem [6; p. 15] we get that f has a fixed point. ∎

Let E be a Banach space, I an interval $[a,b]$ in $\mathbb{R}$, and $F = C[I,E]$ the strong continuous functions from I to E. F is a Banach space with norm $||x|| = \sup_{t \in I} ||x(t)||$.

Remark. Let $C \subset F$ be nonvoid, closed, bounded, and convex and let $T : C \to C$. In section 3 we apply Theorem 1 to show the existence of a fixed point for such a T. In order to apply Theorem 1 we must show that T is β-condensing and weakly continuous. Our method of attack for β-condensing requires relating the β measure of weak noncompactness of subsets of C with the β measure of weak noncompactness of the range of that subset of C, i.e., if $H \subset C$ we want to be able to relate $\beta(H)$ with $\beta(H(t))$ and $\beta(H(I))$. These relationships are developed in Theorem 2. The remainder of the section is devoted to technical lemmas needed to establish the weak continuity of such a T.

Let $V_1, \ldots, V_n$ be open sets such that $I = \bigcup_{i=1}^{n} V_i$. A partition of unity of I subordinate to the cover $\{V_i\}_{i=1}^{n}$ is a collection of functions $\{h_i\}_{i=1}^{n}$ such that $0 \leq h_i(t) \leq 1$ for every $t \in I$, the support of h_i is in V_i, and $\sum_{i=1}^{n} h_i(t) = 1$ for every $t \in I$. A partition of unity subordinate to a cover $\{V_i\}_{i=1}^{n}$ always exists (see [5; p. 41]).

Lemma 1.3. Let H be an equicontinuous subset of F and $\varepsilon > 0$. Then there is a partition $\{t_i\}_{i=0}^{M}$ of I and non-negative, real-valued functions $\{h_i\}_{i=1}^{M}$ on I such that

i) $0 \leq h_i(t) \leq 1$ for every $i = 1, \ldots, M$ and $t \in I$,

ii) $\sum_{i=1}^{M} h_i(t) = 1$ for every $t \in I$,

iii) for every f in H, $||f - \sum_{i=0}^{M} f(t_i)h_i|| < \varepsilon$.

Proof. Let δ be such that if $|s - t| < \delta$ then $||f(s) - f(t)|| < \varepsilon$ for every $f \in H$. Let $\{t_i\}_{i=0}^{M}$ be a partition of I such that $0 < t_{i+1} - t_i < \frac{\delta}{2}$. Let $n \in Z^+$ be large enough so that $t_i - \frac{1}{n} > t_{i-1} + \frac{1}{n}$ for $i = 1,\ldots,M$. Define

$$V_i = \begin{cases} [a, t_i + \frac{1}{n}) & \text{if } i = 1 \\ (t_{i-1} - \frac{1}{n}, t_i + \frac{1}{n}) & \text{if } i = 2,\ldots,M-1 \\ (t_{M-1} - \frac{1}{n}, b] & \text{if } i = M\ . \end{cases}$$

Note that at most two (which are in fact adjacent) V_i's overlap, that the V_i's are open, and that $I = \bigcup_{i=1}^{M} V_i$. Let $\{h_i\}_{i=1}^{M}$ be a partition of unity on I subordinate to the cover $\{V_i\}_{i=1}^{M}$. To see these are the h_i's required, notice they clearly satisfy i) and ii). To see that iii) is satisfied, let $f \in H$ and $t \in I$. Then t is in one of the following sets

1) $t \in V_1 - V_2 = [a, t_1 - \frac{1}{n}]$

2) $t \in V_i \cap V_{i+1} = (t_i - \frac{1}{n}, t_i + \frac{1}{n})$ $i = 2,3,\ldots,M-1$

3) $t \in V_i - (V_{i-1} \cup V_{i+1}) = [t_i + \frac{1}{n}, t_{i+1} - \frac{1}{n}]$ $i=2,3,\ldots,M-1$

4) $t \in V_M - V_{M-1} = [t_{M-1} + \frac{1}{n}, t_M]$

If $t \in V_1 - V_2$ then $|t - a| < \frac{\delta}{2} < \delta$, $h_1(t) = 1$, $h_j(t) = 0$ for $2 \leq j \leq M - 1$ and

$$||f(t) - \sum_{j=1}^{M} f(t_{j-1})h_j(t)|| = ||f(t) - f(a_j)|| < \varepsilon.$$

If $t \in V_i \cap V_{i+1}$ then $|t - t_{i-1}| < \delta$, $|t - t_i| < \frac{\delta}{2} < \delta$ $h_j(t) = 0$ if $j \neq i, i+1$, $h_i(t) + h_{i+1}(t) = 1$ and

$$||f(t) - \sum_{j=1}^{M} f(t_{j-1})h_j(t)|| = ||f(t) - h_i(t)f(t_{i-1}) - h_{i+1}(t)f(t_i)||$$

$$\leq h_i(t)||f(t) - f(t_{i-1})|| + h_{i+1}(t)||f(t) - f(t_i)||$$
$$< \varepsilon.$$

If $t \in V_i - (V_{i-1} \cup V_{i+1})$, then $|t - t_{i-1}| < \frac{\delta}{2} < \delta$, $h_j(t) = 0$ if $j \neq i$, $h_i(t) = 1$ and

$$||f(t) - \sum_{j=1}^{M} f(t_{j-1})h_j(t)|| = ||f(t) - f(t_{i-1})|| < \varepsilon.$$

Finally if $t \in V_M - V_{M-1}$ then $|t - t_{M-1}| < \frac{\delta}{2} < \delta$, $h_j(t) = 0$ if $j \neq M$, $h_M(t) = 1$, and

$$||f(t) - \sum_{j=1}^{M} f(t_{j-1})h_j(t)|| = ||f(t) - f(t_{M-1})|| < \varepsilon.$$

For every case, if $t \in I$ then

$$||f(t) - \sum_{j=1}^{M} f(t_{j-1})h_j(t)|| < \varepsilon \text{ and so}$$

it follows that

$$||f - \sum_{j=1}^{M} f(t_{j-1})h_j|| < \varepsilon. \qquad \blacksquare$$

For the sake of clarity we introduce the following notation for basic weak neighborhoods of $x \in E$ (or F). For $\varepsilon > 0$ and $\phi_1,\ldots,\phi_n \in E^*$ (or F^*) the notation

$$N(x,\phi_i,\varepsilon,M) = \{y \in E \text{ (or } F) : |\phi_i(y-x)| < \varepsilon,\ i=1,\ldots,M\},$$

and $N(x,\phi,\varepsilon) = N(x,\phi,\varepsilon,1)$ will be used.

When we use this notation, it should be clear whether we mean weak neighborhoods in E or in F.

Lemma 1.4. Let K be a weakly compact subset of F. Then $K(t)$ is a weakly compact subset of E for each $t \in I$.

Proof. Let $t_0 \in I$ and $\{N(x_\alpha,\phi_{\alpha,j},\varepsilon_\alpha,M_\alpha)\}_{\alpha \in \Gamma}$ be a cover for $K(t_0)$. Define $\psi_{\alpha,j} \in F^*$ by $\psi_{\alpha,j}f = \phi_{\alpha,j}f(t_0)$ and let $f_\alpha \in F$ be such that $f_\alpha(t_0) = x_\alpha$. Then $\{N(f_\alpha,\psi_{\alpha,j},\varepsilon_\alpha,M_\alpha)\}_{\alpha \in \Gamma}$ is a cover for K from which we can get a finite subcover, say

$\{N(f_i, \psi_{ij}, \varepsilon_i, M_i)\}_{i=1}^N$. Clearly $\{N(f_i(t_0), \phi_{ij}, \varepsilon_i, M_i)\}_{i=1}^N$ is a cover for $K(t_0)$. Thus $K(t)$ is compact. ■

Lemma 1.5. Let K be a weakly compact set in E. Let $g \in C[I, \mathbb{R}]$ and $\tilde{K} = \{xg : x \in K\}$. Then $\tilde{K}$ is a weakly compact subset of F.

Proof. Let $\{N(x_\alpha g, \psi_{\alpha,j}, \varepsilon_\alpha, M_\alpha)\}_{\alpha \in \Gamma}$ be a cover for $\tilde{K}$. Define $\phi_{\alpha,j} \in E^*$ by $\phi_{\alpha,j}(x) = \psi_{\alpha,j}(xg)$. Then $\{N(x_\alpha, \phi_{\alpha,j}, \varepsilon_\alpha, M_\alpha)\}_{\alpha \in \Gamma}$ is a cover of K from which we can get a finite subcover, say $\{N(x_i, \phi_{ij}, \varepsilon_i, M_i)\}_{i=1}^N$. Clearly $\{N(x_i g, \psi_{ij}, \varepsilon_i, M_i)\}_{i=1}^N$ is a finite cover of $\tilde{K}$. ■

The following lemma is a corollary to Lemma 1.5.

Lemma 1.6. Let K be a weakly compact subset of E. Let $g_i \in C[I, \mathbb{R}]$ $(i = 1, \ldots, N)$ and define $\tilde{K} = \{\sum_{i=1}^N x_i g_i : x_i \in K\}$. Then $\tilde{K}$ is weakly compact in F.

The next theorem gives the relationship, referred to earlier between $\beta(H)$, $\beta(H(t))$ and $\beta(H(I))$ for bounded subsets H of F. This theorem parallels a theorem of Ambrosetti's [1].

Theorem 2. Let H be a bounded, equicontinuous subset of F. Then $\beta(H) = \sup_{t \in I} \beta(H(t)) = \beta(H(I))$.

Proof. For clarity, we denote the closed unit ball in E by B_E and the closed unit ball in F by B_F.

a) If $H \subset K + tB_F$ then $H(t) \subset K(t) + tB_E$. It follows that $\beta(H(t)) \leq \beta(H)$ and therefore, $\sup_{t \in I} \beta(H(t)) \leq \beta(H)$.

b) Let $\varepsilon > 0$. By Lemma 1.3 there is a partition $\{t_i\}_{i=1}^M$ of I and continuous, non-negative, real-valued functions h_i $(i = 1, \ldots, M)$ on I such that

1) $\sum_{i=1}^M h_i(t) = 1$ for every $t \in I$

11) for every $f \in H$, $\|f - \sum_{i=0}^M f(t_i) h_i\| < \frac{\varepsilon}{2}$.

Let $d = \beta(H(I))$ and let K be a weakly compact subset of E

such that $H(I) \subset K + (d + \frac{\varepsilon}{2})B_E$. Define $\tilde{K} \subset F$ by $\tilde{K} = \{\sum_{j=1}^{M} x_i h_i : x_i \in K\}$. By Lemma 1.6 $\tilde{K}$ is weakly compact. We now show that $H \subset \tilde{K} + (d + \varepsilon)B_F$. Let $f \in H$. Let $x_i \in K$ be such that $f(t_i) - x_i \in (d + \frac{\varepsilon}{2})B_E$. Then $\sum_{i=1}^{M} x_i h_i \in \tilde{K}$ and

$$||f(t) - \sum_{i=1}^{M} x_i h_i(t)||_E$$

$$\leq ||f(t) - \sum_{i=1}^{M} f(t_i)h_i(t)||_E + ||\sum_{i=1}^{M} (f(t_i) - x_i)h_i(t)||_E$$

$$\leq \frac{\varepsilon}{2} + \sum_{i=1}^{M} h_i(t)||f(t_i) - x_i||_E \leq \frac{\varepsilon}{2} + (d + \frac{\varepsilon}{2}) = d + \varepsilon$$

Thus $||f - \sum_{i=1}^{M} x_i h_i||_F \leq d + \varepsilon$ and this implies that $f - \sum_{i=1}^{M} x_i h_i \in (d + \varepsilon)B_F$ or $f \in \tilde{K} + (d + \varepsilon)B_F$. Since f was an arbitrary element of H we get $H \subset \tilde{K} + (d + \varepsilon)B_F$. Since $\varepsilon > 0$ was arbitrary we have $\beta(H) \leq d = \beta(H(I))$.

c) From a) we know $\sup_{t\in I} \beta(H(t))$ exists. Let $d = \sup_{t\in I} \beta(H(t))$ and let $\varepsilon > 0$. Let $\delta > 0$ be such that if $|s - t| < \delta$ then $||f(s) - f(t)|| < \frac{\varepsilon}{2}$ for all $f \in H$. Let $\{t_i\}_{i=0}^{M}$ be a partition of I such that $t_{i+1} - t_i < \delta$. Let K_i be a weakly compact subset of E be such that $H(t_i) \subset K_i + (d + \frac{\varepsilon}{2})B_E$. We now show that $H(I) \subset (\bigcup_{i=0}^{M} K_i) + (d + \varepsilon)B_E$. Let $x \in H(I)$. Let $f \in H$ and $t \in I$ be such that $x = f(t)$. If $t = t_i$ for some $i = 0,\ldots,M$ then $x \in K_i + (d + \varepsilon)B_E$ and there is nothing more to show. We assume $t \in (t_i, t_{i+1})$ for some $i = 0,\ldots,M$. Since $f(t_i) \in K_i + (d + \frac{\varepsilon}{2})B_E$ and $||f(t) - f(t_i)|| < \frac{\varepsilon}{2}$ then

$$x = f(t_i) + (f(t) - f(t_i)) \in K_i + (d + \frac{\varepsilon}{2})B + \frac{\varepsilon}{2}B_E$$

$$= K_i + (d + \varepsilon)B_E$$

Since x was an arbitrary element of $H(I)$ it follows that $H(I) \subset \bigcup_{i=0}^{M} K_i + (d + \varepsilon)B$. Since $\varepsilon > 0$ was arbitrary and since $\bigcup_{i=0}^{M} K_i$ is weakly compact then $\beta(H(I)) \leq d = \sup_{t\in I} \beta(H(t))$. From a), b), and c) we have

$$\sup_{t\in I} \beta(H(t)) \leq \beta(H) \leq \beta(H(I)) \leq \sup_{t\in I} \beta(H(t))$$ ■

The following three technical lemmas essentially give us a way to approximate the weak-topology on F by the weak topology on E.

Lemma 1.7. Let $\{h_i\}_{i=1}^{M}$ be non-negative real-valued continuous functions on I such that $\sum_{i=1}^{M} h_i(t) = 1$ for every $t \in I$. Let $\phi \in F^*$. Define $\phi_i : E \to \mathbb{R}$ by $\phi_i x = \phi(xh_i)$. Then ϕ_i is a continuous linear functional on E and $\sum_{i=1}^{M} ||\phi_i|| \leq ||\phi||$.

Proof. That $\phi_i \in E^*$ is clear. To see that $\sum_{i=1}^{M} ||\phi_i|| \leq ||\phi||$ suppose this is not true. Suppose there is an $\varepsilon > 0$ such that $\sum_{i=1}^{M} ||\phi_i|| > ||\phi|| + \varepsilon$. Let $x_i \in E$ be such that $||x_i|| = 1$ and $\phi_i(x_i) > ||\phi_i|| - \frac{\varepsilon}{2M}$. Define $f(t) = \sum_{i=1}^{M} x_i h_i(t)$. Then $f \in F$ and $||f|| \leq 1$. Now,

$$\begin{aligned}
\phi f = \phi\left(\sum_{i=1}^{M} x_i h_i\right) &= \sum_{i=1}^{M} \phi(x_i h_i) = \sum_{i=1}^{M} \phi_i x_i \\
&\geq \sum_{i=1}^{M} \left[||\phi_i|| - \frac{\varepsilon}{2M}\right] = \Sigma||\phi_i|| - \frac{\varepsilon}{2} \\
&> ||\phi|| + \varepsilon - \frac{\varepsilon}{2} = ||\phi|| + \frac{\varepsilon}{2}
\end{aligned}$$

Thus we have $||\phi|| + \frac{\varepsilon}{2} < \phi f \leq ||\phi||$ which is not possible. It follows that $\sum_{i=1}^{M} ||\phi_i|| \leq ||\phi||$. ■

Let $\phi \in F^*$. Then ϕ is said to be a *point functional* if there is a $\psi \in E^*$ and a $t \in I$ such that $\phi f = \psi f(t)$ for all $f \in F$.

Lemma 1.8. Let H be an equicontinuous subset of F and

$\phi \in F^*$. Then there is a sequence of finite sums of point functionals which converge uniformly on H to ϕ, i.e., given $\varepsilon > 0$ there are a finite number of point functionals ϕ_i $(i = 1,\dots,M)$ such that $|\phi x - \sum_{i=1}^{M} \phi_i x| < \varepsilon$ for every $x \in H$, moreover, $\sum_{i=1}^{M} ||\phi_i|| \leq ||\phi||$.

Proof. Clearly we can assume $\phi \neq 0$. Let $\varepsilon > 0$. Let $\{t_i\}_{i=0}^{M}$ be a partition of I and $h_i (i=1,\dots,M)$ be non-negative, real valued, continuous functions on I be such that (see Lemma 1.4)

i) $\sum_{i=1}^{M} h_i(t) = 1$ for every $t \in I$

ii) for every $f \in H$, $||f - \sum_{i=1}^{M} f(t_i)h_i|| < \frac{\varepsilon}{||\phi||}$

Define $\psi_i \in E^*$ by $\psi_i x = \phi(xh_i)$ (as in Lemma 1.7). Define $\phi_i \in F^*$ by $\phi_i(f) = \psi_i f(t_i)$ and notice that $||\phi_i||_{F^*} = ||\psi_i||_{E^*}$ and so $\sum_{i=1}^{M} ||\phi_i|| \leq ||\phi||$. Now for $f \in H$

$$
\begin{aligned}
|\phi f - \sum_{i=1}^{M} \phi_i f| &= |\phi f - \sum_{i=1}^{M} \psi_i f(t_i)| \\
&= |\phi f - \sum_{i=1}^{M} \phi(f(t_i)h_i)| \\
&= |\phi(f - \sum_{i=1}^{M} f(t_i)h_i)| \\
&\leq ||\phi||\,||f - \sum_{i=1}^{M} f(t_i)h_i|| \\
&< \varepsilon
\end{aligned}
$$

■

The next lemma is a topological version of Lemma 1.8.

Lemma 1.9. Let H be an equicontinuous subset of F. Let $\phi \in F^*$ and $\varepsilon > 0$. Then there exists a finite number of point functionals $\phi_i \in F^*$ $(i = 1,\dots,N)$ and a $\delta > 0$ such that

$$H \cap N(x, \sum_{i=1}^{N} \phi_i, \delta) \subset H \cap N(x,\phi,\varepsilon)$$

for every $x \in H$. Moreover, $\sum_{i=1}^{N} ||\phi_i|| \leq ||\phi||$.

Proof. Let $\phi_i \in F^*$ be point functionals such that

$|\phi x - \sum_{i=1}^{N} \phi_i x| < \frac{\varepsilon}{3}$ for every $x \in H$ and $\sum_{i=1}^{N} ||\phi_i|| \leq ||\phi||$ (Lemma 1.8). Let $\delta = \frac{\varepsilon}{3}$. Then if $y \in H \cap N(x, \sum_{i=1}^{N} \phi_i, \delta)$ we have

$$\begin{aligned} |\phi(x - y)| &\leq |\phi x - \sum_{i=1}^{N} \phi_i x| + |\sum_{i=1}^{N} \phi_i (x - y)| + |\sum_{i=1}^{N} \phi_i y - \phi y| \\ &< \frac{\varepsilon}{3} + \frac{\varepsilon}{3} + \frac{\varepsilon}{3} \\ &= \varepsilon. \end{aligned}$$

Thus $y \in H \cap N(x, \phi, \varepsilon)$ and the proof follows. ∎

Let $f : I \times E \to E$ be weakly continuous. Let $x \in F$, $\phi \in E^*$ and $\varepsilon > 0$. We would like to be able to assert that there is a weak neighborhood about x in F, $N(x, \psi_i, \delta, M)$, such that if $y \in N(x, \psi_i, \delta, M)$ then $f(t, y(t)) \in N(f(t, x(t)), \phi, \varepsilon)$ for every $t \in I$. The next lemma provides a sufficient set of hypothesis which insures that such neighborhood about x exists and for which the ψ_i's are in fact point functionals.

Lemma 1.10. Let $R_0 \subset E$ and $f : I \times R_0 \to E$ be weakly continuous. Let $H \subset Lip_K(I, R_0) \subset F$. Let $x \in H$, $\phi \in E^*$, and $\varepsilon > 0$. Then there are a finite number of point funtionals, $\psi_i \in F^*$ $(i = 1, \ldots, M)$ and a $\delta > 0$ such that if $y \in N(x, \psi_i, \delta, M) \cap H$ then $|\phi(f(t, x(t)) - f(t, y(t)))| < \varepsilon$ for all $t \in I$.

Proof. For each $t \in I$, let

$$U(t, \delta) = \begin{cases} (t - \delta,\ t + \delta) & \text{if } t \neq a, b \\ [t,\ t + \delta) & \text{if } t = a \\ (t - \delta, t] & \text{if } t = b \end{cases}$$

For each $t \in I$, let $\gamma(t) > 0$ and $\phi_{t,i} \in E^*$ $(i = 1, \ldots, M_t)$, $||\phi_{t,i}|| = 1$, be such that if $(\hat{t}, \hat{y}) \in U(t, \gamma(t)) \times N(x(t), \phi_{t,i}, \gamma(t), M_t)$ then $|\phi(f(t, x(t)) - f(\hat{t}, \hat{y}))| < \frac{\varepsilon}{2}$. For each $t \in I$, let $\delta(t) = \frac{\gamma(t)}{2K}$. Let $a = t_0 < t_1 < \ldots < t_N = b$ be such that

$I = \bigcup_{i=0}^{N} U(t_i,\delta(t_i))$. Define $\psi_j \in F^*$ by $\psi_j f = \phi_{t_i,k} f(t_i)$ where $j = k + M_{t_0} + M_{t_1} + \dots + M_{t_{i-1}}$. Note that the ψ_j's are point functionals. Let $M = M_0 + M_1 + \dots + M_N$ and $\delta = min\{\frac{\gamma(t_i)}{2}\}_{i=0}^{N}$.

Let $y \in N(x,\psi_i,\delta,M) \cap H$ and let $t \in I$. Let j be such that $t \in U(t_j,\delta(t_j))$. Since $y \in N(x,\psi_i,\delta,M)$ then $|\phi_{t_{j,m}}(x(t_j) - y(t_j))| < \frac{\gamma(t_i)}{2}$ for $m = 1,\dots,M_j$ and

$$
\begin{aligned}
&|\phi_{t_{j,m}}(x(t_j) - y(t))| \\
&\quad \le |\phi_{t_{j,m}}(x(t_j) - y(t_j))| + |\phi_{t_{j,m}}(y(t_j) - y(t))| \\
&\quad < \frac{\gamma(t_j)}{2} + ||y(t_j) - y(t)|| \\
&\quad \le \frac{\gamma(t_j)}{2} + K|t_j - t| \\
&\quad < \frac{\gamma(t_j)}{2} + K\frac{\gamma(t_j)}{2K} \\
&\quad = \gamma(t_j) .
\end{aligned}
$$

Thus, $(t,y(t)) \in U(t_j,\gamma(t_j)) \times N(x(t_j),\phi_{t_{j,m}},\gamma(t_j),M_j)$ and so $|\phi(f(t_j,x(t_j)) - f(t,y(t))| < \frac{\varepsilon}{2}$. Also, since $|\phi_{t_{j,m}}(x(t) - x(t_j))| \le ||x(t) - x(t_j)|| \le K|t_j - t| < \frac{\gamma(t_j)}{2}$, then $(t,x(t)) \in U(t_j,\gamma(t_j)) \times N(x(t_j),\phi_{t_{j,m}},\gamma(t_j),M_j)$, and so $|\phi(f(t_j x(t_j) - f(t,x(t))| < \frac{\varepsilon}{2}$. Then, if $y \in N(x,\psi_i,\delta,M) \cap H$ and $t \in U(t_j,\delta(t_j))$ then

$$
\begin{aligned}
&|\phi(f(t,x(t)) - f(t,y(t))| \\
&\quad \le |\phi(f(t,x(t))+f(t_j,x(t_j))|+|\phi(f(t_j,x(t_j))-f(t,y(t)))| \\
&\quad < \frac{\varepsilon}{2} + \frac{\varepsilon}{2} = \varepsilon.
\end{aligned}
$$

Since $I = \bigcup_{j=0}^{N} U(t_j, \delta(t_j))$ then this result is true for every $t \in I$. ∎

II. INTEGRATION AND DIFFERENTIATION

In this section we introduce the concepts of integration and differentiation which will be needed in Section 3. Let E be a Banach space and $I = [a,b]$ be an interval in $\mathbb{R}$.

Definition. Let $g\colon I \to E$. If the Rieman integral, $\int_a^b \phi g(s)ds$ exists for every $\phi \in E^*$ and if there is a $y \in E$ such that $\phi y = \int_a^b \phi g(s)ds$ for every $\phi \in E^*$ then we say that g is integrable on I and write $y = \int_a^b g(s)ds.$

Note that since E^* is a separating family of linear functionals on E then $\int_a^b g(s)ds,$ if it exists, is unique. The following lemma asserts the existence of $\int_a^b g(s)ds$ for the functions g in which we are interested.

Lemma 2.1. If $g\colon I \to E$ is weakly continuous then g is integrable and $\int_a^b g(s)ds \in (b - a)\,\overline{co}\, g(I).$

Proof: Since I is compact we know that $g(I)$ is weakly compact and Lemma 1.1 implies that $\overline{co}\, g(I)$ is weakly compact. This lemma then follows directly from a known theorm [4; p. 74]. ∎

Lemma 2.2. If $g\colon I \to E$ is weakly continuous then $\sup_{t\in I} ||g(t)|| = ||g||_\infty$ is finite and $||\int_a^b g(s)ds|| \le ||g||_\infty (b-a).$

Proof. Since g is weakly continuous then $g(I)$ is weakly compact and hence g is bounded, i.e., $\sup_{t\in I} ||g(t)|| = ||g||_\infty$ is finite. Since $|\phi g(s)| \le ||\phi||\ ||g||_\infty$ for every $\phi \in E^*$ then

$$\left|\phi \int_a^b g(s)ds\right| = \left|\int_a^b \phi g(s)ds\right|$$

$$\leq \int_a^b |\phi g(s)|ds$$

$$\leq ||\phi|| \, ||g||_\infty (b - a).$$

It follows that

$$|| \int_a^b g(s)ds|| \leq ||g||_\infty (b - a).$$ ■

Definition. Let $g : I \to E$. If $\phi \circ g$ is differentiable for every $\phi \in E^*$ and if there is a function $y : I \to E$ such that $(\phi g)'(t) = \phi y(t)$ for every $\phi \in E^*$ and $t \in I$ then g is weakly differentiable and we write $g'(t) = y(t)$.

Remark 1. If E is weakly sequentially complete and $\phi \circ g$ is differentiable for every $\phi \in E^*$ then g is weakly differentiable [2].

Remark 2. If E is a complex Banach space and $\phi \circ g$ is differentiable for every $\phi \in E^*$ then g is strongly differentiable [4; p. 79].

Lemmas 2.3 and 2.4 follow easily from the definitions and the Fundamental Theorem of Calculus.

Lemma 2.3. If $g : I \to E$ is weakly continuous, $x_0 \in E$, and $x(t) = x_0 + \int_a^t g(s)ds$ for $t \in I$, then x is weakly differentiable and $x'(t) = g(t)$.

Lemma 2.4. If $x : I \to E$ is weakly differentiable and if x' is weakly continuous then $x(t) = x(a) + \int_a^t x'(s)ds$ for $t \in I$.

Lemma 2.5. If $x : I \to E$ is weakly differentiable and if x' is weakly continuous then there is a constant $K > 0$ such that $||x'(t)|| \leq K$ for every $t \in I$ and x is Lipschitz with Lipschitz constant K.

Proof. Since x' is weakly continuous then $x'(I)$ is weakly compact and hence bounded. Therefore there is a K such that $||x'(t)|| \leq K$ for every $t \in I$. From Lemmas 2.2 and 2.4 we get for $a \leq t_1 < t_2 \leq b$

$$||x(t_2) - x(t_1)|| = ||\int_{t_1}^{t_2} x'(s)ds|| \leq ||x'||_\infty(t_2 - t_1)$$
$$\leq K(t_2 - t_1).$$ ∎

III. EXISTENCE OF WEAK SOLUTIONS TO THE ABSTRACT CAUCHY PROBLEM

Let E be a Banach space and $I = [t_0, t_0 + a]$ be an interval in $\mathbb{R}$. Let $x_0 \in E$, $L \in \mathbb{R}^+$, and $R_0 = I \times S_L(x_0)$ where $S_L(x_0) = \{x \in E : ||x - x_0|| \leq L\}$. The function $x : I \to E$ is a weak solution to the Abastract Cauchy Problem (ACP)

$$x' = f(t,x) \qquad x(t_0) = x_0, \text{ where } f : R_0 \to E,$$

if

a) x is weakly differentiable;
b) $x(t_0) = x_0$;
c) $x(t) \in S_L(x_0)$ for every $t \in I$;
d) $x'(t) = f(t,x(t))$ for every $t \in I$.

The next lemma follows easily from Lemmas 2.3.

Lemma 3.1. If f is weakly continuous then x is a weak solution to the ACP iff $x(t) \in S_L(x_0)$ for every $t \in I$ and $x(t) = x_0 + \int_{t_0}^{t} f(s,x(s))ds$.

The following theorem appears in Bronson [2]. What distinguishes our result from Bronson's is that our proof uses a fixed point theorem.

Theorem 3. Suppose f has the properties,

i) f is weakly continuous;
ii) $||f(t,x)|| \leq M$ for every $t \in I$, $x \in R_0$;
iii) there exists $k \in \mathbb{R}^+$ such that for any $H \subset S_L(x_0)$, $\beta(f(I \times H)) \leq k\beta(H)$,

where $ak < 1$ and $Ma \leq L$. Then the ACP has a solution.

Proof. From ii) and Lemma 1.4 we know that if x is a solution to the ACP then $x \in Lip_M(I,R_0)$. Hence, this is where we will look for solutions. Remember that $F = C[I,E]$ is a Banach space with sup norm. Let $C = Lip_M(I,R_0) \subset F$. Note that C is nonvoid, closed, bounded and convex. Define $T : C \to C$ by

$$(Tx)(t) = x_0 + \int_{t_0}^{t} f(s,x(s))ds$$

Using the fact that $||f(s,x(s))|| \le M$ for every $S \in I$ and Lemma 2.2 we get that Tx is Lipschitz with Lipschitz constant M and the fact that $Ma \le L$ and Lemma 2.2 gives us that $(Tx)(t) \in S_L(x_0)$ for every $t \in I$. Thus T does map C into C. Note that, as a result of Lemma 3.1, a fixed point of T is a solution to the ACP. We intend to show that T has a fixed point by showing that T is weakly continuous, that T is β-condensing and then applying Theorem 1 (we have already pointed out that C is nonvoid closed bounded and convex).

To see that T is weakly continuous let x C and let $N(Tx,\phi,\varepsilon)$ be a weak neighborhood of Tx. By Lemma 1.9 it is sufficient to assume that $\phi = \sum_{i=1}^{N} \phi_i$ where the ϕ_i's are point functionals. We will assume ϕ has this form. Since ϕ_i is a point functional then there is a $\psi_i \in E^*$ and a $t_i \in I$ such that $\phi_i x = \psi_i x(t_i)$ for every $x \in F$. By Lemma 1.10, there are, for each ψ_i, a finite number of point functionals $\Lambda_{ij} \in F^*$ $(j = 1,\ldots,M_i)$ and a $\delta_i > 0$ such that if $y \in N(x,\Lambda_{ij},\delta_i,M_i)$ then $|\psi_i(f(t,y(t)) - f(t,x(t)))| < \frac{\varepsilon}{Na}$ for every $t \in I$. Then if $y \in C \cap \left[\bigcap_{i=1}^{N} N(x,\Lambda_{ij},\delta_i,M_i)\right]$ we have

$$\begin{aligned}
|\phi(Tx - Ty)| &= \left|\sum_{i=1}^{N} \phi_i(Tx - Ty)\right| \\
&= \left|\sum_{i=1}^{N} \psi_i \int_{t_0}^{t_i} (f(s,x(s)) - f(s,y(s)))ds\right| \\
&= \left|\sum_{i=1}^{N} \int_{t_0}^{t_i} \psi_i(f(s,x(s)) - f(s,y(s)))ds\right|
\end{aligned}$$

$$\leq \sum_{i=1}^{N} \int_{t_0}^{t_i} |\psi_i(f(s,x(s)) - f(s,y(s)))|ds$$

$$\leq \sum_{i=1}^{N} \frac{\varepsilon}{Na} a$$

$$= \varepsilon.$$

Therefore $Ty \in N(Tx,\phi,\varepsilon)$ for every $y \in C \cap [\bigcap_{i=1}^{N} N(x,\Lambda_{ij},\delta_i,M_i)]$. It follows that T is weakly continuous.

To see that T is β-condensing let $H \subset C$. Then

$$\beta(T(H)) = \sup_{t\in I} \beta(T(H(t)))$$

$$= \sup_{t\in I} \beta\left(\left\{ \int_{t_0}^{t} f(s,x(s))ds : x \in H \right\} \right)$$

$$\leq \sup_{t\in I} \left[\beta\{(t - t_0)\, \overline{co}\, f([t_0,t] \times H([t_0,t]))\} \right]$$

$$\leq \sup_{t\in I} (t - t_0)\beta[f(I \times H(I))]$$

$$= a\beta[f(I \times H(I))]$$

$$\leq ak\ \beta(H(I))$$

$$= ak\ \beta(H).$$

The key steps in the above sequence of inequalities follow from Theorem 1 and Lemma 1.1. Since $ak < 1$, then T is β-condensing.

By Theorem 1, T has a fixed point. By Lemma 3.1, the ACP has a solution. ∎

We get the following result, due to Szep [7], as a corollary.

Corollary. Let E be reflexive and let f be weakly continuous. Then there is a $\alpha \leq a$ such that the ACP has a solution on $[t_0, t_0 + \alpha]$.

Proof. Since $S_L(x_0)$ is bounded in E then $S_L(x_0)$ is weakly compact. Since f is weakly continuous then $f(R_0)$ is weakly compact, hence there is an $M \in \mathbb{R}^+$ such that $||f(t,x)|| \leq M$ for every $(t,x) \in R_0$. Let $\alpha > 0$ be such that $M\alpha \leq L$. Let $k > 0$ be scuh that $\alpha k < 1$. All that we need in order to satisfy the hypothesis of Theorem 3 is that $\beta(f(I \times H)) \leq k\beta(H)$ for every $H \subset S_L(x_0)$. But $f(I \times H) \subset f(R_0)$ which is weakly compact, hence $\beta(f(I \times H)) = 0$. Since $H \subset S_L(x_0)$ which is weakly compact then $\beta(H) = 0$. Thus $\beta(f(I \times H)) = k\beta(H)$ for every $H \subset R_0$. The hypothesis of Theorem 3 is satisfied, therefore the ACP has a solution on $[t_0, t_0 + \alpha]$. ∎

IV. REFERENCES

[1] Ambrossetti, A., *UN Teorema Di Existenza Per Le Equazioni Differenzial: Negli Spazi Di Banach*, Rend. Sem. Mat. Univ. Padove 39 (1967), pp. 348-361.

[2] Bronson, E. J., Lakshmikantham, V. and Mitchell, A. R., *On the Existence of Weak Solutions of Differential Equations in Nonreflexive Banach Spaces*, To appear in Journal of Nonlinear Analysis.

[3] De Blasi, F. S., *On a Property of the Unit Sphere in a Banach Space*, (to appear).

[4] Rudin, W., *Functional Analysis*, McGraw-Hill, 1973.

[5] Rudin, W., *Real and Complex Analysis*, McGraw-Hill, 1974.

[6] Smart, D. R., *Fixed Point Theorems*, Cambridge University Press, 1974.

[7] Szep, A., *Existence Theorem for Weak Solutions of Ordinary Differential Equations in Reflexive Banach Spaces*, Studia Scientiarum Mathematicariem Hungarica 6 (1971), pp. 197-203.

MONOTONICITY AND ALTERNATIVE METHODS

Kent Nagle
University of South Florida

I. INTRODUCTION

Let us consider the equation

(1) $$Lx + Nx = 0$$

where $x \in H$, a real Hilbert space, $L: D(L) \to H$, a linear operator on $D(L) \subset D(N) \subset H$ and $N: D(N) \to H$ a nonlinear operator. When L^{-1} exists, equation (1) may be rewritten as a Hammerstein equation $x + L^{-1}Nx = 0$. Hammerstein equations have been extensively studied and existence of solutions is known under very general assumptions on L and N (see Amann [1], Dolph and Minty [7]). However, when L^{-1} does not exist, the question of existence is more difficult. We will use a type of Lyapunov-Schmidt [12, 19] method to study the question of existence of solutions to (1) when L^{-1} does not exist. In particular, we will use an alternative scheme developed by Cesari [3] for the case when L is selfadjoint.

We will consider equations involving large nonlinearities N under monotonicity assumptions. However, we will not require L to be selfadjoint. Our specific assumptions on L and N will be given in Section II.

Let us first consider the equation

(2) $$Ex + Nx = 0$$

N is the same as in equation (1) and $E = L$ in equation (1) is assumed to be selfadjoint. In terms of the alternative method of

ISBN 0-12-434160-8

Cesari, let $P: H \to H$ be an orthogonal projection with range $PH = H_0 \supset \ker E$ and let $H_1 = (I - P)H$. Let $K: H_1 \to H_1 \cap D(E)$ denote a bounded linear operator. We assume the following relations hold: For $x \in D(E)$.

(k_1) $K(I - P)Ex = (I - P)x$

(k_2) $PEx = EPx$

(k_3) $EK(I - P)x = (I - P)x$.

We will refer to (k_1), (k_2), and (k_3) collectively by (k). Under assumptions (k), equation (1) is equivalent to the system

$$x + K(I - P)Nx = x_0 \tag{3}$$

$$P(Ex + Nx) = 0 \tag{4}$$

where $x_0 \in H_0$ (see Cesari [3]).

The idea of the alternative method is to solve equation (3) uniquely for each x_0 in H_0 and hence reduce the problem of solving equation (2) to solving the equation

$$P(Ex^*(x_0) + Nx^*(x_0)) = 0 \tag{5}$$

where $x^*(x_0)$ is the unique solution to (3) for a fixed x_0. Equation (5) is referred to as the alternative problem.

Lyapunov-Schmidt or alternative methods have been used by several authors to study equation (1) under monotonicity assumptions on N. Gustafson and Sather [8] have reduced (1) to an alternative problem when L is selfadjoint. They allow L to have a continuous spectrum and do not require $D(N) = H$. When L is selfadjoint and has eigenvalues approaching ∞, Cesari and Kannan [6] have given sufficient conditions for equation (1) to have a solution. However, they require $D(N) = H$. For non-selfadjoint operators generated by a coercive bilinear form with compact resolvent, Osborn and Sather [18] have shown equation (1) can always be reduced to an alternative problem. They do not require $D(N) = H$. Nagle [15] has extended the results of Cesari and Kannan to nonselfadjoint operators with P a finite rank projection and $D(N) = H$.

Success in applying any of the above results depends upon our knowledge of the linear operator. Consequently, it is to our advantage to work mainly with selfadjoint operators. With this in mind we make the following observation. If L is a nonself-adjoint bounded linear operator, then $L = E + A$ where $E = \frac{1}{2}(L + L^*)$ is a selfadjoint operator and $A = \frac{1}{2}(L - L^*)$ is an antisymmetric operator i.e., $(Ax,x) = 0$. If we assume L has such a decomposition, then we may consider A as part of our nonlinearity and N and $N + A$ will have the same monotonicity properties. We may now apply Cesari's alternative scheme to the equation $Ex + (N + A)x = 0$. While not every non-selfadjoint operator admits such a decomposition, the class of operators which do include many of the operators associated with boundary value problems for both ordinary and partial differential equation. (see Nagle [16] and Nagle and Pothoven [17]).

For the remainder of this paper we will assume:

(L) $L = E + A$ where E is a selfadjoint operator and A an antisymmetric operator.

In Section II we will give sufficient conditions for equation (1) to be reduced to an equivalent alternative problem.

In Section III we will give sufficient conditions for the existence of solutions to the alternative problem or equivalently equation (1). We will not require L to have a discrete spectrum nor $D(N) = H$. In Section IV we will discuss the results of Sections II and III and our assumptions. For applications of these results see Nagle [16] for elliptic boundary value problem on bounded domains and Nagle and Pothoven [17] for the case with unbounded domains.

II. REDUCTION TO AN ALTERNATIVE PROBLEM

For $L = E + A$ let assumptions (k) be satisfied for E

the selfadjoint part of L. In addition we assume:

(E1) $E = S + B$ where S is a positive operator and B a symmetric operator

(E2) On $H_1 \cap D(E)$, E is strongly monotone with constant $m > 0$ i.e., $(Ex,x) \geqslant m \|x\|^2$.

(N1) Let N be semicontinuous with $D(N) \supset D(S^{1/2})$ and quasimonotone with constant $n \geqslant 0$ i.e., $(Nu - Nv, \ u - v) > -n\|u - v\|^2$.

Assumption (E1) is mainly a notational convenience. If E is a selfadjoint operator whose spectrum is bounded below by b, then $E = E + bI + (-bI)$ is the decomposition referred to in (E1). Assumptions (E2) and (N1) are standard types of monotonicity assumption where the assumption $D(N) \supset D(S^{1/2})$ allows us to obtain existence using basic results from monotone operator theory (see Browder [2] or Minty [14]).

The next two theorems give sufficient conditions for equation (1) to be reduced to an alternative problem. The proofs were motivated by techniques found in the papers by Gustafson and Sather [8], Osborn and Sather [18], and the alternative method of Cesari [3].

Theorem 1. Let conditions (L), (E1), (E2), (N1), and (k) hold. If $D(A) \supset D(S^{1/2})$ and $0 \leqslant n < m$, then the equation

$$x + K(I - P)\ (N + A)x = x_0 \tag{6}$$

has a unique solution $x^* = x^*(x_0)$ for each $x_0 \in H_0$. Hence equation (1) is equivalent to the alternative problem

$$P(Ex^* + (N + A)x^*) = 0. \tag{7}$$

Proof of Theorem 1: Since K maps H_1 into $D(E) \cap H_1$, a dense subspace of H_1, it follows from a corollary to the closed graph theorem (e.g., Kato [11, p. 167]) that SK is bounded, i.e. $\|Sx\| \leqslant d\|Ex\|$ for $x \in D(E) \cap H_1$, $d > 0$. Now K is a positive bounded selfadjoint operator with a unique positive selfadjoint square root $K^{1/2}$. It now follows from the spectral theorem (E1), and a Theorem of Heinz [9], that $R(K^{1/2}) \subset D(S^{1/2}) \cap H_1$, so $R(K^{1/2}) \subset D(N) \cap H_1$. Now since K and $K^{1/2}$ are *1-1* solving (6) is equivalent to solving

$$v + K^{1/2}(I - P)N(x_0 + K^{1/2}v) + K^{1/2}(I - P)A(x_0 + K^{1/2}v) = 0,$$

where $x = x_0 + x_1$ and $x_1 = K^{1/2}v$.

Define $Fv = v + K^{1/2}(I - P)\{N(x_0 + K^{1/2}v) + A(K^{1/2}v + x_0)\}$. $F\colon H_1 \to H_1$ and is hemicontinuous since N is hemicontinuous. For u, $v \in H_1$, it follows from the definition of P, (L), (E2), (N1), and our assumption on $D(A)$, that

$$\begin{aligned}(Fu - Fv, u - v) &= (u - v, u - v) + (K^{\frac{1}{2}}(I - P)N(x_0 + K^{\frac{1}{2}}u) \\ &\quad - K^{\frac{1}{2}}(I - P)N(x_0 + K^{\frac{1}{2}}v), u - v) \\ &\quad + (K^{\frac{1}{2}}(I - P)A(x_0 + K^{\frac{1}{2}}u) - K^{\frac{1}{2}}(I - P)A(x_0 + K^{\frac{1}{2}}v), u - v) \\ &= \|u - v\|^2 + (N(x_0 + K^{\frac{1}{2}}u) - N(x_0 + K^{\frac{1}{2}}v), x_0 \\ &\quad + K^{\frac{1}{2}}u - x_0 - K^{\frac{1}{2}}v) \\ &\quad + (A(x_0 + K^{\frac{1}{2}}u) - A(x_0 + K^{\frac{1}{2}}v), x_0 + K^{\frac{1}{2}}u - x_0 - K^{\frac{1}{2}}v) \\ &\geqslant \|u - v\|^2 - n\|K^{\frac{1}{2}}u - K^{\frac{1}{2}}v\|^2 \geqslant (1 - n/m)\|u - v\|^2.\end{aligned}$$

Hence F is a hemicontinuous, strongly monotone operator defined on all of H and it follows from a result from monotone operator theory (Minty [14]), that $Fv = 0$ has a unique solution for each $x_0 \in H_0$. Thus, equation (6) has a unique solution for each $x_0 \in H_0$. *QED.*

The assumption in Theorem 1 that $D(A) \supset D(S^{1/2})$ does not seem to be a necessary assumption for the Theorem to be valid, but a pecularity of the method of proof. The assumption that the $D(A) \supset D(S^{1/2})$ may be replaced by the apparently stronger assumption that $A = S^{1/2}\ T\ S^{1/2}$ where T is antisymmetric and $D(T) \supset R(S^{1/2})$. In practice, A often has such a decomposition. This will be discussed in Section IV.

Theorem 2. Let conditions (L), (E1), (E2), (N1), and (k) hold. If $A = S^{1/2} T S^{1/2}$ where T is antisymmetric with $D(T) \supset R(S^{1/2})$, then equation (6) has a unique solution $x^*(x_0)$ for each $x_0 \in H_0$. Hence equation (1) is equivalent to the alternative problem (7).

Proof: The proof is the same as the proof of Theorem 1 with the exception of the following calculation. For $u,\ v \in H$,

$$(A(x_0 + K^{1/2}u) - A(x_0 + K^{1/2}v),\ x_0 + K^{1/2}u - x_0 - K^{1/2}v)$$
$$= (T(S^{1/2}x_0 + S^{1/2}K^{1/2}u) - T(S^{1/2}x_0 + S^{1/2}K^{1/2}v),$$
$$S^{1/2}x_0 + S^{1/2}K^{1/2}u - S^{1/2}x_0 - S^{1/2}K^{1/2}v)$$
$$\geqslant 0,$$

since $D(S^{1/2}) \supset R(K^{1/2})$.

III. SOLVING THE ALTERNATIVE PROBLEM

Several techniques have been used to study the alternative problem. When H_0 is finite dimensional, techniques involving topological degree or the implicit function theorem are popular. For a survey of the techniques and results see Cesari [4, 5]. In this Section we will use monotone operator methods similar to those used by Cesari and Kannan [6] and Nagle [15] to give sufficient conditions for the solution of the alternative problem.

When trying to solve the alternative problem it is helpful to know that x^* depends continuously on x_0 in H_0. When N is continuous on $D(N)$ with the $D(N) = H$, it follows from a result of Minty [13], that x^* depends continuous on x_0. However, N continuous on H is too strong an assumption to make.

We assume:

(N2) There is a constant k, $0 \leqslant k < 1$ and a nondecreasing function ϕ: $[0,\infty) \to [0,\infty)$ such that for all $x \in D(L)$,

$$\|Nx\| < k\|Lx\| + \phi(\|x\|).$$

Assumption (N2) has been used by Gustafson and Sather [8] and Osborn and Sather [18] and includes the assumptions made by Cesari and Kannan [6] and Nagle [15] that N is a bounded map. The next theorem is essentially due to Osborn and Sather (see theorem 2 in [18]). Their proof needs only slight modifications to allow N to be quasimonotone instead of monotone.

Theorem 3. Under either the assumptions of Theorem 1 or Theorem 2 and the additional assumption (N2), then x^* depends continuously on $x_0 \in H_0$.

When more is known about the operation $K^{1/2}$, defined in the proof of Theorem 1, then the continuous dependence of x^* on x_0 may be shown when N is assumed to be continuous when taken as a map from a suitable Hilbert space H^0, H^0 imbedded in H, into H instead of hemicontinuous on H.

We now give the main result of this paper.

Theorem 4. Let conditions (L), (E1), (E2), (N1), (N2), and (k) hold. If $n = 0$, $m > 0$, E is monotone on all of $D(E)$ and N is coercive, then equation (1) has at least one solution provided either $D(A) \supset D(S^{1/2})$ or $A = S^{1/2}\, TS^{1/2}$ where T is antisymmetric with $D(T) \supset R(S^{1/2})$.

Proof: Since the assumptions of Theorem 1 or Theorem 2 hold, it suffices to solve the alternative problem $P(Ex^* + (N+A)x^*) = 0$. Since $PEx^* = EPx^* = Ex_0$, the alternative problem has the form $Tx_0 = 0$ where $Tx_0 = Ex_0 + P(N+A)(I+K(I-P)(N+A))^{-1}x_0$. T maps H_0 into H_0. Since E and A are continuous, N hemicontinuous, and by Theorem 3 $(I+K(I-P)(N+A))^{-1}$ is continuous, it follows that T is hemicontinuous.

Let $x, y \in H_0$, and let $u = [I + K(I-P)(N+A)]^{-1}x$ and $v = [I + K(I-P)(N+A)]^{-1}y$. Then by (k), (E1), (N1), (L), and the monotonicity of E,

$$(Tx - Ty, x - y) = (Ex - Ey, x - y) + (P(N+A)u - P(N+A)v, x - y)$$

$$\geq ((N+A)u - (N+A)v, x - y) = (Nu - Nv, u - v) + (Au - Av, u - v)$$

$$+ ((N+A) u - (N+A)v, K(I-P)(N+A)u - K(I-P)(N+A)v)$$

$$\geq ((I-P)(N+A)u - (I-P)(N+A)v, K(I-P)(N+A)u$$

$$- K(I-P)(N+A)v) \geq 0.$$

Hence, T is monotone. To show T is coercive, let $x_0 \in H_0$ and $x = [I + I(I-P)(N+A)]^{-1}x_0$. Since $Px = x_0$ and $\|x_0\| \leq \|x\|$, it follows that $\|x_0\| \to \infty$ implies $\|x\| \to \infty$. Now by (k), (E1), (L) and the monotonicity of E.

$$(Tx_0, x_0) = (Ex_0, x_0) + (P(N+A)x, x_0) \geq ((N+A)x, x_0)$$

$$\geq ((N+A)x, x) + ((N+A)x, K(I-P)(N+A)x)$$

$$\geq (Nx,x) + ((I-P)(N+A)x, K(I-P)(N+A)x) \geq (Nx,x).$$

Hence, $\|x\|^{-1}(Tx_0,x_0) \geq \|x\|^{-1}(Nx,x) \geq \|x\|^{-1}(Nx,x)$. Since N is coercive, T is coercive. Now T is a monotone, hemicontinuous, coercive map from H_0 into H_0 hence T is onto and $Tx_0 = 0$ has at least one solution.

IV. DISCUSSION OF THE RESULTS

The alternative scheme of Cesari has been successfully applied to a variety of problem. Its success requires that the projection operator P be correctly chosen in order to allow the existence of the operator K and the verifications of assumptions (k). We will make a few additional comments on the choice of P and refer the reader to any Cesari's papers [3, 4, 5]. Since E is selfadjoint, we may choose P to be one of the spectral projections associated with E. If P is "correctly" chosen from the spectral projections, it is possible to use the spectral theorem to show that (k_2), and (E2) are satisfied and hence on H_1 we can define K so that (k_1) and (k_3) hold provided K is bounded. If both the graph and range of E are closed, then K will be bounded. (See Cesari [4]).

Our assumption (L) requires some comment. It is always true that $L \supset E_0 + A$ where $E_0 = \frac{1}{2}(L + L^*)$ is symmetric. All that is necessary for the applications of Theorem 4 to show existence of solutions to equation (1) is E_0 to have a selfadjoint extension E_1 and $L \supset E_1 + A$, since a solution to $(E_1 + A + N)x = 0$ will then be a solution to $Lx + Nx = 0$.

While it is true that a decomposition $L = E + A$ may not exist, in many applications the decomposition may be obvious. For example, $Lx = -x'' + x'$, with either Dirichlet or periodic boundary conditions, has $Ex = -x''$ and $Ax = x'$. More importantly, if L is generated by a coercive bilinear form, then $L = E + A$ and $A = S^{1/2}\, TS^{1/2}$ as assumed in Theorems 2, 3, and 4. This representation was first used by Osborn and Sather [18] and is used by Nagle [16] and Nagle and Pothoven [17].

The technique of decomposing K into $K^{1/2}\, K^{1/2}$ has been generalized by Kannan and Locker [10]. In studying nonlinear Hammerstein equations they replaced the square root decompositions by $K = J^*J$ and obtained similar results under these more general assumptions. The proof of Theorem 2 needs little modification if we assume $S = U^*U$, $K = J^*J$, $D(N) \supset D(U) \supset R(J)$ and either $D(A) \supset D(U)$ or $A = U^*TU$.

Finally, the conclusions of Theorem 4 hold if N is assumed to be strongly monotone with constant ℓ instead of quasimonotone and coercive. Moreover, if E satisfies (E1) and is quasimonotone with constant h on $D(E)$ and $\ell > h$, then the proof of Theorem 4 needs only slight modification to reach the same conclusions.

V. REFERENCES

[1] Amann, H., *Existence theorems for equations of Hammerstein type*, Applicable Analysis 2, (1973), pp. 385-397.

[2] Browder, F., *Nonlinear elliptic boundary value problems*, Bull. Amer. Math. Soc. 69 (1963), pp. 862-874.

[3] Cesari, L., *Functional analysis and Galerkin's method*, Mich. Math. J. 11 (1964), pp. 385-414.

[4] Cesari, L., *Alternative methods in nonlinear analysis*, in International conference on differential equations, ed. H. Antosieivicz, Academic Press (1975), pp. 95-148.

[5] Cesari, L., *Nonlinear oscillations under hyperbolic systems*, An International Conference, Providence, R. I. Dynamical Systems (Cesari, Hale, LaSalle, eds.), Academic Press, Vol. 1 (1975), pp. 251-261.

[6] Cesari, L. and Kannan, R., *Functional analysis and nonlinear differential equations*, Bull. Amer. Math. Soc. 79 (1973), pp. 1216-1219.

[7] Dolph, C. L. and Minty, G., *On nonlinear integral equations of the Hammerstein type*, Integral Equations, Madison, University of Wisconsin Press (1964), pp. 99-154.

[8] Gustafson, K. and Sather, D., *Large nonlinearities and monotonicity*, Arch. Rat. Mech. Anal. 48 (1972), pp. 109-122.

[9] Heinz, E., *Beiträge zur Störungstheorie der Spectralzerlegung*, Math. Ann. 123 (1951), pp. 415-438.

[10] Kannan, R. and Locker, J., *Operators* J^*J *and nonlinear Hammerstein equations*, J. Math. Anal. Appl. 53 (1976), pp. 1-7.

[11] Kato, T., *Demicontinuity, hemicontinuity, and monotonicity*, Bull. Amer. Math. Soc. 70 (1964), pp. 548-555.

[12] Lyapunov, A. M., *Sur les Figures d'equilibre peu Différeutes des ellipsoids d'une masse Liquide Homogene Douée d'un Mouvement de Rotation*, Zap. Akad. Nauk., St. Petersburg 1, (1906), pp. 1-225.

[13] Minty, G., *Monotone (Nonlinear) operators in Hilbert Space*, Duke Math. J. 29 (1962), pp. 341-345.

[14] Minty, G., *On a "monotonicity" method for the solution of nonlinear equations in Banach Spaces*, Proc. Nat. Acad. Sci. U.S.A. 50 (1963), pp. 1038-1041.

[15] Nagle, K., *Monotonicity and alternative methods for non-linear boundary value problems*, Pacific J. Math. (to appear).

[16] Nagle, K., *Nonlinear Equations at Resonance*, (to appear).

[17] Nagle, K. and Pothoven, K., *Nonlinear equations on unbounded domains*, (to appear).

[18] Osborn, J. and Sather, D., *Alternative problems and monotonicity*, J. Differential Equations 18 (1975), pp. 393-410.

[19] Schmidt, E., *Zur Theorieder Linearen und Nichtlinearen Integralgleichungen und der Verzweigung ihrer Lösungen*, Math. Ann. 65 (1908), pp. 370-399.

THE OLP[1] METHOD OF NON-LINEAR STABILITY ANALYSIS OF TURBULENCE IN NEWTONIAN FLUIDS[2]

Fred R. Payne
The University of Texas at Arlington

I. INTRODUCTION

Paper treats a method of flow stability analysis specialized to Newtonian flow in the turbulent as opposed to laminar mode. In order to study turbulence, one really should solve either the Louiville equation in phase or probability space or its consequence, the Boltzmann equation whose continuum limit, under suitable assumption, yields Navier-Stokes equation as the conservation of linear momentum (CLM) equation; other continuum limits of moments of the Boltzmann equation yield fluid continuity, i.e., the conservation of mass relationship (CM), conservation of angular momentum (CAM) and conservation of energy relationship (CE) where energy is here understood to be the sum of kinetic or mechanical and internal or thermal energy (CE = CKE + CIE). All the aforementioned relations from Louiville's "master equation" are special cases of a Generalized Conservation Principle (GCP) in an Eulerian coordinate frame. A slightly more general formulation is the General Balance Equation (GBE) formulated in Lagrangian coordinates or, to use continuum mechanics terminology, "material-embedded" coordinate system.

---

[1] OLP=Orr (1907), Lumley (1966), Payne (1968)

[2] Supported, in part, by NASA/Ames Grant NSG-2077, Dr. Morris W. Rubesin, Technical Monitor.

ISBN 0-12-434160-8

II. DEFINITION OF TURBULENCE

Purists recognize seven characteristics of turbulent flow as opposed to the laminar mode. The most distinctive difference in the two modes is that of randomness or, to use a less imprecise term, "stochastic" processes. The other six characteristics of turbulence can be exhibited by certain classes of laminar flows and include 2) vorticity; 3) a full three-dimensionality of the fluctuating flow field; 4) an inherent non-linearity in the process; this non-linearity takes two forms: one occurs in the partial differential or integral operator; the second is in the form of the "viscosity" coefficient of the highest order derivative where said coefficient is a fundamentally non-linear function of the solution itself; 5) the diffusion property of turbulence; 6) the dissipative quality of turbulence (at least, characteristic of any viscous flow in the presence of solid boundaries and relative fluid motion thereto, i.e., "shear flows" wherein transverse gradients exist; and 7) the flow can be considered not as a aggregate of discrete particles (for example, the hard-sphere molecules of classical statistical mechanics); rather, turbulence is a continuum process whereby field properties such as mass density, defined as the limit of the amount of mass contained in a volume, Δv, as $\Delta v \to 0$, are everywhere defined and continuous to arbitrary order. A more concise definition of turbulence is due to Stewart (1972) who defines a "Turbulence Syndrome"; this syndrome exhibits three basic symptoms: 1) a fully three-dimensional vorticity field, 2) inherent disorder, and 3) a wide range of scales which contribute to efficient transport and mixing. Finally, a most descriptive incapsulization of turbulence is contained in a ditty.

Big whirls have little whirls
  That feed on their velocity;
Little whirls have lessor whirls,
  and so on to viscosity
In the molecular sense.

-- Richardson (?), Circa 1921

III. PRIOR METHODS OF FLOW STABILITY ANALYSES

Orr (1907) first formulated the "energy method" of flow stability analysis; Sommerfeld (1908) derived a second equation, usually termed the Orr-Sommerfeld equation which is restricted to 2-dimensional, parallel shear flows with small disturbance i.e., linearized. The energy method fell into disuse early in the 1930's when the second method became more successful in predicting values of, specifically, the Reynolds' number stability parameter; the "Orr-Energy" method tended to predict values of critical Reynolds' number lower than those actually observed in experimental flows. This second, Orr-Sommerfeld method is the "linear theory" of flow stability analysis (Heisenberg, 1924; Tollmein, 1929; Schlichting, 1933, 1935; the full list of workers read like a "Who's Who" in applied mathematics and "natural philosophy" from Lord Rayleigh (1880) onward). The linear theory came to fruition in the 1940's and early 1950's through the work of C. C. Lin (1955) and others (See Betchov and Criminale, 1967 but beware mathematical misprints therein). The linear theory since Lin's work has become the predominant method of flow stability analysis; one advantage is the huge number of methodologies and algorithms that have been developed in the past decade for machine solution of the Orr-Sommerfeld equation and extraction of its eigenvalues, i.e. combinations of wave speed, Reynolds' number, and wave lengths arising in solutions to the Orr-Sommerfeld equations.

Lumley (1966), Elswick (1967), and Payne (1968) extended the "Orr-Energy" method whose basic advantage is applicability not only to two-dimensional, parallel shear flows with linearized disturbances, but also to fully three-dimensional turbulence with arbitrary, finite disturbances. Lumley's seminal paper (1966) has never appeared in the open literature primarily due to an earlier paper by Serrin (1959). Serrin's maximization principle is quite similar to Lumley's extremum approach but was formulated for strictly laminar flows and the variation of molecular viscosity. Lumley applied an extremum principle to the turbulent "eddy viscosity" coefficient; this parameter arises in turbulence due to an "engineering" approximation for the Reynolds' stress tensor in the Reynolds' averaged (1894) Navier-Stokes equations. (Sec. Vd)

IV. BRIEF OVERVIEW OF THE OLP METHODOLOGY

This methodology involves conceptually three separate stages of development. The first stage is due to Orr (1907) who integrated the mechanical energy equation over the entire flow region and then assumed stationarity of the disturbance energy on a global basis. This implies that the flow disturbances may gain energy locally and lose energy at another space point; however, summation over the entire flow is such that at an instant in time the global sum of mechanical energy contained in the perturbations to the basic motion remains constant. Stationarity of global energy for the case of the mechanical energy equation reduces to the simple statement that the rate of global production of disturbance energy balances the rate of total dissipation of that energy by viscous actions.

Stage two in the development of OLP is due to Lumley (1966) whose extremum principle is somewhat analogous to that of Serrin (1959). Lumley assumes that the viscosity parameter is independent of time and extremizes some representative value; this value may be that occurring in a wall region, or at a point of maximum shear gradient or other characteristic region of the flow. This leads to an eigenvalue/vector problem as the solution of his homogeneous, partial differential equations.

Stage three of OLP development is due to Elswick (1967) and Payne (1968); for computational simplicity, at least with the digital machinery then available, homogeneity is assumed for any space direction in which the statistics of the turbulence field are invariant under translation. Homogeneity in space is mathematically equivalent to stationarity in time; that is, homogeneity (stationarity) implies that two-point statistics in that space (time) direction are not functions of those two points in space, (time), i.e. the position of a "fixed probe" and "movable probe". Rather, such statistics are functions only of differences in position of the two probes, namely their separation vector. Homogeneity permits a Fourier integral transformation or Fourier-Stieltjes transformation if one requires more generality. The transformation from laboratory coordinates into a Fourier, or wave-number $\underline{k}$, space reduces the number of independent variables by one for each such homogeneity (stationarity) assumption. For example, assuming ergodicity so that one can convert ensemble averages to meaningful temporal averages, if the flow may be assumed homogeneous in the x_1 direction, then Fourier transformation to k_1-space is legitimate. The Lumley eigenvalue problem may then be solved in this wavenumber space for selected parametric values of k_1 and, in principle, inverse Fourier transformation back to laboratory coordinates is possible. For all applications to date of the OLP method two-dimensional homogeneity has been assumed; this implies that eigen-solutions are functions of a single space coordinate, y, and a two-dimensional parametric wavenumber pair, $\underline{k} = (k_1, 0, k_3)$. The resultant savings in computational complexity are obvious and considerable.

The Elswick-Payne methodology constructs Green's functions for the "reduced" operator under homogeneity assumptions (Fourier integral transforms). As a result, past applications of OLP (Elswick, 1967; Payne, 1968) each convert a system of three differential equations into a pair of coupled integral equations; the eigen-vectors are functions of three independent space variables, namely $\underline{x} = (x_1, x_2, x_3)$; this "real" space is transformed into a "mixed space", (k_1, y, k_3), wherein the partial differential equations are reduced to ordinary differential equations and converted, by use of Green's functions, to integral equations.

The solutions are then given as functions at point y on a two-dimensional, parametric wavenumber grid $\underline{k} = (k_1, k_3)$.

V. FUNDAMENTAL PHYSICAL AND MATHEMATICAL QUESTIONS

a) OLP avoids the usual "closure" hypothesis required in the predominant approaches to the turbulence problem; this avoidance is at the cost of, however ...

b) "Lock-in" to a complex, wavenumber space, $\underline{k} = (k_1, k_3)$. The difficulty arises from the necessity to inversely transform from the "mixed" space of (k_1, y, k_3) back to "real" space (x_1, y, x_3); the phase relation ambiguity across $\underline{k}$-space, due to complex eigen-vectors, is related to the criteria for the inverse transformation, i.e. return to laboratory space, of the eigen-solutions of the Lumley equations obtained in the "mixed" space of $(\underline{k}, y)$.

c) OLP should provide improved Reynolds stress models for digital or analog algorithms. (Rubesin, 1975; Payne, 1969, 1973, 1975, 1976, 1977a-e).

d) Is the loss of information due to temporal averaging, under the usual ergodic hypothesis, critical to the analysis/prediction of turbulence? Note that this difficulty is probably non-removable, in principle, due to the random, i.e., stochastic nature of turbulence. This stochastism is definitely not Guassian nor other well-known statistical distribution, at least for the "first order" turbulent statistics, the simplest of which is two-point co-variances. For "higher order" statistics of a turbulent field, namely, three-point co-variances or co-variances of velocity moments or derivatives, the departure from known statistical distributions becomes even more dramatic. A fundamental characteristic of turbulence, as pointed out by many authors (e.g. Townsend, 1956, 1976; Lumley, 1970), is the "intermittency" in two areas. These areas occur in the high frequency components which are enhanced in the statistics of derivatives and in the random convolutions of the interfacial boundary between turbulent and non-turbulent flow regions. The second area of departure from known statistical distribution involves the so-called "closure" problem in turbulence which arises fundamentally from the linear Reynolds' decomposition (1894) of turbulent signals into an average plus deviation from that average. This linear decomposition inserted into non-linear Navier-Stokes or Prandtl equations, which arise from Navier-Stokes under the usual "boundary layer" or "thin shear layer" approximation, results in a homo/heterodyning which produces the Reynolds' stress tensor. This new quantity $\underline{\underline{R}}(\underline{x})$, arising from the linear decomposition of a non-linear process, is the heart of the turbulence "closure" problem. See below for details (over bar denotes temporal average):

Incompressible Navier-Stokes
$$\frac{D\underline{u}}{Dt} = \frac{\partial \underline{u}}{\partial t} + (\underline{u}\cdot\nabla)\ \underline{u} = \nu\nabla^2\underline{u} - \frac{1}{\rho}\nabla P$$

Under Reynolds decomposition: $\underline{u} = \underline{\overline{U}} + \underline{u}',\ \overline{\underline{u}'} = 0$ becomes the Reynolds averaged equation:

$$\frac{\partial \underline{\overline{U}}}{\partial t} + (\underline{\overline{U}}\cdot\nabla)\ \underline{\overline{U}} = \nu\nabla^2\underline{\overline{U}} - \overline{\underline{u}'\cdot\nabla\underline{u}'} - \frac{1}{\rho}\nabla\overline{P}$$

and eqs for Reynolds' stress involve velocity triples and so on to ∞. $\underline{\underline{R}}(\underline{x})$, the Reynolds' stress tensor is defined by

$$\underline{\underline{R}}(\underline{x}) = \overline{\underline{u}'(\underline{x})\underline{u}'(\underline{x})};\ \therefore\ \overline{\underline{u}'\cdot\nabla\underline{u}'} \to \nabla\cdot\underline{\underline{R}}$$

and ν_T is defined via $\nu_T\nabla\underline{U} \equiv -\underline{\underline{R}}(x)$, analogous to molecular ν.

e) OLP utilizes "dynamical" information arising from the mathematical form of the model equations of motion as opposed to the PODT-SAS methodology (lumley, 1967; Payne, 1966) which is independent upon the "dynamics" of the flow except in the "the large". Hence, results of OLP depend directly upon the mathematical equations used for the process in question. A fundamental difficulty in turbulence even with the recent advent of huge vector processors is the shear magnitude of the calculational problem with present approximations most of which are ad hoc and semi-empiric at best.

f) OLP avoids the "constitutive relation" guessing game even under such objective principles used by continuum mechani-eians as "material independence", "observer independence", and other quite general principles as elucidated by Trusedell and Toupin (1960) and others.

g) Even so, OLP places severe requirements upon computing machinery; however, this severity is considerably less than that of any "complete simulation" of turbulence.

h) OLP and PODT-SAS (see Payne, 1966) eigen-values and eigen-vectors have somewhat different physical interpretation;

	λ, eigenvalue	ψ_i, eigenvector
OLP (Prediction)	Stability Parameter	Unstable Modes (of Turbulent Profile)
PODT-SAS (Experimental)	Mean Square Energy	Strong, "Large" Modes (of Turbulence)

As seen from the chart above, the extraction from experiment method (PODT-SAS) is straight-forward, in a both physical and mathematical sense, in contrast to the predictive, stability approach (OLP), or to use a more precise term, "bifurcation theory".

i) Are the three-dimensional OLP differential operators self-adjoint?

j) Do there exist more efficient and/or optimum calculational schemes for the eigensolution of the Lumley equations? (than used to date).

VI. OLP DERIVATION

(A) The "Orr-Energy" method of flow stability analysis involves the following steps:

1) *Given:* The kinetic, i.e., mechanical energy equation (CKE)

2) *Integrate:* CKE over the global flow region: This operation (Orr, 1907) implies that the CKE equation becomes:

$$\frac{\partial}{\partial t}\iiint_{\Omega} \rho E d\sigma + \oint\!\!\oint_{Bd(\Omega)} \rho\, E(\underline{V}\cdot\underline{n})dA = \iiint_{\Omega} \underline{V}\cdot(\nabla\cdot\underline{\underline{T}})^{t} d\sigma \qquad (1)$$

Where ρ = mass density; $E = (\frac{1}{2})\ \underline{v}\cdot\underline{v}\cdot$, the kinetic energy per unit mass; $\underline{n}$ is the unit normal, positive direction taken outward from the "control volume"; $\underline{\underline{T}}$ = the stress tensor including both isotropic hydrostatic and viscous terms; Ω = a "control volume" encompassing the entire flow region. $\underline{V}$ = the vector velocity relative to the boundary of Ω.

3) *Reynolds decomposition:* $V_i = \overline{U}_i + u_i,\ \overline{u}_i = 0,$ (2)
wherein the overbar denotes the usual temporal average under the ergodic hypothesis.

4) *Time average* eq. (1) with (2) inserted and

5) *Assume incompressibility* (for simplicity), $\nabla\cdot\underline{V} = 0.$

6) *Assume either* solid boundaries or cyclic boundary conditions on the flow field; this implies that the non-linear transport ("inertia") terms integrate to zero via the divergence theorem of Gauss.

7) $\rho E = \rho\overline{E} + \overline{e},\quad e = \frac{1}{2}\rho\,\underline{u}\cdot\underline{u}$ (By eq. (2))
(where $\overline{E}$ = mean flow KE, e = disturbance KE)

8) The result for time averaged disturbance energy from eq. (1), (2), with assumptions above is:

$$\frac{\partial}{\partial t}\int \rho\bar{e}d\sigma = -\int\left(\overline{\frac{\partial u_i}{\partial x_j}t_{ij}} + \rho\,\overline{u_i u_j}\,\frac{\partial U_i}{\partial x_j}\right)d\sigma \tag{3}$$

where $t_{ij} = T_{ij} - \overline{T}_{ij}$, the usual Reynolds decomposition wherein $t_{ij} = t_{ij}(\underline{x}, t)$ is the instantaneous disturbance stress tensor.

9) *Assume incompressible Newtonian* fluid constitutive relation is descriptive of the disturbance:

i.e., $\underline{\underline{t}} \equiv -P\underline{\underline{I}} + 2\mu\underline{\underline{s}}$, $\underline{\underline{s}} \equiv \frac{1}{2}[\nabla\underline{u} + (\nabla\underline{u})^t]$

This in turn implies that eq. (3) becomes eq. (4) below:

$$\frac{\partial}{\partial t}\int_\Omega \bar{e}d\sigma = -\rho\int_\Omega \overline{u_i u_j}\,S_{ij}\,d\sigma - \int_\Omega \overline{\mu u_{i,j}(u_{i,j} + u_{j,i})}\,d\sigma \tag{4}$$

where $S \equiv \frac{1}{2}[\nabla\underline{u} + (\nabla\underline{u})^t]$, mean strain-rate tensor

10) Orr's extremum (1907) assumes instantaneous stationarity of global kinetic energy:

Orr's Extremum Principle: $$\frac{\partial}{\partial t}\int_\Omega \bar{e}d\sigma = 0 \tag{5}$$

In words: "Orr ≡ Stationarity of global disturbance kinetic energy"

TOTAL PRODUCTION	=	TOTAL DISSIPATION

(rates of disturbance energy)

(B) *Lumley's (1966) Extremum Principle:*

1) Take the spatial variation of viscosity as fixed in eq. (4) and

2) Extremize some representative value, ν_t, which leads, via standard variational calculus, to . . .

(C) *The General Lumley Equations:* Lumley (1967) obtains

$$S_{ij}u_j = \phi,_i + \frac{\partial}{\partial x_j}[\nu_T(u_{i,j} + u_{j,i})], \quad \nabla \cdot \underline{u} = 0 \tag{7}$$

where S_{ij} = mean rate of strain, and ν_T = "eddy viscosity" and "ϕ is essentially a Lagrange multiplier which satisfies the constraint of incompressibility or (may be) a pressure, depending upon point of view" - Lumley (1966).

VII. PRIOR APPLICATIONS OF OLP

(A) *Wall Turbulence* (Elswick, 1967) To compare with experiment Elswick assumed:

1) Steady, 2-*D* mean flow, $\overline{U} = f(x,y)$, f given and

2) $\nu_t = g(y)$, g given, piecewise smooth.
Hence, the Lumley eqs. (7) reduce to:

$$\begin{aligned} S_{11}u_1 + S_{12}u_2 &= \phi,_1 + \frac{\partial}{\partial x_j} g(u_{1,j} + u_{j,1}) \\ S_{21}u_2 + S_{22}u_2 &= \phi,_2 + \frac{\partial}{\partial x_j} g(u_{2,j} + u_{j,2}) \end{aligned} \tag{8}$$

By use of stream function: $u_k \equiv -\varepsilon_{1jk}\frac{\partial\psi}{\partial x_j}$

Eq. (8) $\Longrightarrow$

$$\left.\begin{aligned} \frac{ik_3}{2}\frac{U'}{\nu}\hat{\psi} &= \left[\nabla^2 + \frac{\nu'}{\nu}D\right]\hat{u}_1 \\ \frac{ik_3}{2}\frac{U'}{\nu}\hat{u}_1 &= \left[\nabla^4 + 2\frac{\nu'}{\nu}\nabla^2 D + \frac{\nu''}{\nu}[D^2 + k_3^2]\right]\hat{\psi} \end{aligned}\right\} \tag{9}$$

$$D = \frac{d}{dt}, \quad \hat{u}_1 = F.T.(u_1), \quad \hat{\psi} = F.T.(\psi)$$

which Elswick solved via Green's function method and was able to predict, in good correlation with experiment,

1) The dominant frequencies observed and

2) The *2-D* "eddy" structure, u_1, u_2.

(B) *Turbulent Wake of a Circular Cylinder* (Payne, 1968) To the author's knowledge this is the only 3-dimensional application of the PODT-SAS, much less the OLP, methodology to date. However, Reed (1977) is currently applying the PODT-SAS methodology to a circular jet experiment.

The motivation is to compare OLP results with those of the extractive method, PODT-SAS, (Payne, 1966; Payne & Lumley, 1967) for the large eddy structure from experimental two-point velocity covariances measured by hotwire anemometry (Grant, 1958). To do so, one assumes steady, fully developed, two-dimensional mean flow, $\underline{U} = \underline{f}(y/d)$. In addition, ν_T is taken as constant throughout the flow; this assumption is easily verified on a empirical basis for the central 90% of the wake. Under these assumptions Lumley's equations become $(\eta = y/d)$.

$$
\begin{aligned}
\frac{1}{2}\frac{dU}{d\eta}u_2 &= \phi,_1 + \nu_T \nabla^2 u_1 \\
\frac{1}{2}\frac{dU}{d\eta}u_1 &= \phi,_2 + \nu_T \nabla^2 u_2 \\
0 &= \phi,_3 + \nu_T \nabla^2 u_3
\end{aligned}
\tag{10}
$$

In eq. (10) cross differentiation eliminates the quantity ϕ, yielding:

$$
\begin{aligned}
\frac{1}{2}\frac{dU}{d\eta}u_{2,3} &= \nu_T \nabla^2 (u_{1,3} - u_{3,1}) \\
\frac{1}{2}\frac{dU}{d\eta}u_{1,3} &= \nu_T \nabla^2 (u_{2,3} - u_{3,2})
\end{aligned}
\tag{11}
$$

If one non-dimensionalizes eq. (11) by the appropriate velocity and length scales, U_0 and d, then the form of eq. (11) is unchanged except that the eddy viscosity coefficient, ν_T, becomes a reciprocal, turbulent Reynolds number, $\nu_T \to R_T^{-1}$. One can also assume, for the experiment in question, homogeneity in the plane of the wake; this validates the usual Fourier transform:

$(\tilde{\underline{x}} = \underline{x}/d)$

$$\underline{u}(\tilde{x}) = (2\pi)^{-2} \iint_{-\infty}^{\infty} e^{i\underline{k}\cdot\tilde{\underline{x}}} \, \underline{\phi}(\tilde{x}_2, \underline{k}) \, dk \tag{12}$$

where $\underline{k} = (k_1, 0, k_3)$. Note that eq. (12) can be generalized, if necessary, to a Fourier-Stieltjes transform wherein the orthogonal increments, $d\underline{z}$ are given by $\phi_i d\underline{k} \to dz_i(x_2, \underline{k})$ in eq. (12). Under the assumption of homogeneity, then, (see Appendix B; Payne, 1968) the Lumley equations (11) become eq. (13) below:

$$\left.\begin{aligned} k^2\, \nabla^2\, \psi_1 &= \left(ik_1\, \nabla^2 D + \tfrac{1}{2}\, k_3^2\, R_T U'\right) \psi_2 \\ \nabla^2\left(k_3^2 - D^2\right) \psi_2 &= \left(ik_1\, \nabla^2 D + \tfrac{1}{2}\, k_3^2\, R_T U'\right) \psi_1 \end{aligned}\right\} \tag{13}$$

Where $\nabla^2 \equiv D^2 - k^2$, $D \equiv d/dy$.

One immediately notes that eq. (13) is of the form of

$$\begin{aligned} L_1(\psi_1) &= M(\psi_2) \\ L_2(\psi_2) &= M(\psi_1) \end{aligned} \tag{14}$$

where L_1, L_2, and M are linear operators defined in the obvious way from eq. (13). Note that L_1 is second order in D, L_2 fourth order in D whereas the M operator is third order in D.

One can cross-differentiate eq. (13) to try and partly untangle the $\psi_1 - \psi_2$ couplings. (See Appendix C, Payne 1968 for details). Of course, one immediately raises the question as to whether spurious roots are introduced or not.

$$\left.\begin{aligned} \nabla^4\, \psi_1 &= -\tfrac{1}{2}\, R_T \left[ik_1 D(U'\psi_1) + \left(k_3^2 - D^2\right)\left(U'\psi_2\right)\right] \\ \nabla^4\, \psi_2 &= -\tfrac{1}{2}\, R_T \left[ik_1 D(U'\psi_2) + k^2(U'\psi_1)\right] \end{aligned}\right\} \tag{15}$$

<u>Note</u> form of (15): $\{\nabla^4\psi_\alpha = L_0(U'\psi_\alpha) + L_{\alpha\beta}(U'\psi_\beta)$, $\alpha \neq \beta$, $\alpha, \beta = 1, 2$

Where $L_{\alpha\beta}$ is the "interaction" operator.

Payne (1966) also found that the experimental eigen-functions for the *2-D* wake occurred in pairs:

$$\psi_1(y) = \pm\psi_1(-y), \quad \psi_2(y) = \mp\psi_2(-y) \tag{16}$$

Where the upper signs were taken as the strongest mode throughout Fourier wave-number space for purposes of both interpretation and inverse transformation back to laboratory-space. Hence, $\psi_1(y) = \psi_1(-y)$, $\psi_2(y) = -\psi_2(-y)$ is used: This and $\text{Limit}_{|y|\to\infty} \{\psi_i(y)\} \to 0$ complete the conditions needed to solve (15).

Caution:

The original PODT-SAS work (Payne, 1966) pointed out that there is an unresolved, in principle or theory, problem of the interphase relationships of complex eigen-vectors across the Hilbert space associated with the (parametric) $\underline{k}$-space; correct phase relations are crucial for inverse transformation from (k_1,y,k_3) into (x,y,z)-space, i.e., back to laboratory coordinates.

The Green's function found for eq. (15) are given below where $k = (\underline{k}\cdot\underline{k})^{1/2}$:

$$\left.\begin{aligned}
H_1(y^*,y) &= -\frac{1}{4k^2} N_1(k,y^*,y) \\
H_2(y^*,y) &= -\frac{1}{4k^2} N_2(k,y^*,y) \\
\text{where } N_1 &= N + \eta_1, \quad N_2 = N + \eta_2 \\
N &= (-\frac{1}{k} - |y - y^*|) \ \text{Exp}\{-k\,|y - y^*|\} \\
\eta_1 &= (-\frac{1}{k} - y - y^*) \ \text{Exp}\{-k(y + y^*)\} \\
\eta_2 &= (\frac{1}{k} + y + y^*) \ \text{Exp}\{-k(y + y^*)\}
\end{aligned}\right\} \tag{17}$$

Where H_i are subject to identical *B. C.* as ψ_i :

$$\left.\begin{aligned} D\psi_1(0) &= D^3\psi_3(0) = \psi_1(\infty) = D\psi_1(\infty) = 0 \\ \psi_2(0) &= D^2\psi_2(0) = \psi_2(\infty) = D\psi_2(\infty) = 0 \end{aligned}\right\}$$

Then the eigen-solutions, under the given conditions to eq. (15) may be written in the integral formulation given below:

$$\begin{aligned} \psi_1(y^*,\underline{k}) &= \frac{1}{2} R_T \int_0^\infty \left[-i_{k_1} U'\psi_1 \, DH_1 + \psi_2 U'(k_3^2 H_1 - D^2 H_1) \right] dy \\ \psi_2(y^*,\underline{k}) &= \frac{1}{2} R_T \int_0^\infty \left[-i_{k_1}\psi_2 \, U'DH_2 - \psi_1 U' k^2 H_2 \right] dy \end{aligned} \tag{18}$$

Numeric methodology for solution of eq. (18) can be written as

$$\psi_i(\underline{k}, y^*) = \frac{1}{2} R_T \, K_{ij}\psi_j, \quad K_{ij}\psi_j = \lambda\psi_i, \quad \lambda \equiv 2/R_T \tag{19}$$

where K_{ij}, defined by eq. (19), generates a classical eigenvalue problem for which, since it was derived via a maximization principle. Lumley (1970) formulated a rather general iterative scheme:

$$\begin{aligned} K_{ij}\,\psi_j^{(0)} &= \psi_i^{(1)} \\ &\vdots \\ K_{ij}\,\psi_j^{(n)} &= \psi_i^{(n+1)}, \qquad n = 0, 1, 2 \ldots \\ &\vdots \end{aligned} \tag{20}$$

where $\lim_{n\to\infty} \left\{ \dfrac{\psi_i(n+1)}{\psi_i(n)} \right\} \longrightarrow \lambda_1$ uniformly in y

and $\lambda_1 \geqslant \lambda_2 \geqslant \ldots \geqslant o$, ψ_i are orthonormal subject only to K_{ij} Hermitian.

Cautions:

1) Payne (1968) was unable to prove with mathematical

rigor that K_{ij} is a self-adjoint matrix but ...

2) The physical problem insures that R_T is real and non-negative and

3) Green's functions of K_{ij} are real and symmetric. It should be noted that Elswick (1967) showed that K_{ij} was self-adjoint for his special, two-dimensional case. See Payne (1968, 1977d) for details of the computational scheme and results of the OLP prediction which include favorable comparisons with the PODT-SAS extracted (Payne, 1966) eigen-structures of turbulence experimental data (Grant, 1958). Comparisons there (Payne, 1968) are two-fold; phase relations of $\underline{k}$-space generally agree to within 3-10% of a full cycle and amplitude variation with y, distance from the wake central plane of symmetry, is qualitatively excellent and quantitatively satisfactory.

VIII. FORMULATION OF OLP FOR THE BOUNDARY LAYER

(A) *General Lumley Equations:*

$$S_{ij}u_j = \phi,_i + \frac{\partial}{\partial x_j}\left[\nu_T(u_{i,j} + u_{j,i})\right], \quad \nabla \cdot \underline{u} = 0 \quad \text{-(7R)}$$

(B) *3-D, Newtonian Fluid Flow* (in the average): No restriction

(C) 2-D Mean Flow, 3-D Disturbance: $S_{i3} = S_{3i} = 0$

(D) $\nabla P = 0$; No restriction other than kinematic on $\underline{\underline{S}}$

(E) *NSG-2077 Grant by NASA/Ames:* No new restriction except that ν_T must be allowed to vary with y which is like Elswick, 1967 and more complex than the wake (Payne, 1966).

Remark: Here is a major new complexity for which "appeal" is made.

(F) *Implications for Flow Simulation/Calculation:*

1) Rubesin (1976, private communication): "Essential to input spectral information into the Reynolds' stess models". And as of now (June 1977) there appears to be no other rational scheme (current author's claim).

2) *Schematic of "application" of OLP:* (Payne, 1969)

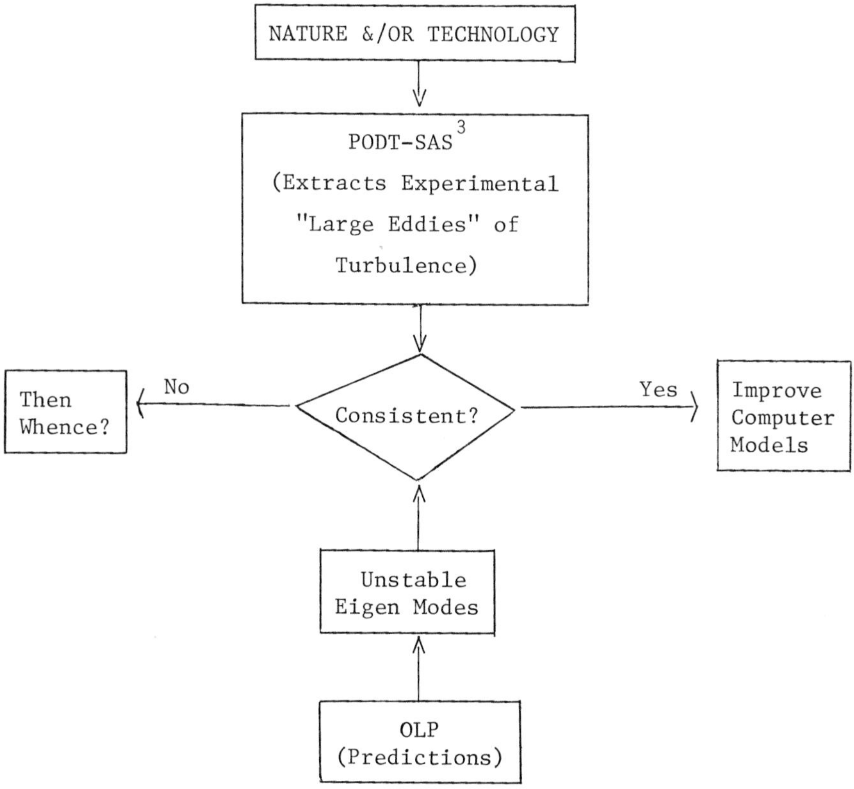

---

[3] PODT-SAS ≡ Proper Orthogonal Decomposition Theorem -- Structural Analysis System (Lumley 1967, Payne 1966, Payne and Lumley 1967, Payne 1968, 1969, 1973-77; Lemmerman 1976, Lemmerman and Payne, 1977).

IX. SUMMARY AND RECAPITULATION:

(A) Problem Re-visited

1) OLP has been successful for *2-D* wake (Payne, 1968, 1977d)

2) OLP boundary layer in progress (Payne, 1977c)

3) OLP needs to be applied to other flow proto-types (Jets, ∇P, etc.)

(B) Current Status of boundary Layer on Flat-Plate (NSG-2077)

1) OLP specialized to boundary layer $(\nabla P = 0)$

2) ν_T model (must come from experiment as of now) but sample calculations could be done for "constitutive relations for turbulence"

(C) Appeal for Assistance from the "Mathematics Community"

1) Phase relationship (across $\underline{k}$-space) of complex eigen-vectors for "Reconstitution" in laboratory coordinates?

2) Study of "covariances"

3) Is information loss critical (mostly of phase, details of local flow field) due to averaging?

4) $\exists$ optimal calculational schemes?

(D) *Conclusions:* OLP has had good success for one flow proto-type, the *2-D* wake, and a second application to the *2-D* boundary layer is being initiated but will require, due to presence of solid boundary, considerably more ingenuity than the first case. OLP predictions are complementary to the PODT-SAS analysis of experimental data for the case of statistical turbulence and hold promise of improving our knowledge (decreasing our ignorance) of the challenging phenomena of turbulence.

X. REFERENCES CITED

[1] Betchov, R. and Criminale, W. O., *Stability of Parallel Flows*, Academic Press, NY (1967).

[2] Elswick, R. C., *MSAE Thesis*, Penn State University, and rep. to U.S.N./ONR under (G)-00043-65 (1967).

[3] Grant, H. L., *Journal Fluid Mechanics*, (1958), p. 149.

[4] Heisenber, W., *Ann. d. Phys.*, 74 (1924), pp. 577-627.

[5] Lemmerman, L. A., *Ph.D. Dissertation*, University of Texas at Arlington (1976).

[6] Lemmerman, L. A. and Payne, F. R., "*Extracted Large Eddy Structure of a Turbulent Boundary Layer*," AIAA Paper 77-717, 10th Fluid & Plasma-dynamics Conference, Albuquerque (1977).

[7] Lin, C. C., *The Theory of Hydronamic Stability*, Cambridge U. Press, London (1955).

[8] Lumley, J. L., "*Large Disturbances to the Steady Motion of a Liquid*," Memo/Ordnance Res. Lab., Penn State, 22 August (1966).

[9] Lumley, J. L., "*The Structure of Inhomogeneous Turbulent Flows*," Paper presented in 1966 at Moscow and printed in Doklady Akad. Nauk SSSR, Moscow (1967).

[10] Lumley, J. L., *Stochastic Tools in Turbulence*, Academic Press, New York and London (1970).

[11] Orr, W., "*The Stability ... of steady Motions of a Perfect Fluid and a Viscous Fluid*," Proc. Royal Irish Acad., Sec. A. Vol. XXVII, (1907), p. 69.

[12] Payne, F. R., *Ph.D. Thesis*, Penn State Univ. and rep. to U.S.N./ONR under Nonr 656(33), (1966).

[13] Payne, F. R. and Lumley, J. L., *Phys. Fluids*, SII (1967), p. S194.

[14] Payne, F. R., *Predicted Large Eddy Structure of a Turbulent Wake*, rep. to U.S.N/ONR under Nonr 656(33), (1968).

[15] Payne, F. R., "*Analysis of Large Eddy Structure of Turbulent Boundary Layers*," ARR-13, General Dynamics, Fort Worth

(1969), pp. 50-54.

[16] Payne, F. R., "*Toward a Better Eddy Viscosity Model of Turbulence*," First AIAA-Mini-Symposium, UTA (1973).

[17] Payne, F. R., "*In Search of Big Eddies*," 14th Midwestern Mech. Conf., Norman, Oklahoma (1975).

[18] Payne, F. R., "*Large Eddy Structure of a Flat-Plate Boundary -Layer*," 4th AIAA Mini-Symposium, UTA (1976).

[19] Payne, F. R., "*Large Eddy Structure of Turbulent Coundary Layers*," paper at 5th AIAA Mini-Symposium, U.T. Arlington, February and 15th Mid-Western Mech. Conf., Chicago (1977a).

[20] Payne, F. R., "*Extracted Large Eddy Structure of a Flat-Plate Boundary Layer*," Symposium on Turbulent Shear Flows, Penn State U., April (1977b).

[21] Payne, F. R., "*Extraction and Prediction of the Large Scale Structures in Turbulent Shear Flows*," RCAS-sponsored Conference on Industrial Mathematics (CIM), UTA, May (1977c), (Current article expands Sec. IV of that paper.)

[22] Payne, F. R., "*Comparison of PODT-SAS Extractive with OLP-Predictive Eigen-Structures in a Turbulent Wake*," SIAM Fall 1977 meeting (abstract submitted) (1977d).

[23] Payne, F. R., "*Large-Scale Eigen-Structures of Turbulent-Shear-Flows: A Dual Strategy for Prediction and Measurement of Stochastic Fields*," SIAM Fall 1977 meeting (abstract submitted) (1977e).

[24] Rayleigh, Lord, *Sci. Papers I*, (1880), pp. 474-487.

[25] Reed, X. B., Jr., et al, *Proc. Symposium on Turbulent Shear Flows*, Penn State, April 18-20 (1977), pp. 2.23-2.32.

[26] Reynolds, O., "*On the Dynamical Theory ... and the Determination of the Criterion*," Philos. Trans., A. CLXXXVI, (1894) p. 123.

[27] Rubesin, M. W., "*Subgrid or Reynolds Stress Modeling for Three Dimensional Turbulent Computations*," NASA SP-347 (1975).

[28] Schlichting, H., *Nachr. Ges. Wiss. Gottingen, Math. Phys.*

Klasse, (1933), pp. 182-208, (also (ZAMM, 13, pp. 171-174).

[29] Schlichting, H., *NGWGMPK, Fach Gruppe* I, 1, (1935), pp. 47-78.

[30] Serrin, J., "*On the Stability of Viscous Fluid Motions*," Arch. Rat. Mech. Anal., 1, e, (1959), p. 1ff.

[31] Sommerfeld, A., *Atti del 4. Congr. Int'l dei Mat., III*, (1908), pp. 116-124, Roma.

[32] Stewart, R. W., "*Turbulence*," Illustrated Exp. in Fluid Mech., MIT Press, Cambridge (1972), pp. 82-88.

[33] Tollmein, W., (1929), Engl. Trans. NACA TM 609 (1931).

[34] Townsend, A. A., *The Structure of Turbulent Shear Flow*, Cambridge University Press (1956).

[35] Townsend, A. A., *Ibid*, (1976), 2nd Edition.

[36] Tritton, D. J., *Journal Fluid Mechanics*, (1967), p. 439.

[37] Trusedell, C. and Toupin, R., "*The Classical Field Theories*", Ency. of Physics, III/1, Springer-Verlag (1960).

GENERALIZED CONTRACTIONS AND SEQUENCE OF ITERATES

K. L. Singh
Texas A&M

I. INTRODUCTION

The main aim of this paper is to study some properties of generalized contraction mappings. In Section I we have shown that if T is a generalized contraction mapping of a closed, bounded and convex subset of a uniformly convex Banach space into itself with nonempty fixed points set, then the mapping T_λ defined by $T_\lambda = \lambda I + (1 - \lambda)T$, for any λ such that $0 < \lambda < 1$ is asymptotically regular. In Section II, it is shown that if T is a generalized contraction self mapping of a closed, convex subset of a Hilbert space with nonempty fixed points set, then the mapping T_λ defined as above is a reasonable wanderer with the same fixed point as T. In Sections III and IV we have obtained some results for the weak and strong convergence of sequence of iterates of such kind of mappings. Finally in Section V, we prove some results for the convergence of sequence of iterates of this kind of mappings for infinite summability matrices.

Definition 1.1. Let X be a normed linear space and C be a nonempty subset of X. A mapping $T\colon\ C \to C$ is said to be quasi-nonexpansive provided T has at least one fixed point in C, and if $p \in C$ is any fixed point of T then $\|Tx - p\| \leqslant \|x - p\|$ for all $x \in C$.

ISBN 0-12-434160-8

Remark 1.1. A nonexpansive mapping $T\colon C \to C$ with at least one fixed point $\in C$ is quasi-nonexpansive. A linear quasi-nonexpansive mapping on a subspace is nonexpansive on that subspace; but there exist continuous and discontinuous nonlinearquasi-nonexpansive mappings which are not nonexpansive. For example, the mapping $T\colon R \to R$ (real) defined by $T(x) = 2x/3 \sin(1/x)$, $x \neq 0$, $T(0) = 0$ is quasi-nonexpansive but not nonexpansive. Indeed, the only fixed point of T is zero, since if $x \neq 0$ and $x = Tx$, then $x = 2x/3 \sin(1/x)$, or $3/2 = \sin 1/x$; which is impossible. We claim that T is quasi-nonexpansive, since if $y \in R$, $p = 0$, then $|Ty - p| = |Ty - 0| = |2y/3 \sin(1/y)| = |2y/3|$ $|\sin 1/y| \leq 2/3|y| < |y| = |y - p|$. However, T is not a nonexpansive mapping. Indeed, let $x = 2/\pi$ and $y = 2/3\pi$, then $|Tx - Ty| = |\frac{4}{3\pi} \sin \pi/2 - \frac{4}{9\pi} \sin \frac{3\pi}{2}| = |\frac{4}{3\pi} + \frac{4}{9\pi}| = \frac{16}{9\pi}$, whereas $|x - y| = |\frac{2}{\pi} - \frac{2}{3\pi}| = \frac{4}{3\pi}$.

Definition 1.2. Let X be a normed linear space and C be a nonempty subset of x. A mapping $T\colon C \to C$ is said to satisfy condition (α) if

$$\|Tx - Ty\| \leq a\|x - y\| + b(\|x - Tx\| + \|y - Ty\|) + c(\|x - Ty\| + \|y - Tx\|) \tag{1}$$

for all x, y in C, where a, b, c are nonnegative and $a + 2b + 2c \leq 1$.

Remark 1.2. Letting $b = c = 0$ and $a = 1$, in (1) it follows that T is nonexpansive. Letting $a = c = 0$ in (1), we get the mapping of Kannan [10], and letting $c = 0$ in (1), we get the mapping of Yadav [31].

Remark 1.3. It has been shown by Petryshyn and Williamson Jr. [16] that a mapping satisfying condition (α) is quasi-nonexpansive. However, the following simple example shows that the quasi-nonexpansive mappings and the mappings satisfying condition (α) are independent.

Example 1.1. Let $X = [0,1]$, with the usual metric. Define T

by $T(x) = 0$, $0 \leq x \leq \frac{1}{2}$, $T(x) = \frac{1}{2}$, $\frac{1}{2} < x \leq 1$. Then T is quasi-nonexpansive but T does not satisfy condition (α). On the other hand, if $T(x) = \frac{1}{4}$, $0 \leq x \leq \frac{1}{3}$, $T(x) = 0$, $\frac{1}{3} < x \leq 1$. Then T satisfies the condition (α), however T is not quasi-nonexpansive.

In [27] and [28] we studied the following mapping.

Definition 1.3. Let X be a normed linear space and C be a nonempty subset of X. A mapping $T\colon C \to C$ is said to satisfy condition (B) if

$$\|Tx - Ty\| \leq \max\{\|x - y\|;\ \tfrac{1}{2}(\|x - Tx\| + \|y - Ty\|);\ \tfrac{1}{2}(\|x - Ty\| + \|y - Tx\|)\}$$

for all x,y in C.

Definition 1.4. Let x be a normed linear space and c be a nonempty subset of x. A mapping $T\colon c \to c$ is called generalized contraction if

$$\|Tx - Ty\| \leq \max\{\|x - y\|;\ \|x - Ty\|,\ \|y - Ty\|;\ \tfrac{1}{2}(\|x-Tx\| + \|y - Ty\|);\ \tfrac{1}{2}(\|x - Ty\| + \|y - Tx\|)\}$$

for all x,y in C.

Remark 1.4. It is clear that any mapping which satisfies condition (α) or condition (β) is a generalized contraction. The following example shows that a generalized contraction need not satisfy either condition (α) or condition (β).

Example 1.2. Let

$$M_1 = \{\, m/n\colon\ m = 0, 1, 3, \ldots, n = 1, 4, \ldots, 3k+1, \ldots\}$$

$$M_2 = \{\, m/n\colon\ m = 1, 3, 9, \ldots, n = 2, 5, \ldots, 3k+2, \ldots\}$$

and let $M = M_1 \cup M_2$ with the usual metric. Define the mapping $T\colon M \to M$ by

$$T(x) = \begin{cases} 4x/5 & \text{for } x \in M_1 \\ x/3 & \text{for } x \in M_2. \end{cases}$$

The mapping T is a generalized contraction mapping.

Indeed, if both x and y are in M_1 or in M_2, then $|Tx - Ty| \leq 4/5|x - y| < |x - y|$. Now consider, for example $x \in M_1$ and $y \in M_2$. Then $x > (5/12)y$ implies $|Tx - Ty| = 4/5(x - (5/12)y) < 4/5(x - y/3) = 4/5|x - Ty|$. $x < (5/12)y$ implies $|Tx - Ty| = 4/5(5y/12 - x) \leq 4/5(y - x)$ $= 4/5(y - x)$. Therefore T on M is a generalized contraction. To show T does not satisfy either condition (α) or condition β, let $x = 1$, $y = 1/2$. Then we have $|Tx - Ty| = 4/5 - 1/6 = 19/30$. However,

$$a(x-y) + b(|x - Tx| + |y - Ty|) + c(|x - Ty| + |y - Tx|)$$

$$= a \cdot \frac{1}{2} + b(\frac{1}{5} + \frac{1}{3}) + c(\frac{5}{6} + \frac{3}{10}) = a \cdot \frac{1}{2} + \frac{8b}{15} + \frac{34c}{30}$$

$$= a \cdot \frac{1}{2} + \frac{16b}{30} + \frac{34c}{30} < (a + 2b + 2c)\frac{34}{60} \leq \frac{17}{30} < \frac{19}{30} = |Tx - Ty|.$$

Hence the condition (α) is not satisfied. Moreover,

$$\max\{|x - y|;\ \frac{1}{2}(|x - Tx| + |y - Ty|);\ \frac{1}{2}(|x - Ty| + |y - Tx|)\}$$

$$= \max\{1/2,\ 8/30,\ 17/30\} = 17/30 < 19/30 = |Tx - Ty|.$$ Thus the condition (β) is not satisfied.

Definition 1.5. A Banach space X is said to be uniformly convex (or uniformly rotund) if given $\varepsilon > 0$ there exists $\delta(\varepsilon) > 0$ such that $\|x - y\| \geq \varepsilon$ for $\|x\| \leq 1$ and $\|y\| \leq 1$ implies $\|(x + y)/2\| \leq 1 - \delta(\varepsilon)$.

Definition 1.6. Let X be a Banach space and C be a closed, convex subset of X. A mapping $T\colon C \to C$ is called asymptotically regular at x is and only if $\|T^n x - T^{n+1}x\| \to 0$ as $n \to \infty$.

Theorem 1.1. Let D be a nonempty, bounded, closed and convex subset of a uniformly convex Banach space X. Let $T\colon D \to D$ be a generalized contraction mapping. Let us assume that $F = \{x \in D\colon Tx = x\}$ is nonempty. Then the mapping T_λ defined by $T_\lambda = \lambda T + (1 - \lambda)I$ for any λ such that $0 < \lambda < 1$ is asymptotically regular with the same fixed point as T.

Proof. It is clear that $F(T) = F(T_\lambda)$, where $F(T)$ and $F(T_\lambda)$ are fixed point sets of T and T_λ respectively. Indeed, $x \in F(T)$ implies $Tx = x$. Thus $T_\lambda x = \lambda x + (1 - \lambda)x = x$, hence $x \in F(T_\lambda)$, i.e. $F(T) \subset F(T_\lambda)$. Conversely $y \in F(T_\lambda)$ implies $T_\lambda y = y \quad = \lambda Ty + (1 - \lambda)y$, which implies $\lambda Ty = \lambda y$, or $Ty = y$. Thus $F(T_\lambda) \subset F(T)$. Hence $F(T_\lambda) = F(T)$.

Let x_0 in $D, x_{n+1} = T_\lambda(x_n)$, $n = 0, 1, 2, \ldots$. Since $T_\lambda x - x = \lambda(x - Tx)$ for x in D, it is enough to show that $\|x_n - Tx_n\| \to 0$ as $n \to \infty$. Now let x in D and y in $F(T)$, hence in $F(T_\lambda)$, then

$$\begin{aligned}(1)\qquad \|Tx - y\| = \|Tx - Ty\| &\leqslant \max\{\|x - y\|;\ \|x - y\|;\ \|y - Ty\|;\\ &\quad \tfrac{1}{2}(\|x - Tx\| + \|y - Ty\|);\ \tfrac{1}{2}(\|x - Ty\| + \|y - Tx\|)\}\\ &= \max\{\|x - y\|;\ \tfrac{1}{2}\|x - Tx\|;\ \tfrac{1}{2}(\|x - y\| + \|y - Tx\|)\}.\end{aligned}$$

Now

$$\|Tx - y\| \leqslant \tfrac{1}{2}\|x - Tx\| \leqslant \tfrac{1}{2}(\|x - y\| + \|y - Tx\|) \quad \text{implies}$$

$$\|Tx - y\| \leqslant \|x - y\|.$$

Thus from (1) we obtain

$$(2)\qquad \|Tx - y\| \leqslant \|x - y\|.$$

Also,

$$\begin{aligned}(3)\qquad \|y - T_\lambda x\| &= \|y - (1 - \lambda)x - \lambda Tx\|\\ &= \|(1 - \lambda)(y - x) + \lambda(y - Tx)\|\\ &\leqslant (1 - \lambda)\|y - x\| + \lambda\|y - x\| = \|y - x\|\end{aligned}$$

So the sequence $\{\|y - x_n\|\}$ is bounded by $M = \|y - x_0\|$. If $y = x_n$ for some n, then from (3), $\{x_n\}$ converges to y and the proof is complete. So we may assume that $y \neq x_n$ for all $n = 0, 1, 2, \ldots$ Suppose $\lambda \leqslant 1/2$. Now

(4) $$\|y - x_{n+1}\| = \|\lambda(y - x_n + y - Tx_n) + (1 - 2\lambda)(y - x_n)\|$$

$$\leqslant \lambda\|y - x_n + y - Tx_n\| + (1 - 2\lambda)\|y - x_n\|$$

$$= (\|y - x_n\|)\lambda\left\|\frac{(y - x_n + y - Tx_n)}{\|y - x_n\|}\right\| + (1 - 2\lambda)\|y - x_n\|.$$

(5) Let $a = \dfrac{y - x_n}{\|y - x_n\|}$ and $b = \dfrac{y - Tx_n}{\|y - x_n\|}$, then

$$\|a + b\| = \frac{\|y - x_n + y - Tx_n\|}{\|y - x_n\|} \text{ and } \|a\| \leqslant 1, \quad \|b\| \leqslant 1.$$

Thus from (4) we get

(6) $$\|y - x_{n+1}\| \leqslant 2\lambda\|(a+b)/2\|\,\|y - x_n\| + (1 - 2\lambda)\|y - x_n\|$$

$$= \{2\lambda\|(a+b)/2\| + (1 - 2\lambda)\}\|y - x_n\|.$$

Since X is uniformly convex

$$\delta(\varepsilon) = \inf\{1 - \|(x+y)/2\| : \|x\| \leqslant 1, \|y\| \leqslant 1, \|x - y\| \geqslant \varepsilon\}$$

is positive for $\varepsilon \in (0,2]$. Also $\delta(0) = 0$.

From (5) we have

$$\|(a + b)/2\| \leqslant 1 - \delta(\|x_n - Tx_n\|/\|y - x_n\|).$$

Since δ is monotonically nondecreasing on $[0,2]$

(7) $$\|(a + b)/2\| \leqslant 1 - \delta(\|x_n - Tx_n\|/M).$$

Thus from (6) and (7) we have

(8) $$\|y - x_{n+1}\| \leqslant 2\lambda\|y - x_n\|(1 - \delta(\|x_n - Tx_n\|/M) + (1 - 2\lambda)\|y - x_n\|$$

$$= \{2\lambda - 2\lambda\delta(\|x_n - Tx_n\|/M) + (1 - 2\lambda)\}\|y - x_n\|$$

$$= \{1 - 2\lambda\delta(\|x_n - Tx_n\|/M)\}\|y - x_n\|.$$

From (8) and induction we obtain

(9) $$\|y - x_{n+1}\| \leqslant \prod_{j=0}^{n} 1 - 2\lambda\delta(\|x_j - Tx_j\|/M\} M.$$

Suppose $\{\|x_n - Tx_n\|\}$ does not converge to zero. Then there exists a subsequence $\{x_{k(n)}\}$ of $\{x_n\}$ such that $\{\|x_{k(n)} - Tx_{k(n)}\|\}$ converges to some constant α in $(0,\infty)$. Since δ is monotonically nondecreasing and $1 - 2\lambda\delta(\|x_j - Tx_j\|/M)$ belongs to $[0,1]$ for each j, from (9) for sufficiently large n we have

$$\|y - x_{k(n+1)}\| \leqslant 1 - 2\lambda\delta(\alpha/2M)^n M.$$

So $\{x_{k(n)}\}$ converges to y, but then from (2) we get the convergence of $\{Tx_{k(n)}\}$ to y. Therefore $\{\|x_{k(n)} - Tx_{k(n)}\|\}$ converges to zero, a contradiction to the choice of α. If $\lambda \geqslant 1/2$, then $1 - \lambda \leqslant 1/2$, we can apply the same kind of argument as above by replacing (5) as

$$\begin{aligned}\|y - x_{n+1}\| &= \|(1 - \lambda)(y - x_n + y - Tx_n) + (2\lambda - 1)(y - Tx_n)\| \\ &\leqslant (1 - \lambda)\|y - x_n + y - Tx_n\| + (2\lambda - 1)\|y - Tx_n\| \\ &= (1 - \lambda)\|y - x_n\|\,\|(a + b)/2\| + (2\lambda - 1)\|y - Tx_n\|.\end{aligned}$$

By interchanging the roles of λ and $1 - \lambda$ we can obtain as earlier a contradiction. Thus T_λ is asymptotically regular.

Remark 1.5. A theorem similar to our Theorem 1.1. for nonexpansive mappings was proved by Schaefer [21] and for mappings satisfying condition (β) by the present author [27].

Definition 2.1. Let H be a Hilbert space and C be a closed, convex and nonempty subset of H. A mapping $T: C \to C$ is said to be reasonable wanderer in C if starting at any point x_0 in C, its successive steps $x_n = Tx_0^n$ $(n = 1, 2, 3, \ldots)$ are such that the sum of squares of their lengths is finite, i.e.

$$\sum_{n=0}^{\infty} \|x_{n+1} - x_n\|^2 < \infty.$$

Theorem 2.1. Let H be a Hilbert space and C be a nonempty, closed and convex subset of H. Let $T\colon C \to C$ be a generalized contraction mapping. Suppose F, the fixed points set of T in C is nonempty. Let $T_\lambda = \lambda I + (1-\lambda)T$ for any given λ with $0 < \lambda < 1$ be a mapping of C into itself. Then T_λ is a reasonable wanderer with the same fixed points as T.

Proof. For any x in C, set $x_n = T_\lambda^n x$ and let y be a fixed point and, hence of T_λ. Then

(10) $$x_{n+1} - y = \lambda x_n + (1-\lambda)Tx_n - \lambda y - (1-\lambda)y$$
$$= \lambda(x_n - y) + (1-\lambda)(Tx_n - y).$$

For any constant a, we have

(11) $$a(x_n - Tx_n) = a(x_n - y + y - Tx_n) = a(x_n - y) - a(Tx_n - y).$$

Since T is a generalized contraction for y in F we have

(12) $$\|Tx_n - y\| \leqslant \|x_n - y\|.$$

Indeed (12) follows from (2) replacing x by x_n. Using (10) and (12) we have

(13) $$\|x_{n+} - y\|^2 = \lambda^2\|x_n - y\|^2 + (1-\lambda)^2\|Tx_n - y\|^2$$
$$+ 2\lambda(1-\lambda)(Tx_n - y, x_n - y)$$
$$\leqslant \{\lambda^2 + (1-\lambda)^2\}\|x_n - y\|^2 + 2\lambda(1-\lambda)(Tx_n - y, x_n - y).$$

Using (11) and (12) we obtain

(14) $$a^2\|x_n - Tx_n\|^2 = a^2\|x_n - y\|^2 + a^2\|Tx_n - y\|^2$$
$$- 2a^2(Tx_n - y,\ x_n - y)$$
$$\leqslant 2a^2\|x_n - y\|^2 - 2a^2(Tx_n - y,\ x_n - y).$$

Adding (13) and (14) we obtain

$$(15)\quad \|x_{n+1} - y\|^2 + a^2\|x_n - Tx_n\|^2 \leqslant \{2a^2 + \lambda^2 + (1 - \lambda)^2\}$$
$$\|x_n - y\|^2 + \{2\lambda(1 - \lambda) - a^2\}(Tx_n - y, x_n - y).$$

If we assume that a is such that $a^2 \leqslant \lambda(1 - \lambda)$, then from (12), (15) and using Cauchy-Schwarz inequality we get

$$(16)\quad \|x_{n+1} - y\|^2 + a^2\|x_n - Tx_n\|^2 \leqslant \{2a^2 + \lambda^2 + (1 - \lambda)^2\}$$
$$\|x_n - y\|^2 + \{2\lambda(1 - \lambda) - 2a^2\}\|x_n - y\|^2 = \|x_n - y\|^2.$$

Letting $a^2 = \lambda(1 - \lambda) > 0$ and summing up (16) from $n = 0$ to $n = N$ we get

$$\lambda(1 - \lambda) \sum_{n=0}^{N} \|x_n - Tx_n\|^2 \leqslant \sum_{n=0}^{N} \{\|x_n - y\|^2 - \|x_{n+1} - y\|^2\}$$
$$\leqslant \|x_0 - y\|^2 - \|x_{N+1} - y\|^2 \leqslant \|x_0 - y\|^2.$$

Hence $\sum_{n=0}^{\infty} \|x_n - Tx_n\|^2 < \infty$. Since $x_{n+1} - x_n = (1 - \lambda)(Tx_n - x_n)$, we obtain

$$\lambda(1 - \lambda) \sum_{n=0}^{\infty} \|x_n - Tx_n\|^2 = \lambda(1 - \lambda) \sum_{n=0}^{\infty} \frac{1}{(1 - \lambda)^2} \|x_n - x_{n+1}\|^2$$
$$= \frac{\lambda}{1 - \lambda} \sum_{n=0}^{\infty} \|x_{n+1} - x_n\|^2 \leqslant \|x_0 - y\|^2.$$

Hence

$$\sum_{n=0}^{\infty} \|x_{n+} - x_n\|^2 \leqslant \frac{(1 - \lambda)}{\lambda} \|x - y\|^2 < \infty,$$

that is T_λ is a reasonable wanderer.

Remark 2.1. A similar theorem for nonexpansive mappings was proved by Browder and Petryshyn [2] and for mappings satisfying condition (β) by the present author [27].

Lemma 3.1. Let H be a Hilbert space and C be a nonempty, convex subset of H. Let $T\colon C \to C$ be a generalized contraction mapping. Suppose F, the fixed points set of T in C is non-empty, then F is convex.

Proof. We may assume that F consists of more than one point; otherwise the result is proved. Let x, y be in F. It is enough to show that $z = \lambda x + (1 - \lambda)y$, $0 < \lambda < 1$ belongs to F. Since C is convex z belongs to C. Since T is generalized contraction, from (2) we have

$$\|Tz - x\| \leqslant \|z - x\| \text{ and } \|Tz - y\| \leqslant \|z - y\|.$$

Now $z - x = -(1 - \lambda)(x - y)$. Hence $x - z = (1 - \lambda)(x - y)$. Thus we obtain

$$\|x - y\| \leqslant \|x - Tz\| + \|Tz - y\| \leqslant \|x - z\| + \|z - y\|$$
$$= (1 - \lambda)\|x - y\| + \lambda\|x - y\| = \|x - y\|.$$

Hence

$$\|x - Tz\| + \|Tz - y\| = \|x - Tz + Tz - y\|.$$

If $x - Tz = 0$, then $\|Tz - y\| = \|x - y\| \leqslant \|z - y\| = \lambda\|x - y\|$, whence $1 \leqslant \lambda$, which is not true. Similarly $Tz - y = 0$ implies $1 \leqslant 1 - \lambda$, whence $\lambda \leqslant 0$, which is not true. Since H is strictly convex, there exists $\alpha > 0$ such that $Tz - x = \alpha(y - Tz)$, whence $Tz = (1 - \beta)x + \beta y$, where $\beta = (\alpha/1 + \alpha)$. We have $Tz - x = \beta(y - x)$ and so $\beta\|y - x\| = \|Tz - x\| \leqslant \|z - x\| = (1 - \lambda)\|x - y\|$, which gives $\beta \leqslant 1 - \lambda$. Using $Tz - y = (1 - \beta)(x - y)$, a similar argument gives $\beta \geqslant 1 - \lambda$. Thus $\beta = 1 - \lambda$ and so $Tz = \lambda x + (1 - \lambda)y = z$, i.e. $z \in F$.

Lemma 3.2. [4, Proposition 2.5., pp. 53]. Let X be a Banach space and g a convex continuous real valued function on X. Then g is weakly lower semicontinuous.

Lemma 3.3. [4, Proposition 1.4., pp. 32]. Let X be a topological space and C be a compact subset of it. Let $g\colon X \to R$ be a lower semicontinuous function in X. Then there exists x_0 in C such that $g(x_0) = \inf_{x \in C} g(x)$.

Definition 3.1. Let X be a Banach space. A mapping $T\colon X \to X$ is said to be demiclosed if for any sequence $\{x_n\}$ such that $x_n \rightharpoonup x$ (i.e. x_n converges weakly to x) and $Tx_n \to y$, then $y = Tx$.

Theorem 3.1. Let H be a Hilbert space and T be a generalized contraction asymptotically regular mapping of H into itself. Suppose T is continuous and $I - T$ is demiclosed. Let F, the fixed points set of T in H be nonempty. Then for each x_0 in H, the sequence of iterates $\{T^n x_0\}$ converges weakly to a point of F.

Proof: Since F is nonempty we see that a ball B about some fixed point and containing x_0 is mapped into itself by T; consequently B contains the sequence of iterates $T^n x_0$. So we restrict ourselves to a mapping of ball into itself. It follows from Lemma 3.1. that F is convex. The continuity of T implies that F is closed. Thus F being closed and convex is weakly compact. Let us observe that for any y in F it follows from the definition of T that

$$\|T^n x_0 - y\| \leq \|T^{n-1} x_0 - y\| \leq \ldots \leq \|Tx_0 - y\|.$$

So the sequence $\{\|T^{n-1} x_0 - y\|\}$ is non-increasing.

Define in F the following mapping

$$g\colon\ F \to R^+ \quad (R^+ = \text{nonnegative real numbers})$$

(17) $$g(y) = \inf_n \|T^n x_0 - y\| = \lim_{n\to\infty} \|T^n x_0 - y\|.$$

(In (17), lim = inf, because the sequence $\{\|T^n x_0 - y\|\}$ is non-increasing). The mapping g so defined is continuous. Indeed,

$$g(z) = \lim \|T^n x_0 - z\| \leq \lim \|T^n x_0 - y\| + \|y - z\| = g(y) + \|y - z\|$$

from this inequality it follows that $|g(y) - g(z)| \leq \|y - z\|$, hence g is continuous. Moreover g is convex. In fact

$$g(\alpha y + (1-\alpha)z) = \lim \|T^n x_0 - (\alpha y + (1-\alpha)z)\|$$

$$= \lim \|\alpha T^n x_0 - \alpha y + (1-\alpha)(T^n x_0 - z)\|$$

$$\leqslant \alpha \lim \|T^n x_0 - y\| + (1-\alpha) \lim \|T^n x_0 - z\|$$

$$= \alpha g(y) + (1-\alpha) g(z).$$

So, by Lemma 3.2 we conclude that g is weakly lower semi-continuous. Applying Lemma 3.3 we obtain the existence of a point u in F such that $g(u) = d = \inf_{y \in F} g(y)$.

We claim that u is unique. In fact, suppose this is not so, i.e. there exists another point v in F such that $g(v) = d$. Since g is convex for $0 \leqslant \lambda \leqslant 1$ we have

$$g(\lambda u + (1-\lambda)v) \leqslant \lambda g(u) + (1-\lambda) g(v) = \lambda d + (1-\lambda) d = d.$$

Thus

$$d \geqslant g(\lambda u + (1-\lambda)v) = \inf \|T^n x_0 - (\lambda u + (1-\lambda)v)\|$$

$$= \inf \|\lambda(T^n x_0 - u) + (1-\lambda)(T^n x_0 - u)\|$$

$$\geqslant \lambda \inf \|T^n x_0 - u\| + (1-\lambda) \inf \|T^n x_0 - u\|$$

$$= \lambda g(u) + (1-\lambda) g(v) = \lambda d + (1-\lambda) d = d.$$

Hence, $g(\lambda u + (1-\lambda)v)) = d$. Since $u, v \in F$ and T is a generalized contraction, it follows that $\|T^n x_0 - u\| \leqslant \|T^{n-1} x_0 - u\|$ and $\|T^n x_0 - v\| \leqslant \|T^{n-1} x_0 - v\|$. So, the sequences $\{\|T^n x_0 - u\|\}$ and $\{\|T^n x_0 - v\|\}$ are non-increasing. Therefore $\|x_n - u\| = \|T^n x_0 - u\| \to d$ and $\|x_n - v\| = \|T^n x_0 - v\| \to d$. Thus from uniform convexity of H, we conclude that $\|(x_n - u) - (x_n - v)\| \to 0$, i.e. $u = v$.

It remains to show that the sequence $\{T^n x_0\}$ converges weakly to u. Suppose not, then by the reflexivity of H and the boundedness of the sequence $\{T^n x_0\}$, there exists a convergent subsequence $\{T^{n(j)} x_0\}$ of $\{T^n x_0\}$ whose limit say z is different from u. Since T is asymptotically regular, it follows that the sequence $\{(I - T)(T^{n(j)}x_0)\}$ tends to zero as $n \to \infty$. Since by hypothesis $I - T$ is demiclosed, $(I - T)z = 0$, i.e. z is a fixed point of T. We claim that $z = u$. Indeed, we have

$$\|T^{n(j)}x_0 - u\|^2 = \|T^{n(j)}x_0 - z + z - u\|^2$$
$$= \|T^{n(j)}x_0 - z\|^2 + \|z - u\|^2 + 2\mathrm{Re}\,(T^{n(j)}x_0 - z,\ z - u).$$

Taking limits we obtain

$$g(u) = g(z) + \|z - u\|$$

which is possible only if $z = u$. Thus the theorem.

Theorem 3.2. Let X be a reflexive Banach space and T be an asymptotically regular generalized contraction mapping from X into itself. Suppose T is continuous and $I - T$ is demiclosed. Let F, the fixed points set of T in X be nonempty. Then, for each x_0 in X, every subsequence of $\{T^n x_0\}$ contains a further subsequence which converges weakly to a fixed point of T. In particular, if F consists of precisely one point then the whole sequence $\{T^n x_0\}$ converges to this point.

Proof: Let y in F. Since T is a generalized contraction, it follows that $\|T^n x_0 - y\| \leqslant \|x_0 - y\|$ for all n. So the sequence $\{T^n x_0\}$ bounded. Thus it follows from the reflexivity of X that every subsequence $\{T^{n(j)}x_0\}$ of $\{T^n x_0\}$ contains a further subsequence, which we again denote by $\{T^{n(j)}x_0\}$ such that $T^{n(j)}x_0 \to z$. Now we show that z is a fixed point of T. Indeed, since $T^{n(j)}x_0 \to z$, it follows that $(I - T)T^{n(j)}x_0 \to (I - T)z$. On the other hand since T is asymptotically regular it follows that $(I - T)T^{n(j)}x_0 = T^{n(j)}x_0 - T^{n(j)+1}x_0 \to 0$ as $n \to \infty$. Thus $(I - T)z = 0$, i.e. z is a fixed point of T. If F consists of only one point y then the whole sequence must converge to y.

Remark 3.1. A theorem similar to our Theorem 3.1. for nonexpansive mappings was proved by Opial [13], and a theorem similar to our Theorem 3.2 was obtained by Browder and Petryshyn [1].

In the sequel we will prove some theorems for the strong convergence of sequence of iterates for the generalized contraction mappings.

Theorem 4.1. Let X be a Banach space and T be a continuous generalized contractive asymptotically regular mapping of X into itself. Suppose F, the fixed points set of T in X is non-empty. Let us further assume that T satisfies the following condition:

(A) $I - T$ maps bounded closed sets into closed sets.

Then, for any x_0 in X, the sequence $\{T^n x_0\}$ converges strongly to some point in F.

Proof: Let y be a fixed point of T. Since T is a generalized contraction, it follows that $\|T^{n+1} x_0 - y\| \leqslant \|T^n x_0 - y\|$, $n = 1, 2, 3, \ldots$. So the sequence $\{T^n x_0\}$ is bounded. Let D be the strong closure of $\{T^n x_0\}$. By condition (A) it follows that $(I - T)(D)$ is closed. This together with the fact that T is asymptotically regular implies that $0 \in (I - T)(D)$. So there exists a $z \in D$ such that $(I - T)z = 0$. But this implies that either $z = T^n x_0$ for some n, or there exists a subsequence $\{T^{n(j)} x_0\}$ converging to z. Since z is a fixed point of T, we can then conclude that in either case the whole sequence $\{T^n x_0\}$ converges to z.

Corollary 4.1. Let X be a uniformly convex Banach space and T be a continuous generalized contraction mapping of X into itself. Suppose F, the fixed points set of T in X is non-empty and T satisfies the condition (A). Then, for each point x_0 in X, the sequence $\{x_n\}$ defined by

$$x_{n+1} = \lambda x_n + (1 - \lambda) T x_n, \quad 0 < \lambda < 1$$

converges strongly to a fixed point of T.

Proof: Let λ be such that $0 < \lambda < 1$. Let $T_\lambda = \lambda I + (1 - \lambda)T$.

It follows from Theorem 1.1 that T_λ is asymptotically regular. T satisfies condition (A) if and only if T_λ also does. Indeed, we just observe that $I - T_\lambda = (1 - \lambda)(I - T)$. Let us observe that T_λ is not a generalized contraction, however, for any $y \in F$ it follows from (3) that $\|T_\lambda^n x - y\| \leqslant \|x - y\|$. From this we conclude that the sequence $\{T_\lambda^n x_0\}$ is bounded, hence the corollary follows from Theorem 4.1.

Definition 4.1. A continuous mapping T from a Banacn space X into itself is said to be demicompact if for every bounded sequence $\{x_n\}$ such that $\{(I - T)(x_n)\}$ converges strongly, contains a strongly convergent subsequence $\{x_{n(j)}\}$.

Remark 4.1. It follows from Proposition II.4 [4, pp. 47] that a demicompact mapping T of a Banach space into itself satisfies condition (A). Thus we have the following corollary.

Corollary 4.2. Let X be a uniformly convex Banach space. Let T be a generalized contractive demicompact mapping of X into itself. Suppose F, the fixed points set of T in X is non-empty. Then, for each point x_0 in X, the sequence $\{x_n\}$ defined by

$$x_{n+1} = \lambda x_n + (1 - \lambda)Tx_n, \quad 0 < \lambda < 1$$

converges strongly to a fixed point of T.

Remark 4.2. Theorem 4.1, Corollary 4.1 and Corollary 4.2 for non-expansive mappings were proved by Browder and Petryshyn [1] and for mappings satisfying condition (β) by the present author [27].

Definition 4.2. A Banach space X is said to be strictly convex (rotund) if for any pair of vectors x, y in X from $\|x + y\| = \|x\| + \|y\|$ it follows that $x = \lambda y$, $\lambda > 0$ (or, in trivial case, $y = 0$).

Remark 4.3. Every uniformly convex Banach space is strictly convex [14, Proposition 1.1, pp. 5] but the converse is not true, a counter example may be found in Wilansky [30, pp. 111, problem 9].

Theorem 4.2. Let X be a strictly convex Banach space, and D be a compact convex subset of X. Let $T\colon D \to D$ be a continuous generalized contraction. Then F, the fixed points set of T is nonempty and compact, moreover for any x_0 in D and any λ such that $0 < \lambda < 1$, $\{T_\lambda^n x\}$ converges to a fixed point of T, where $T_\lambda x = (1 - \lambda)x + \lambda Tx$, x in X.

Proof: By the continuity of T and the Schauder-Tychonoff theorem, it follows that F, the fixed points set of T is compact and nonempty. Let $n \geqslant 0$, $x_n = T_\lambda^n x_0$. Since D is compact, $\{x_n\}$ has a convergent subsequence $\{x_{k(n)}\}$ which converges to some point in D. We need to show that x is a fixed point of T. From (2) it follows that $\{\|x_n - y\|\}$, where y is a fixed point of T is monotonically non-increasing. So by the continuity of norm $\|\ \|$ and T_λ we have

$$\text{(18)}\qquad \begin{aligned}\|x - y\| &= \lim_{n\to\infty} \|x_{k(n+1)} - y\| \leqslant \lim_{n\to\infty} \|x_{k(n)+1} - y\| \\ &= \lim_{n\to\infty} \|T(x_{k(n)}) - y\| = \|T_\lambda x - y\|.\end{aligned}$$

By (18) and (3) we obtain

$$\text{(19)}\qquad \|T_\lambda x - y\| = \|x - y\|.$$

Moreover,

$$\text{(20)}\qquad \begin{aligned}\|T_\lambda x - y\| &= \|(1 - \lambda)(x - y) + \lambda(Tx - y)\| \\ &\leqslant (1 - \lambda)\|x - y\| + \lambda\|Tx - y\| \\ &\leqslant (1 - \lambda)\|x - y\| + \lambda\|x - y\| = \|x - y\|.\end{aligned}$$

Combining (19) and (20) we conclude that all inequalities in (20) are equalities. So

$$\text{(21)}\qquad \|(1 - \lambda)(x - y) + \lambda(Tx - y)\| = (1 - \lambda)\|x - y\| + \lambda\|Tx - y\|$$

and

$$\text{(22)}\qquad \|Tx - y\| = \|x - y\|.$$

By (21) and strict convexity of X, either $x = y$, or $Tx - y = t(x - y)$ for some $t > 0$. From (22) it follows that $t = 1$. Thus $Tx - y = x - y$ or $x = Tx$. Hence x is a fixed point T. It follows from (20) that the sequence $\{x_n\}$ converges to x.

Remark 4.4. The above result was proved by Edelstein [6] for nonexpansive mappings and by the present author [27] for mappings satisfying condition (β).

Following Reinermann [17] we define a summability matrix A by

$$\text{(I)} \qquad \begin{aligned} a_{n_k} &= c_k \prod_{j=k+1}^{n} (1 - cj) \qquad k < n \\ &= c_n \qquad k = n, \\ &= 0 \qquad k > n, \end{aligned}$$

where the real sequence $\{c_n\}$ satisfies (i) $c_0 = 1$, (ii) $0 < c_n < 1$ for $n \geqslant 1$ and (iii) $\sum_k (1 - c_k) c_k$ diverges. Let X be a normed linear space, C a nonempty, closed, bounded and convex subset of X. Let $T\colon C \to C$ be a mapping for any given x_0 in C we define the iteration scheme as follows: $\overline{x_0} = x_0$ and $x_{n+1} = \sum_{k=0}^{n} a_{n_k} T(x_k)$, which can be written as

$$\text{(II)} \qquad x_{n+1} = (1 - c_n) x_n + c_n T(x_n).$$

We denote the scheme (II), where the real sequence $\{c_n\}$ satisfies (i), (ii), and (iii) by $M(x_0, A, T)$.

Lemma 5.1. [9]. Let X be a uniformly convex Banach space. Suppose x, y in X and $\|x\| \leqslant 1$, $\|y\| \leqslant 1$. Then for $0 \leqslant \lambda \leqslant 1$ we have $\|\lambda x + (1 - \lambda) y\| \leqslant 1 - 2\lambda(1 - \lambda)\delta(\varepsilon)$.

Theorem 5.1. Let X be a uniformly convex Banach space. Let D be a nonempty, convex subset of X. Let $T\colon D \to D$ be a generalized contraction mapping with at least one fixed point. Then the sequence $\{(I - T)x_n\}$ converges strongly to zero for each x_0 in D.

Proof: Let p be a fixed point of T. For any x in D

$$(23)\qquad \|x_{n+1} - p\| = \|(1 - c_n)x_n + c_n Tx_n - p\|$$
$$= \|(1 - c_n)(x_n - p) + c_n(Tx_n - p)\|$$
$$\leqslant (1 - c_n)\|x_n - p\| + c_n\|Tx_n - p\|.$$

Since T is a generalized contraction, for p in F from (2) we have (taking $x_n = x$) $\|Tx_n - p\| \leqslant \|x_n - p\|$. Thus we can write (23) as

$$(24)\quad \|x_{n+1} - p\| \leqslant (1 - c_n)\|x_n - p\| + c_n\|x_n - p\| = \|x_n - p\|.$$

Thus the sequence $\{\|x_n - p\|\}$ is non-increasing for all n. Also $\|x_n - Tx_n\| \leqslant \|x_n - p\| + \|p - Tx_n\| \leqslant 2\|x_n - p\|$.

Suppose $\{\|x_n - Tx_n\|\}$ does not converge to zero. Then there exists a $\alpha > 0$ such that $\|x_n - Tx_n\| \geqslant \alpha$ for all n. Now using Lemma 5.1 and (24) we have

$$(25)\qquad \|x_{n+1} - p\| \leqslant \|x_n - p\| - \|x_n - p\|c_n(1 - c_n)b,$$

where $b = 2\delta(\varepsilon/\|x_0 - p\|)$.

Also

$$(26)\quad \|x_n - p\| \leqslant \|x_{n-1} - p\| - \|x_{n-1} - p\|(1 - c_{n-1})c_{n-1}.b.$$

Substituting the values from (26) into (25) we obtain

$$(27)\quad \|x_{n+1} - p\| \leqslant \|x_{n-1} - p\| - \|x_{n-1} - p\|c_{n-1}(1 - c_{n-1})b$$
$$- \|x_n - p\|c_n(1 - c_n)b.$$

Now $\|x_n - p\| \leqslant \|x_{n-1} - p\|$ implies $-\|x_n - p\| \geqslant -\|x_{n-1} - p\|$.

Thus we can write (27) as

$$\|x_{n+1} - p\| \leqslant \|x_{n-1} - p\| - \|x_n - p\|c_{n-1}(1 - c_{n-1})b - \|x_n - p\|c_n(1 - c_n)b$$
$$= \|x_{n-1} - p\| - \|x_n - p\| \{c_n(1 - c_n)b + c_{n-1}(1 - c_{n-1})b\}.$$

By induction we have

$$\alpha \leqslant \|x_{n+1} - p\| \leqslant \|x_0 - p\| - \|x_n - p\| b \sum_{k=1}^{n} c_k(1 - c_k).$$

Therefore;

$$\alpha + \|x_n - p\| \; b \sum_{k=1}^{n} c_k(1 - c_k) \leqslant \|x_0 - p\|.$$

But by assumption, $\|x_n - p\| \geqslant \alpha$, hence

$$\alpha + \alpha b \sum_{k=1}^{n} c_k(1 - c_k) \leqslant \|x_0 - p\|$$

or

$$\alpha(1 + b \sum_{k=1}^{n} c_k(1 - c_k) \leqslant \|x_0 - p\|,$$

a contradiction, since the series on the left diverges.

Corollary 5.1. Let X be a uniformly convex Banach space. Let C be a closed and convex subset of X. Let $T\colon\ C \to C$ be a generalized contraction mapping of C into itself with at least one fixed point. Suppose T satisfies condition (A). Then $M(x_0,A,T)$ converges strongly to a fixed point of T for each x_0 in C.

Proof: Let D be the strong closure of $\{x_n\}$. Since $(I - T)(D)$ is closed, we conclude from Theorem 5.1 that zero belongs to $(I - T)(D)$. Thus there exists a subsequence $\{x_{n_k}\}$ of $\{x_n\}$ which converges strongly to y, where $(I - T)y = 0$. Therefore by (24) it follows that $\{x_n\}$ converges to y.

Corollary 5.2. Let C be a uniformly convex Banach space. Let C be a closed and convex subset of X. Let $T\colon\ C \to C$ be a demicompact generalized contraction with at least one fixed point. Then $M(x_0,A,T)$ converges strongly to a fixed point of T for each x_0 in C.

Definition 5.1. Let X be a real Banach space. Let D be a nonempty, bounded subset of X. A continuous mapping $T\colon\ D \to D$ is called densifying if for every bounded subset C of D with $\gamma(C) > 0$ we have $\gamma(T(C)) < \gamma(C)$.

Remark 5.1. It follows from Lemma 1 [12, pp. 80] that a densifying mapping satisfies condition (A). Thus we have the following corollary.

Corollary 5.3. Let X be a real Banach space. Let D be a nonempty, bounded, closed and convex subset of X. Let $T: D \to D$ be densifying and generalized contraction mapping. Then $M(x_0,A,T)$ converges strongly to a fixed point of T for all x_0 in D.

Proof: It follows from a Theorem [8] that F, the fixed points set of T in D is nonempty. Thus using Remark 5.1 we see that Corollary 5.3 follows from Corollary 5.1.

Definition 5.2. Let X,Y be two Banach spaces. Let C be bounded subset of X. A continuous mapping $T: C \to Y$ is said to be compact if $T(C)$ is relatively compact, i.e. the closure of $T(C)$ is compact.

Corollary 5.4. Let X be a uniformly convex Banach space. Let C be a closed, bounded and convex subset of X. Let $T: C \to C$ be a compact generalized contraction mapping. Then $M(x_0,A,T)$ converges strongly to a fixed point of T for each x_0 in C.

Proof: Since T is compact mapping, it follows from Schauder Tychonoff theorem that F, the fixed points set of T in C is nonempty. Moreover T being compact is demicompact [14, Proposition 2, pp. 39]. Thus Corollary 5.4 follows from Corollary 5.2.

Definition. Let C be a subset of a Banach space X, $T: C \to X$ be a continuous mapping. We shall say that T is LANE (locally almost nonexpansive) (locally in the weak topology) if and only if given x in C and $\varepsilon > 0$, there exists a weak neighborhood N_x of x in C such that for u,v in N_x,

$$\|Tu - Tv\| \leqslant \|u - v\| + \varepsilon$$

Corollary 5.5. Let X be a uniformly convex Banach space, C be a closed, bounded and convex subset of X. Let $T: C \to C$ be a generalized contraction LANE mapping. Then $M(x_0,A,T)$ converges strongly to a fixed point T for each x_0 in C.

Proof: It follows from [12, Lemma 3, pp. 139] that $I - T$ satisfies condition (A). Moreover we conclude from the Theorem [12, pp. 141] that F, the fixed points set of T in C is non-empty. Thus Corollary 5.5 follows from Corollary 5.1.

Definition. Let X be a Banach space, X^* be its dual space. A mapping $T\colon X \to X^*$ is said to be duality mapping if for any x in X

$$\|T(x)\| = \|x\|$$

$$(Tx,x) = \|Tx\|\,\|x\| - \|x\|^2$$

Definition. Let X, Y be two Banach spaces. A mapping $T\colon X \to Y$ is called weakly continuous if $x_n \rightharpoonup x$ in X implies $Tx_n \rightharpoonup Tx$ in Y for x_n and x in X; where $\rightharpoonup$ denotes the weak convergence.

Corollary 5.6. If X has a weakly continuous duality mapping (in particular if X is a Hilbert space) and if T is a generalized contraction mapping of a closed and convex subset C of X into itself with at least one fixed point. Then $M(x_0,A,T)$ converges weakly to a fixed point of T for each x_0 in C.

II. REFERENCES

[1] Browder, F. E. and Petryshyn, W. V., *The solution by iteration of nonlinear functional equations in Banach spaces*, Bull. Amer. Math. Soc. 72 (1966, pp. 571-575.

[2] Browder, F. E. and Petryshyn, W. V., *Construction of fixed points of nonlinear mappings in Hilbert space*, Jour. Math. Anal. and Appl. 20 (1967), pp. 197-228.

[3] Ćirić, Lj. B., *Fixed point theorems for mappings with a generalized contractive iterates at a point*, Publ. Inst. Math. (Beograd), 13 (27), (1972), pp. 11-16.

[4] deFigueiredo, D. G., *Topics in nonlinear functional analysis*, Lecture series No. 48, University of Maryland (1967).

[5] Doston, W. G. Jr., *On the Mann iterstes process*, Trans. Amer. Math. Soc. 149 (1970), pp. 65-73.

[6] Edelstein, M., *A remark on a theorem of M. A. Krasnoselskii*, Amer. Math. Monthly, 73 (1966), pp. 509-510.

[7] Eisenfeld, J. and Lakshmikantham, V., *On measure of nonconvexity and applications*, Yokohama Math. Jour. Vol. XXIV, Nos. 1-2, (1976), pp. 133-140.

[8] Furi, M. and Vignoli, A., *On* α*-nonexpansive mappings and fixed points*, Acad. Naz. Dei Lincei, 48 (1970), pp. 195-198.

[9] Groetsch, C. W., *A note on segmenting Mann iterates*, Jour. Math. Anal. and Appl. 40 (1972), pp. 369-372.

[10] Kannan, R., *Fixed point theorems in reflexive Banach spaces*, Proc. Amer. Math. Soc. 38 (1973), pp. 111-118.

[11] Kannan, R., *Construction of fixed points of a class of nonlinear mappings*, Jour. Math. Anal. and Appl. 41 (1973), pp. 430-438.

[12] Nussbaum, R. D., *k-set contractions and degree theory*, Ph.D. Dissertation, University of Chicago (1969).

[13] Opial, Z., *Weak convergence of the sequence of successive approximations for nonexpansive mappings*, Bull. Amer. Math. Soc. 73 (1967), pp. 591-597.

[14] Opial, Z., *Nonexpansive and monotone mappings in Banach spaces*, Lecture series No. 1, January 1967, Brown University.

[15] Petryshyn, W. V., *Construction of fixed points of demicompact mappings in Hilbert spaces*, Jour. Math. Anal. and Appl. 14 (1966), pp. 276-284.

[16] Petryshyn, W. V. and Williamson, T. E., Jr., *Strong and weak convergence of the sequence of successive approximations for quasi-nonexpansive mappings*, Jour. Math. Anal. and Appl. 43 (1973), pp. 459-497.

[17] Reinermann, J., *über Toeplitzsche iteration verfahren und einige ihre anwendungen in der konstruktiven fixpunktheorie*, Studia, Math. 32 (1969), pp. 209-227.

[18] Rhoades, B. E., *Fixed point iteration using infinite matrices*, Trans. Amer. Math. Soc. 196 (1974), pp. 161-175.

[19] Reich, S., *Some remarks concerning contraction mappings*, Can. Math. Bull. 14 (1971), pp. 121-124.

[20] Reich, S., *Fixed point of contractive functions*, Bull. Uni. Math. Italiano, 4(5), (1972), pp. 26-42.

[21] Schaefer, H., *Über dei methode sukzessive approximationes*, Jbr. Deutch. Math. Verein, 59 (1957), pp. 131-140.

[22] Singh, K. L., *Contraction mappings and fixed point theorems*, Annales de la Soc. Scientifique de Bruxelles, Tome 83 (1968) pp. 34-44.

[23] Singh, K. L., *Some further extensions of Banach's contraction principle*, Riv. di Mat. Univ. Parman, 2 (10), (1969), pp. 139-155.

[24] Singh, K. L., *On some fixed point theorems I*, Riv. Di. Mat. Univ. Parma, 2 (10), (1969), pp. 13-21.

[25] Singh, K. L., *Fixed point theorems for quasi-nonexpansive mappings*, Acad. Naz. Dei Lincei, Vol. LXI(1976),pp.354-363.

[26] Singh, K. L., *Convergence of sequence of iterates of generalized contractions*, Acad. Naz. Dei Lincei Vol. LXII, No. 2 (to appear).

[27] Singh, K. L., *Sequence of iterates of generalized contraction*, Fundamenta Mathematicae (Accepted).

[28] Singh, K. L., *Fixed and common fixed points for generalized contractions*, Bulletin De L'Acaademie Polonaise Des Sciences Vol. XXV, No. 5 (1977), pp. 41-47.

[29] Singh, K. L., *Fixed and common fixed points in convex metric spaces*, Annales Polonici Mathematici (Submitted).

[30] Wilansky, A., *Functional Analysis*, Blaisdell Publishing Co., New York (1964).

[31] Yadav, R. K., *Fixed point theorems in generalized metric spaces*, Banaras Math. Jour. (1969).

CRITERIA FOR THE EXISTENCE AND COMPARISON OF SOLUTIONS TO NONLINEAR VOLTERRA INTEGRAL EQUATIONS IN BANACH SPACE

R. L. Vaughn
University of Texas at Arlington

I. INTRODUCTION

In this paper criteria for the existence of solutions to nonlinear Volterra integral equations in a Banach space E are given. The equation under consideration has the form

$$x(t) = x_0(t) + \int_{t_0}^{t} K(t,s,x(s))ds, \tag{1.1}$$

where $x_0 \in C[J,\Omega]$, $K \in C[J \times J \times \Omega, \Omega]$, Ω is an open subset of E, and $J = [t_0, t_0 + \alpha] \subset R$. The existence criteria include compactness type conditions, which are in terms of the Kuratowski measure of noncompactness, α. For a bounded subset A of E, $\alpha(A)$ is defined by

$\alpha(A) = \inf \{\varepsilon > 0 \mid A$ can be covered by a finite number of sets each with diameter $\leqslant \varepsilon\}$.

Maximal solutions to (1.1) are also considered. This is accomplished by inducing a partial ordering on the Banach space with respect to a cone $H \subseteq E$ with nonvoid interior, H^0. The orderings of elements $u, v \in E$ are given by the relations

$$u \leqslant v \quad \text{if} \quad v - u \in H,$$

$$u < v \quad \text{if} \quad v - u \in H^0$$

In this setting we say a map $\phi\colon E \to E$ is monotone if $u \leqslant v$ implies $\phi(u) \leqslant \phi(v)$.

ISBN 0-12-434160-8

II. PRELIMINARIES

It is assumed that the reader is familiar with the properties of the Kuratowski measure of noncompactness, α. A list of these properties can be found in Martin [1].

The following extension of the Darbo fixed point theorem will be used in the remainder of the paper; however, in addition to the hypothesis of the following theorem, we shall require that the comparison function ρ is monotone when using the theorem in proving existence of solutions to (1.1).

Theorem 2.1. Let E be a Banach space, and let A be a nonvoid closed bounded convex subset of E. If $T \in C[A,A]$ with $\alpha(TB) \leqslant \rho(\alpha(B))$ for every $B \subseteq A$, where $\rho\colon [0,\infty) \to [0,\infty)$ is a right continuous function with $\rho(r) < r, \forall r > 0$, then T has a fixed point in A.

Proof: Let $S = \{x \in A \mid Tx = x\}$. It suffices to show that $S \neq \phi$. Let $A_1 = A$, $A_2 = \overline{\text{co}}\, TA$, and define recursively $A_n = \overline{\text{co}}\, TA_{n-1}$, for $n \geqslant 2$. Clearly each $A_n \neq \phi$. And $A_2 = \overline{\text{co}}\, TA \subset \overline{\text{co}}\, A_1 = A_1$, inductively $A_n = \overline{\text{co}}\, TA_{n-1} \subseteq A_{n-1}$. Set $C_n = \alpha(A_n)$ $n = 1, \ldots,$ then

$$C_2 = \alpha(A_2) = \alpha(TA_1) \leqslant \rho(\alpha(A_1)) = \rho(C_1) < C_1 = \alpha(A_1)$$

and inductively

$$C_n = \alpha(A_n) = \alpha(TA_{n-1}) \leqslant \rho(\alpha(A_{n-1})) = \rho(C_{n-1}) < C_{n-1} = \alpha(A_{n-1}).$$

Thus $C_n \downarrow \delta$, by the right continuity of ϕ we have that

$$\delta = \lim_{n\to\infty} C_n \leqslant \lim_{n\to\infty} \rho(C_{n-1}) = \rho(\delta) < \delta, \quad \text{and thus} \quad \delta = 0.$$

Consequently $\alpha(\bigcap_{n=1}^{\infty} A_n) = 0$, and thus $\bigcap_{n=1}^{\infty} A_n$ is a closed bounded nonvoid compact convex subset of E. Further $T\colon \bigcap_{n=1}^{\infty} A_n \to \bigcap_{n=1}^{\infty} A_n$, by Tychonoff fixed point theorem $\exists x \in \bigcap_{n\subset 1}^{\infty} A_n \subset A$ so that $Tx = x$. Thus $S \neq \phi$.

III. EXISTENCE CRITERIA

Using the previous fixed point theorem we now state and

prove an existence theorem for solution of (1.1).

Theorem 3.1. Let $x_0 \in C[J,\Omega]$ and $K \in C[J \times J \times \Omega, \Omega]$ and suppose the following conditions are satisfied:

(A$_1$) $\|K(t,s,x)\| \leqslant M$ for every $(t,s,x) \in J \times J \times \Omega$

(A$_2$) $\lim_{t\to\tau}(\sup\{\int_I \|K(t,s,\psi(s)) - K(\tau,s,\psi(s))\| ds \mid \psi \in C[I,B]\}) = 0$ for each bounded $B \subseteq \Omega$, and for every $I \subseteq J$, $\forall t \in J$.

(A$_3$) $\alpha(K(J \times J \times B)) \leqslant \rho(\alpha(B))$ where $\rho: [0,\infty) \to [0,\infty)$ is a right continuous monotone nondecreasing function with $\rho(r) < r, \forall\, r > 0$.

Then there exists a solution to (1.1) on $[t_0, t_0 + \gamma]$ for some $\gamma > 0$.

Proof: Let $\eta = \sup\{\varepsilon > 0 \mid B_\varepsilon(x_0(t_0)) \subseteq \Omega\}$ where $B_\varepsilon(x) = \{y \mid \|y-x\| < \varepsilon\}$ Set $\gamma = \min(a, \delta, \eta/2M, 1)$ where $\delta > 0$ is so that $|t - t_0| < \delta$ implies $\|x_0(t) - x_0(t_0)\| < \eta/2$. Let $J_0 = [t_0, t_0 + \gamma]$. Define $A \subseteq C[J_0,\Omega]$ by

$$A = \{\psi \in C[J_0,\Omega] \mid \sup_{t\in J_0} \|\psi(t) - x_0(t)\| \leqslant \eta/2\},$$

clearly A is a closed bounded convex subset of the Banach space $C[J\ ,E]$, and $x \in A$. Define $T: C[J_0,\Omega] \to C[J_0,\Omega]$ by

$$(T\phi)(t) = x_0(t) + \int_{t_0}^{t} K(t,s,\phi(s))ds.$$

Note that,

$$\|(T\phi)(t) - x_0(t)\| \leqslant \int_{t_0}^{t} \|K(t,s,\phi(s))\| ds \leqslant (t - t_0)M \leqslant \eta/2,$$

and thus $TA \subset A$ and TA is bounded

If $\{x_n\} \subseteq A$ with $\{x_n\} \to x \in A$ we have that $\|K(t,s,x_n(s))\| \leqslant M$ and $K(t,s,x_n(s)) \to K(t,s,x(s))$, thus an easy application of the Bounded Convergence theorem shows T is continuous. The equicontinuity of TA is established by using the uniform continuity of x_0 on J_0 and assumptions (A$_1$) and (A$_2$).

We now show T satisfies the comparison hypothesis of Theorem 2.1. Let $B \subseteq A$.

$$\alpha((TB)(t)) = \alpha(\{x\ (t) + \int_{t_0}^{t} K(t,s,\psi(s))ds:\ \ \psi \in B\})$$
$$\leqslant \alpha(\{\ \int_{t_0}^{t} K(t,s,\psi(s):\ \ \psi \in B\})$$
$$\leqslant |t - t_0|\ \alpha\ (\overline{\text{co}}K(t,\ [t_0,t],\ B[t_0,t]))$$
$$\leqslant \gamma \cdot \alpha(K(J \times J \times B[J_0])) \leqslant \gamma\rho(\alpha(B(J_0)))$$
$$\leqslant \rho(\alpha(B[J_0])) \leqslant \rho(\alpha(B)).$$

Thus T has a fixed point in A. Clearly such a fixed point is a solution of (1.1).

With $\rho(r) = \beta r$ with $\beta \in (0,1)$, the above existence theorem clearly implies the result contained in [2]. For $\beta \geqslant 1$, theorem 2.1 can be modified by requiring that 0 is the unique fixed point of ρ instead of $\rho(r) < r\ \forall r > 0$.

IV. MAXIMAL SOLUTIONS

Let $H \subseteq E$ be a proper cone with $H^0 \neq \phi$. We shall say a function ϕ: $E \rightarrow E$ is monotone nondecreasing if $u \leqslant v$ implies that $\phi(u) \leqslant \phi(v)$.

The following inequality can be found in [2] and will be used in proving that maximal solutions of (1.1) exist, when $K\ (t,s,u)$ is monotone nondecreasing in u for each $(t,s) \in J \times J$.

Theorem 4.1. Let $g \in C[J \times J \times E,\ E]$ and $x_0,u,v \in C[J,E]$ with $g(t,s,u)$ monotone nondecreasing in u for each $(t,s) \in J \times J$. Then

$$u(t) \leqslant x_0(t) + \int_{t_0}^{t} g(t,s,u(s))ds, \quad \text{and}$$
$$v(t) > x_0(t) + \int_{t_0}^{t} g(t,s,v(s))ds, \quad \text{and}$$
$$u(t_0) < v(t_0)$$

implies $$u(t) \leqslant v(t) \quad \forall\ t \geqslant t_0.$$

We now state the results concerning maximal solutions and the comparison of solutions.

Theorem 4.2. Let K and x_0 be as in Theorem 3.1 and suppose $K(t,s,u)$ is monotone nondecreasing in u for each

$(t,s) \in J \times J$, then $\exists \gamma > 0$ so that the maximal solution to (1.1) exists on $[t_0, t_0 + \gamma]$.

Proof: Let η and δ be as in Theorem 3.1 and set $\gamma = \min(a, \delta, \eta/4, 1)$. Define $A \subseteq C[J_0, \Omega]$ and $T: A \to A$ as in Theorem 3.1. Let $y_0 \in H^0$ with $\|y_0\| \leq \eta/4$, and $y_n = \frac{1}{n} y_0$, $n = 1, \ldots$. Define $T_n: A \to A$ by $T_n\phi = T\phi + y_n$. That $T_n \in C[A,A]$ and that T_nA is equicontinuous is established as in Theorem 3.1.

If $B \subseteq A$, then

$$\alpha(\{T_n\phi: \phi \in B\}) = \alpha(\{T\phi + y_n: \phi \in B\})$$
$$\leq \alpha(\{T\phi: \phi \in B\}) = \alpha(TB) \leq \rho(B),$$

thus each T_n has a fixed point ϕ_n.

Using Theorem 4.1 it is easy to see that

$$\phi_n(t) > \phi_n(t) \ \forall t \in J, \quad \text{if} \quad m > n .$$

It is also easy to establish that $\{\phi_n\}$ is uniformly bounded and equicontinuous. Also

$$\alpha(\{\phi_n(t)\}) = \alpha(\{T\phi_n(t) + y_n\})$$
$$\leq \alpha(\{T\phi_n(t)\}) = \alpha(T\{\phi_n\}(t)).$$

Thus $\alpha(\{\phi_n\}) \leq \alpha(T\{\phi_n\}) \leq \rho(\alpha(\{\phi_n\}))$ and so $\alpha(\{\phi_n\}) = 0$. Thus $\{\phi_n\}$ contains a uniformly convergent subsequence $\{\phi_{nk}\}$ which by the Bounded Convergence Theorem, converges to a fixed point ψ of T.

By applying Theorem 4.1 again we see that if $x(t)$ is any fixed point of T, then $x(t) < \phi_{nk\to\infty}(t)$ $t \in J_0$, and thus $x(t) \leq \lim_{nk\to\infty} \phi_{nk}(t) = \psi(t)$. Thus ψ is the maximal solution of (1.1).

Theorem 4.3. Let the hypothesis of Theorem 4.2 be satisfied. And suppose $m \in C[J, \Omega]$ with $m(t) \leq x_0(t) + \int_{t_0}^{t} K(t,s,m(s))ds$, then $m(t) \leq r(t)$ $\forall t \in [t_0, t_0 + \gamma]$, where $r(t)$ is the maximal solution of (1.1).

The proof of this theorem can be found in [2].

V. REFERENCES

[1] Martin, R. H., *Nonlinear Operators and Differential Equations in Banach Space*, J. Wiley and Sons, 1976.

[2] Vaughn, R. L., *Existence and Comparison Results for Nonlinear Volterra Integral Equations in a Banach Space*, (to appear).

SEMILINEAR BOUNDARY VALUE PROBLEMS IN BANACH SPACE

James R. Ward*
Pan American University

I. INTRODUCTION

In this paper we consider the integral equation

$$\text{(I)} \qquad u(t) = W(t,0)z + \int_0^t W(t,s)F(s,u(s))ds \qquad (t \in J)$$

in a Banach space X together with nonlinear boundary conditions of the form

$$\text{(II)} \qquad Qu = Bu$$

Here $J = [0,T]$ is a real number interval, $\{W(t,s): 0 \leqslant s \leqslant t \leqslant T\}$ is a strongly continuous family of evolution operators on X, and $F: J \times X \to X$ is a continuous function. Q in (II) is a continuous linear operator mapping $C(J,X)$ into X and B is a completely continuous (nonlinear) operator mapping $C(J,X)$ into X.

We offer sufficient conditions for the existence of a solution to the integral equation (I) which also satisfies the boundary condition (II).

If the evolution system $W(t,s)$ has a closed densely defined family of generators $A(t)$ then equation (I) may be seen to be an integrated form of the differential equation

$$\text{(I')} \qquad u'(t) = A(t)u(t) + F(t,u(t)).$$

The approach we take is motivated by the papers of Opial [6],

* This research was supported by a Pan American University Faculty Research Grant.

ISBN 0-12-434160-8

Conti [1], Kartsatos [4], on boundary value problems for ordinary differential equations in R^n, and by the paper of Pazy [7] on initial value problems in Banach space. The present author has considered in [8] equation (I) subject to linear boundary conditions.

Let R denote the set of real numbers and let $J = [0,T]$ be a compact interval of real numbers. Let X be a Banach space with norm $|\cdot|$. By $C = C(J,X)$ we mean the Banach space of continuous X valued functions defined on J, with norm $\|\cdot\|$ defined by

$$\|u\| = \sup_{t \in J} |u(t)|$$

for $u \in C$. The space of bounded linear transformations on X will be denoted by $L(X)$, and norms in $L(X)$ by $|\cdot|$.

We make the following assumptions regarding the linear evolution system $W(t,s)$:

(1.1) $W(t,s) \in L(X)$ whenever $0 \leqslant s \leqslant t \leqslant T$ and for each $x \in X$ the mapping $(t,s) \to W(t,s)x$ is continuous.

(1.2) $W(t,s)W(s,r) = W(t,r)$ whenever $0 \leqslant r \leqslant s \leqslant t \leqslant T$.

(1.3) $W(t,t) = I$, the identity operator on X, for each $t \in J$.

(1.4) $W(t,s)$ is a compact linear operator on X whenever $t - s > 0$ $(0 \leqslant s < t \leqslant T)$.

The reader is referred to [3] for sufficient conditions for (1.1) - (1.3) to hold for evolution systems with generator $A(t)$. If the basic conditions in ([3], p. 108) are satisfied, and if in addition it is true that for each $t \in J$ there is a number λ in the resolvent set of $A(t)$ such that the resolvent $R(\lambda; A(t))$ is compact then the generated evolution system $W(t,s)$ will satisfy (1.4); see Fitzgibbon [2].

The following lemma, proven in [8], will be needed.

Lemma 1.1. Let $W(t,s)$: $0 \leqslant s \leqslant t \leqslant T$ satisfy (1.1) - (1.4); then for each fixed $s \in [0,T)$ the mapping $t \to W(t,s)$

from $(s,T]$ into $L(X)$ is continuous in the uniform operator topology on $L(X)$. Moreover, this continuity is uniform with respect to s in sets bounded away from t; i.e., as long as $t - s \geqslant \beta$ for any fixed $\beta > 0$.

The operator Q in (II) is a continuous linear mapping from $C(J,X)$ into X. Let $\tilde{Q} \in L(X)$ be defined by

(1.5) $\tilde{Q}x = Q[W(\cdot,0)x]$ for each $x \in X$.

Lemma 1.2. Let $W(t,s)$ satisfy (1.1) - (1.3), let $F \in C(J \times X, X)$, and assume $\tilde{Q}$ has a bounded inverse $\tilde{Q}^{-1}$. Then a function $u \in C(J,X)$ solves (I), (II) if and only if u is a solution to the integral equation

(1.6) $u(t) = W(t,0)\tilde{Q}^{-1}\ (Bu - Q\psi(\cdot,u)) + \psi(t,u)$ where $\psi(t,u)$ is defined by

(1.7) $\psi(t,u) = \int_0^t W(t,s)F(s,u(s))ds$ for $t \in J$ and $u \in C$.

The proof of Lemma 2.2 is immediate and will be omitted.

II. MAIN RESULTS

We will assume $F: (0,T) \times X \to X$ is continuous and satisfies

(F1) For each natural number K there is a function $g_k \in L^1(J)$ such that for a.a. $t \in J$

$$\sup_{|x| \leqslant k} |F(t,x)| \leqslant g_k(t)$$

and

$$\lim_{k\to\infty} \inf \frac{1}{k}\int_0^T g_k(s)ds = 0.$$

Let $B: C(J,X) \to X$ be completely continuous; we will use the condition

(B1) $$\lim_{\|u\|\to\infty} |Bu| / \|u\| = 0.$$

Theorem 2.1. Let $W(t,s)$ be an evolution system satisfying (1.1) - (1.4). Assume (F1) and (B1) hold, and that $\tilde{Q} \in L(X)$ has a bounded inverse. Then there is a strongly continuous function

$u: J \to X$ which satisfies (I) and (II).

Proof: We define an operator S from $C(J,X)$ into itself by

$$Su(t) = W(t,0)\tilde{Q}^{-1}[Bu - \psi(\cdot,u)] + \psi(t,u)$$

for each $u \in C = C(J,X)$ and $t \in J$. Here ψ is defined by

$$\psi(t,u) = \int_0^t W(t,s)F(s,u(s))ds.$$

It will be shown that S has a fixed point. It can be easily verified that S maps C into itself continuously.

For each natural number n let $B_n = \{u \in C: \|u\| \leqslant n\}$. Then for some natural number N, $SB_N \subseteq B_N$. If this were not the case, then for each natural number n there would exist a function $u_n \in B_n$ with $Su_n \notin B_n$, i.e., with $\|Su_n\| > n$. Hence

$$1 < \frac{1}{n} \|Su_n\|.$$

But (F1) and (B1) imply

$$\liminf_{n\to\infty} \frac{1}{n}\|Su_n\| = 0.$$

This establishes the existence of a natural number N with $SB_N \subseteq B_N$.

The operator S maps B_N into a compact subset of B_N as the following argument will show.

Let t, $0 < t \leqslant T$ be fixed and ε a real number satisfying $0 < \varepsilon < t$. Define $\psi_\varepsilon(t,\cdot)$ by

$$\psi_\varepsilon(t,u) = \int_0^{t-\varepsilon} W(t,s)F(s,u(s))ds$$

for all $u \in B_N$. By (1.2)

$$\psi_\varepsilon(t,u) = W(t,t-\varepsilon)\int_0^{t-\varepsilon} W(t-\varepsilon,s)F(s,u(s))ds.$$

By hypothesis $W(t,t-\varepsilon)$ is a compact operator and hence the set

$$K_\varepsilon(t) = \{\psi_\varepsilon(t,u): u \in B_N\}$$

is precompact in X. Also,

$$|\psi(t,u) - \psi_\varepsilon(t,u)| \leq \int_{t-\varepsilon}^{t} |W(t,s)F(s,u(s))|ds$$
$$\leq \int_{t-\varepsilon}^{t} |W(t,s)|g_N(s)ds \leq M \int_{t-\varepsilon}^{t} g_N(s)ds$$

where $M = \sup\{|W(t,s)| : 0 \leq s \leq t \leq T\}$ which is finite by the principle of uniform boundedness. Thus there are precompact sets arbitrarily close to the set

$$K_0(t) = \{\psi(t,u) : u \in B_N\}$$

and therefore $K_0(t)$ is precompact.

We next show that ψ maps the functions in B_N into an equicontinuous family of functions. Let $u \in B_N$ and $t, \tau \in J$. Let ε be a positive real number. Then if $0 < \varepsilon < t < \tau$

$$|\psi(t,u) - \psi(\tau,u)| \leq \left|\int_t^\tau W(\tau,s)F(s,u(s))ds\right| + \left|\int_0^{t-\varepsilon} [W(t,s) - W(\tau,s)]F(s,u(s))ds\right| + \left|\int_{t-\varepsilon}^{t} [W(t,s) - W(\tau,s)]F(s,u(s))ds\right|$$
$$= I_1 + I_2 + I_3.$$

We have for estimates

$$I_1 \leq M \int_t^\tau g_N(s)ds$$

$$I_2 \leq \int_0^{t-\varepsilon} |W(t,s) - W(\tau,s)|g_N(s)ds$$

$$I_3 \leq \int_{t-\varepsilon}^{t} 2Mg_N(s)ds$$

The bounds on I_1 and I_3 may be made small by choosing t close to τ and ε sufficiently small. By Lemma 1.1 $W(t,s)$ is continuous in the operator norm, uniformly for $t - s \geq \varepsilon$ and thus $I_2 \to 0$ as $t \to \tau$, or as $\tau \to t$. It is true, then, that ψ maps B_N into an equicontinuous family of functions uniformly bounded on J. By the Arzela-Ascoli theorem ψB_N is precompact in $C(J,X)$. Hence ψ is a completely continuous operator on $C(J,X)$.

Define an operator Φ by $\Phi = S - \psi$, so that

$\Phi u(t) = W(t,0)\tilde{Q}^{-1}\,[Bu - Q\psi(\cdot,u)]$ for $u \in B_N$ and $t \in J$.

The set ΦB_N is clearly bounded in $C(J,X)$. The complete continuity of ψ and B and the continuity of Q, $\tilde{Q}^{-1}$ and $W(t,0)$ $(t \in J)$ insure the continuity of Φ and that the set

$$Z(t) = \{\Phi u(t): \ u \in B_N\}$$

is precompact in X for each $t \in J$. Indeed, the set

$$\{\, y \in X \mid y = \tilde{Q}^{-1}[Bu - Q\psi(\cdot,u)] \ \text{ for some } \ u \in B_N \,\}$$

is precompact in X. The strong continuity of $W(t,0)$ on J now implies that for $\tau \in J$

$$\lim_{t \to \tau} |\Phi u(t) - \Phi u(\tau)| = 0$$

uniformly for $u \in B_N$. Thus the family of functions ΦB_N is pointwise equicontinuous on J, and Φ is a completely continuous operator from B_N into $C(J,X)$.

Finally, because $SB_N \subseteq B_N$ and $S = \Phi + \psi$ we have that S is a completely continuous mapping from B_N into itself. By the Schauder fixed point theorem there is a function $u \in B_N$ with $u = Su$. By Lemma 1.2 the function u is a solution to (I), (II). This completes the proof of the theorem.

III. EXAMPLES

Our first example is to illustrate that classical results in ordinary differential equations are included in our theorem.

Let $f\colon [0,\infty) \times R^n \to R^n$ be a continuous function satisfying, for some $p > 0$ and all $x \in R^n$ and $t \geqslant 0$

$$f(t + p,\, x) = f(t,x)$$

and

$$f(t,-x) = -f(t,x).$$

Also suppose that

$$\lim_{|x| \to \infty} |f(t,x)| \,/\, |x| = 0 \tag{3.1}$$

Lastly, we suppose that solutions to initial value problems for

$$\frac{dx}{dt} = f(t,x) + g(t) \tag{3.2}$$

are unique. Then there exists a $2p$-periodic solution to (3.2) provided g is continuous and $g(t + p) = -g(t)$.

To apply Theorem 2.1 consider the boundary value problem

$$\begin{aligned} u' &= f(t,u) + g(t) \\ u(0) &+ u(p) = 0 \end{aligned} \tag{3.3}$$

This problem may be written in the form

$$u' = Au + f(t,u) + g(t)$$

$$Qu = Bu$$

where $Ax \equiv 0$, $Qu \equiv u(P) + u(1)$ and $Bu \equiv 0$. Because the only solution to

$$u' = Au$$

$$Qu = 0$$

is $u(t) \equiv 0$ the operator $\tilde{Q}$ is invertible. Indeed, $\tilde{Q} = 2$. Here $W(t,s) \equiv I$, the identity operator on R^n. Clearly (F1) holds by (3.1). By Theorem 2.1 there exists a solution $u(t)$ to (3.3). Let $z(t) = -u(t + p)$. Then $z'(t) = f(t,z(t)) + g(t)$ and $z(0) = -u(p) = u(0)$. By uniqueness of solutions we have $u(t) = z(t)$ and $u(t + 2p) = u(t)$.

Example 2

Consider the equation

$$\frac{\partial u}{\partial t} = \frac{\partial^2 u}{\partial x^2} + f(t,x,u(t,x)) \quad (0 < x < 1, \quad o < t < T)$$

with boundary conditions

$$u\ (t,0) = u(t,1) = 0$$

and

$$u(0,x) - u(T,x) = \int_0^1 \int_0^T K(x,s,y)p(u(s,y))dsdy$$

Here $f\colon (0,T) \times [0,1] \times R \to R$ is a continuous function satisfying

$$f(t,0,u) = f(t,1,u) = 0$$

Also assume there is a function $g \in L^1(0,T)$ and a number α, $0 < \alpha < 1$. such that $|f(t,x,u)| \leqslant g(t)(|u|^\alpha + 1)$ for $0 < t < T$, $0 \leqslant x \leqslant 1$ and $u \in R$.

$$K\colon [0,1] \times [0,T] \times [0,1] \to R$$

satisfies $K(0,y,s) = K(1,y,s) = 0$ and $K(x,y,s)$ is continuous.

$P\colon R \to R$ is continuous and satisfies

$$|p(x)| \leqslant M(|x|^\beta + 1).$$

for some $M > 0$ and $\beta \in R$, $0 < \beta < 1$.

Let $C = C([0,1], R)$ and let $X = C_{00} = \{u \in C\colon u(0) = u(1) = 0\}$. Then X is a Banach space with the sup norm. Let $D \subseteq X$ be the set

$$D = \{u \in X\colon u' \in C, \ u'' \in X\}$$

and let $A\colon D \to X$ be defined by $Au = u''$. Define a Nemytskii operator $F\colon (0,T) \times X \to X$ by the rule $F(t,u)(y) = f(t,y,u(t,y))$ for all $(t,u) \in [0,T] \times X$ and $y \in [0,1]$. Define an operator $B\colon C([0,T], X) \to X$ by

$$Bu(y) = \int_0^1 \int_0^T K(y,s,t)g(u(s,t))dsdt$$

where $u(s,\cdot) \in X$ for $s \in [0,T]$.

It is known ([5], p. 312) that A generates an analytic semi-group $S(t)$ on X and $S(t)$ is compact for each $t > 0$. Moreover, there are numbers $M \geqslant 1$ and $\delta > 0$ such that $|S(t)| \leqslant Me^{-\delta t}$ for all $t \geqslant 0$. It follows that the only solution to the problem

$$u' = Au$$

$$u(0) - u(T) = 0$$

is the zero solution. Thus the only solution to the equation

$$\tilde{Q}x \equiv (I - S(T))\, x = 0$$

is $x = 0$ By the Fredholm alternative, $\tilde{Q}^{-1}$ is bounded. It is not difficult to show that the remaining hypotheses of Theorem 2.1 are satisfied, and thus there exists a mild solution to the abstract version of this boundary problem.

IV. REFERENCES

[1] Conti, R., *Recent trends in the theory of boundary value problems for ordinary differential equations*, Boll. U. M. I. (3), Vol. XXII (1967), pp. 135-178.

[2] Fitzgibbon, W. E., *Semilinear functional differential equations in Banach space*, to appear in J. of Diff. Eq.

[3] Friedman, A., *Partial Differential Equations*, Holt, Rinehart, and Winston, New York, 1969.

[4] Kartsatos, A. G., *Locally invertible operators and existence problems in differential systems*, Tohoku Math. J., 28 (1976), pp. 167-176.

[5] Martin, R. H., Jr., *Nonlinear Operators and Differential equations in Banach Spaces*, Wiley-Interscience, 1976.

[6] Opial, Z., *Linear problems for systems of nonlinear differential equations*, J. Diff. Eq. 3 (1967), pp. 580-594.

[7] Pazy, A., *A class of semi-linear equations of evolution*, Israel J. of Math. 20 (1975), pp. 23-36.

[8] Ward, J. R., *Boundary value problems for differential equations in Banach space*, to appear.

POLYNOMIAL PERTURBATIONS TO THE LAPLACIAN L^p

Fred B. Weissler
The University of Texas at Austin

I. INTRODUCTION

This paper is concerned with existence of solutions to the evolution equation

(1) $$u'(t) = \Delta u(t) \pm u(t)^k, \quad u(0) = \phi, \quad t \geq 0$$

and the corresponding integral equation

(2) $$u(t) = e^{t\Delta}\phi \pm \int_0^t e^{(t-s)\Delta} u(s)^k ds,$$

where $k \geq 2$ is an integer and $u(t)$ is a curve in $L^p(R^n)$. In other words, $J\phi = \pm \phi^k$ is considered as a perturbation to Δ. The methods that are described apply in a more general context, but this example illustrates the main ideas. Complete proofs, in a more general setting, and a complete bibliography are found in [4], [5], and [6].

II. FIRST METHOD

We would like to prove existence of solutions to (2) by a contraction mapping argument, i.e. for a curve $u(t)$ in L^p with $u(0) = \phi$ let

(3) $$(Fu)(t) = e^{t\Delta}\phi + \int_0^t e^{(t-s)\Delta} u(s)^k ds.$$

(The sign in (2) is irrelevant for this first method, so for convenience we take the plus sign.) As usual, the idea is to show that F has a fixed point.

The difficulty is that the integral in (3) does not

ISBN 0-12-434160-8

obviously make sense in $L^p(R^n)$. If $u(s)$ is a curve in L^p, then $u(s)^k$ is a curve in $L^{p/k}$. On the other hand, $e^{t\Delta}$ is convolution with a Gaussian kernel

$$g_t(x) = (4\pi t)^{-n/2} \exp(-|x|^2/4t);$$

and so Young's convolution inequality [3, p. 271] implies that for $t > 0$, $e^{t\Delta}$ maps $L^{p/k}$ into L^p with norm less than or equal to

$$Ct^{-n(k-1)/2p},$$

for some C independent of t. (We need to require that $p \geqslant k$ in order for Young's inequality to apply.)

Consequently, for $s < t$ the integrand in (3) is indeed in L^p, and if $p > n(k-1)/2$ the integral converges as a Bochner integral in L^p. Once this has been observed, it is relatively straightforward to show that F is a contraction mapping on a suitable space of curves in L^p, and therefore has a fixed point. In other words, if

(4) $$p \geqslant k$$

(5) $$p > n(k-1)/2,$$

then a local solution $u(t)$ to (2) exists for any initial data ϕ in L^p We remark that (5) can be interpreted to mean that $J\phi = \phi^k$ is "less singular" on L^p than Δ. Also, condition (4) entered as a technicality, and with a fair amount of work one can replace it by $p \geqslant 1$.

Let $W_t\phi = u(t)$, where $u(t)$ is the maximal solution to (2) in L^p with initial data ϕ. (We are assuming, of course, that p satisfies (4) and (5).) The transformations W_t form a <u>local semi-flow</u> on L^p, i.e. $W_{t+s}\phi = W_t W_s \phi$ if either expression makes sense. We mention without proof two properties of the W_t. First, if $\sup\{t\colon W_t\phi \text{ exists}\} = T < \infty$ (i.e. if the maximal solution of (2) is not global), then $\|W_t\phi\|_p \to \infty$ as $t \to T$. Second, the curve $u(t) = W_t\phi$ satisfies the differential equation (1) for $t > 0$ with any initial data ϕ in L^p. Indeed, the transformations W_t exhibit smoothing properties analogous to those of $e^{t\Delta}$.

Finally, we mention that the maps $K_t = e^{t\Delta}J$, $t > 0$, are semi-Lipschitz (Lipschitz on bounded sets) on L^p. This suggests that on an arbitrary Banach space one should formulate the problem in terms of maps K_t satisfying certain conditions. See [6].

III. SECOND METHOD

For the arguments that follow, we need to take the minus sign in (1) and (2). For the moment let us consider the equations on $C_0(R^n)$, the continuous functions vanishing at infinity. $J\phi = -\phi^k$ is semi-Lipschitz, and so there exist local semi-flows V_t and W_t on $C_0(R^n)$ satisfying

(6) $$V_t\phi = \phi - \int_0^t (V_s\phi)^k ds$$

(7) $$W_t\phi = e^{t\Delta}\phi - \int_0^t e^{(t-s)\Delta}(W_s\phi)^k ds.$$

(For the existence of W_t, see [2].)

It follows from (6) that

$$d(V_t\phi(x))/dt = -(V_t\phi(x))^k, \quad t \geqslant 0, \quad x \in R^n.$$

Therefore, if $\phi \geqslant 0$ then $V_t\phi \geqslant 0$; and if $\phi, \psi \geqslant 0$, then

(8) $$|V_t\phi(x) - V_t\psi(x)| \leqslant |\phi(x) - \psi(x)|, \quad t \geqslant 0, \quad x \in R^n.$$

In other words, V_t is a contraction (non-linear) semi-group on the positive cone in $C_0(R^n)$. Moreover, (8) implies that V_t extends to a contraction semi-group on the positive cone in $L^p(R^n)$, $1 \leqslant p < \infty$. Note that $e^{t\Delta}$ is also a contraction semi-group on the positive cones in C_0 and L^p.

The idea now is to combine the information we have about $e^{t\Delta}$ and V_t to gain information about W_t. The vehicle we use is a non-linear Trotter product formula:

(9) $$W_t\phi = C_0(R^n) - \lim_{m\to\infty} (e^{t/m\Delta}V_{t/m})^m\phi$$

for all $\phi \geqslant 0$ in $C_0(R^n)$ and $t \geqslant 0$ for which $W_t\phi$ exists. (See [5] for a proof.) Since $e^{t\Delta}$ and V_t are both contraction semi-groups on the positive cone in $C_0(R^n)$, it follows from (9) that W_t is also a contraction semi-group on the positive cone in $C_0(R^n)$.

As for L^p, one can use the dominated convergence theorem to show that if $\phi \geqslant 0$ is in $L^p \cap C_0$, then the limit in (9) holds in the sense of L^p. Therefore

$$\| W_t\phi - W_t\psi\|_p \leqslant \| \phi - \psi\|_p$$

for $\phi,\psi \geqslant 0$ in $L^p \cap C_0$. W_t thereby extends to a contraction semi-group on the positive cone in L^p.

Let B be the generator of this semi-group, i.e.

$$B\phi = L^p - \lim_{t \downarrow 0} t^{-1}(W_t\phi - \phi),$$

with domain $D_p(B)$ equal to the set of $\phi \geqslant 0$ in L^p for which this limit exists. Then

$$D_p(B) \cap L^{pk} = D_p(\Delta) \cap L^{pk},$$

where $D_p(\Delta)$ is the domain of Δ as the generator of $e^{t\Delta}$ on the positive cone in L^p; and if $\phi \in D_p(\Delta) \cap L^{pk}$, then

$$B\phi = \Delta\phi - \phi^k. \tag{10}$$

Moreover, if $p > 1$, then W_t preserves both $D_p(B)$ and L^{pk}; and so for $\phi \in D_p(\Delta) \cap L^{pk}$ the curve $u(t) = W_t\phi$ satisfies the differential equation (1) for $t \geqslant 0$, at least if $u'(t)$ is interpreted as a right derivative.

We remark that no conditions have been placed on p and k to require that $J\phi = -\phi^k$ be "relatively bounded" with respect to Δ. Also, any positivity preserving strongly continuous contraction semi-group e^{tA} would suffice in place of $e^{t\Delta}$ and $J\phi = -\phi^k$ can be replaced by any non-increasing locally Lipschitz function $J\colon [0,\infty) \to R$ with $J(0) = 0$. For these and further extensions, see [5].

Finally, we would like to mention that much of what can be

proved using this second method is obtainable by the somewhat different techniques developed in Chapters 8 and 9 of [1].

IV. REFERENCES

[1] Martin, R. H., *Nonlinear Operators and Differential Equations in Banach Spaces*, John Wiley and Sons, New York, 1976.

[2] Segal, I., *Non-linear semi-groups*, Ann. of Math., 78(1963), pp. 339-364.

[3] Stein, E. M., *Singular Integrals and Differentiability Properties of Functions*, Princeton, 1970.

[4] Weissler, F. B., *Semi-groups, Sobolev inequalities, and non-linear evolution equations*, Doctoral Dissertation, University of California, Berkeley, 1976.

[5] Weissler, F. B., *Construction of non-linear semi-groups using product formulas*, Israel Jnl. Math., to appear.

[6] Weissler, F. B., *Semilinear evolution equations in Banach spaces*, preprint, 1977.

A
B
C 8
D 9
E 0
F 1
G 2
H 3
I 4
J 5